SULPHUR'S VERSATALITY

A journey through SO_2, SO_3, H_2SO_4 and Oleum

Mahendra Sheth
Aditya Sheth

Book Creation Services

by

BookMyStory

It's time to tell your story

www.bookmystorypublishing.com

A venture of

R&S Enterprises

SULPHUR'S VERSATALITY

A journey through S02, SO3, H2S04 and Oleum

Mahendra Sheth & Aditya Sheth

Published in 2024

© Published by

Qurate
qaratebooks.com

Qurate Books Pvt. Ltd.
Goa 403523, India
www.quratebooks.com
Tel: 1800-210-6527, Email: info@quratebooks.com

ISBN: 978-93-58989-25-0

DEDICATION

This book is a tribute to the memory of my cherished wife, Kanan (Kokila Muni) Sheth, whose unwavering support and encouragement played a pivotal role in shaping this technical volume, 'Versatilities of Sulphur'.

She accompanied me to worldwide Sulphur conferences and exhibitions and encouraged me to share my knowledge, deepening my passion in the field.

In every section, paragraph, and idea presented herein, her influence resonates. Her unwavering love and endless encouragement throughout our journey together significantly contributed to this manuscript. Though her physical presence may be absent, her impact remains palpable, serving as a continual source of inspiration throughout the exploration of Sulphur's various applications.

With each passage, I find reassurance in the enduring presence of her essence within these lines, a testament to our shared aspirations. This publication stands as a tribute to her support and the enduring bond that transcends the limitations of time and space.

In remembrance of my dear wife, whose legacy lives on within these pages,

Mahendra Sheth

ACKNOWLEDGEMENT

Being inspired to write a book is a journey that demands long-term dedication and relies on having the right environment and support.

I want to express my heartfelt gratitude to everyone who has been a part of my life and supported me along the way:

- **My Parents**: Thank you for teaching me the importance of living a simple life.
- **My beloved departed wife, Kanan (Kokila)**, your unwavering support and encouragement.
- **My Children, Ratna and Aditya:** Your constant care and encouragement have kept me going through thick and thin.
- **My Grandchildren, Arjun, Anika, and Nandini:** Your laughter and joy have brought immense happiness into my life.
- **My Daughter-in-law Shivani and Son-in-law Vishal:** Thank you for bringing love and happiness into our family with your wonderful children.
- **Late Brother Bakulesh and Late Sister-in-law Rekha:** Your love and inspiration have meant the world to me, and I will always cherish the memories we shared.
- **My Sister, Pratima, Brothers, Harshad and Kamlesh and Brother-in-law, Jyotindra:** Your companionship and support have been invaluable to me.
- **My EA Shweta, and Shri Sancheti:** Your assistance and feedback have been crucial in completing this book.
- **My entire office staff in Mumbai and Ankleshwar:** For translating my knowledge into tangible reality
- **My Home Staff, Ajay:** Thank you for making my home life comfortable..
- **My Industry Colleagues, Late Shri J. L Thakker, Mr. Christer Anderson, Mr. Passoni and Bonzanni, Dr. Jitov and Nikolay:** Your support has been instrumental in the growth of our company.
- **My Book Editors, Shri Rajnishji and Dr. Rutuja:** Your corrections and guidance have been invaluable in bringing this book to its final form.
- **My Family, Friends, and Colleagues all over the globe:** Your respect and encouragement have been a driving force behind my endeavour

Table of Contents

FOREWORD by Mr. Bimal L. Goculdas 1-3

FOREWORD by Mr. Shiv Shukla 4-6

FOREWORD by Mr. Sharad Chandra Nigam 7-7

PREFACE........ 9-10

INTRODUCTION 11-12

SECTION 1

Chapter 1: Comprehensive Exploration of Sulphur 13-49

HISTORY 13-13

SOURCES OF SULPHUR 15-26

1. Sulphur Deposits 16-20
2. Sulphur Extraction from Sulphide Ore 20-21
3. Sulphur Recovery from Flue Gases 21-22
4. Sulphur Recovery from Gypsum 22-23
5. Extraction of Sulphur from Pyritic Sulphur in Coal 23-24
6. Sulphur Extraction from Hydrogen Sulphide 24-24
7. Sulphur Extraction from Petroleum Refinery 25-25

Summary 25-26

APPLICATION OF SULPHUR 27-47

1. Agriculture: Insecticide, Fungicide, and Soil Conditioner 28-28
2. Styrene Monomer Production 29-29
3. Utilising Sulphur in Sodium Sulphide Production 30-31
4. Halides and Oxyhalides Production 31-32
5. Disulphur Trioxide Production 32-33
6. Carbonyl Sulphide Production 33-33
7. Impact on Malt Production 34-35
8. Ultramarine Blue Pigment Production 35-35
9. Free Machining Steel 35-36
10. Sulphur Cement 36-37
11. Sulphur Monoxide Production 37-38

12. Vulcanizing of Rubber....38-39
13. Poly-Sulphide Elastomer....39-40
14. Tetra Sulphur Tetranitride....40-40
15. Treflan....40-40
16. Sodium Polysulphide....41-42
17. Chromic Oxide....42-43
18. Carbon Disulphide....43-47
SUMMARY:....47-48
REFERENCES:....48-49

Chapter 2: Sulphur Dioxide: A Historical and Chemical Exploration....50-89
HISTORICAL INSIGHTS....50-50
SOURCES OF SULPHUR DIOXIDE....52-55
1. Production of Sulphur Dioxide from Sulphur....53-53
2. Production from Sludge Acid....53-53
3. Production from Hydrogen Sulphide....54-54
4. Production from Smelting of Ores....54-54
5. Production from Iron Pyrite....54-55
APPLICATION OF SULPHUR DIOXIDE SHEET 1....56-68
1. Sodium Bisulphite....57-57
2. Sulphite Pulp....58-59
3. Sugar Refining Enhanced by Sulphur Dioxide....59-60
4. Sulphuryl Chloride Production and Its Derivatives....60-62
5. Corn Oil Production....62-63
6. Production of Starch....63-64
7. Sodium sulphite from Sulphur Dioxide....64-66
8. Utilization in Casein Production....66-67
9. Soybean Protein Preservation....67-67
10. Zinc Hydrosulfite Production....67-68
APPLICATION OF SULPHUR DIOXIDE SHEET 2....69-86
11. Chrome Liquor Production....70-70

12. Glass Production....70-71
13. Decolorization of Beet Sugar Juice....71-72
14. Fumigant....72-73
15. Sulfolane....73-74
16. Vanillin: A Synthetic Approach....74-75
17. Refrigerants....75-75
18. Organic Sulphonates....75-76
19. Sulfones....76-77
20. Thionyl Chloride....77-79
21. Sulphonyl Chloride....79-80
22. Gelatine....81-81
23. Glue....82-83
24. Wine....83-83
25. Magnesium Metallurgy....84-85
26. Preservative....85-86
27. Sulphur Trioxide & Oleum Production....86-87

Summary....87-88

REFERENCES:....88-89

Chapter 3: The Dynamic World of Sulphur Trioxide....90-119

HISTORY:....90-90

SOURCES OF SULPHUR TRIOXIDE....92-93

APPLICATION OF SULPHUR TRIOXIDE- SHEET 1....94-103

1. Thionyl Chloride....95-95
2. Poly Sulphuryl Chloride....96-96
3. Trimethylamine Sulphur Trioxide....97-97
4. Sulphur Trioxide Dioxane....97-98
5. Sodium Dodecylbenzenesulfonate (SDBS) and its Role in Detergent Manufacturing....98-99
6. Sulfonating Agents....99-100

7. Ethylene Sulphate 100-101
8. Methylene Sulphate 101-101
9. Ethionic Acid 102-103
APPLICATION OF SULPHUR TRIOXIDE- SHEET 2 104-119
10. Oleum 105-105
11. Sulfamic Acid 106-107
12. AmidoSulphuric Acid 108-108
13. Pyrosulphuric Acid 109-109
14. Sulphur Trioxide Pyridine Complex 110-110
15. Haloalkyl Chlorosulphates 111-112
16. Dichlorodiphenyl Sulfone 112-113
17. Chloro Silanes 113-113
18. Methyl Hydrogen Sulphate 114-115
19. Dimethyl Sulphate 115-117
Summary 118-118
REFERENCES: 119-119

Chapter 4: The Evolution and Properties of Sulphuric Acid 120-218
HISTORY 120-120
Sources Of Sulphuric Acid 122-123
APPLICATION OF SULPHURIC ACID-SHEET 124-140
1. Ferrous Sulphate 125-125
2. Alum 125-126
3. Phosphoric Acid 126-127
4. Diammonium Phosphate (DAP) Derived from Phosphoric Acid 127-129
5. Ammonium Sulphate (AMS) as Fertilizer 129-130
6. Single super phosphate production 130-131
7. Citric Acid 131-132
8. Bromine 132-133
9. Chloroform: 133-135
10. Glutamic Acid 135-136

11. Butyl Acetate 136-137
12. Oxalic Acid 137-137
13. Acetic Acid in the Presence of Sulphuric Acid 137-138
14. Lactic Acid in the Presence of Sulphuric Acid 138-139
15. Chromic Acid Derived from Sulphuric Acid 139-140

APPLICATION OF SULPHURIC ACID- SHEET 2 141-164

16. Calcium Sulphate 142-142
17. Copper Sulphate 142-143
18. Electroplating 143-144
19. Sodium Bisulphate 144-145
20. Ethyl Alcohol 145-147
21. Benzene 147-148
22. Toluene 148-150
23. Xylene 150-151
24. Dialkyl Phthalates 151-153
25. Naphthalene 153-154
26. Dicalcium Phosphate 154-155
27. β-Naphthol 155-158
28. Anthraquinone 158-160
29. Boric Acid 160-163
30. Chloral 163-164

APPLICATION OF SULPHURIC ACID- SHEET 3 165-180

31. Stearic Acid 166-167
32. Tartaric Acid 167-167
33. Oleic Acid 168-169
34. Cellulose Acetate 169-170
35. Nitrobenzene 170-172
36. Explosive 172-172
37. Ethyl Acetate 173-173
38. Ethers 175-175
39. Formic Acid 176-178

40. Sodium Dichromate 178-180
APPLICATION OF SULPHURIC ACID- SHEET 4 181-196
41. Salicylic Acid 182-182
42. Sulfonic Acid 182-183
43. Titanium Dioxide 183-184
44. Magnesium Sulphate 184-184
45. Zinc Sulphate 185-185
46. Sodium Sulphate 186-188
47. Lithium Carbonate 188-189
48. Isopropyl Alcohol 189-190
49. Ethylene Glycol 190-191
50. Alkylation Gasoline 191-192
51. Fatty Acid 192-192
52. Aluminum Sulphate 193-193
53. Pyridine 193-194
54. Vanilline 194-195
55. Dithionous Acid 195-196
APPLICATION OF SULPHURIC ACID- SHEET 5 197-208
56. Iron & Steel Pickling 198-198
57. Cyclic Sulphates 199-199
58. Chlorine 199-201
59. Cresols 201-203
60. Phenol 203-204
61. Hydrochloric Acid 204-205
62. Nitroglycerin 206-207
63. Tetryl 207-207
64. Hydrogen Peroxide 208-208
APPLICATION OF SULPHURIC ACID- SHEET 6 209-218
65. Hydrofluoric Acid 210-211
66. Hydrogen Cyanide 211-213
67. Fatty Alcohols 213-214

68. Industrial Alcohol....214-215
69. Storage Batteries....215-216
70. Tall Acids....216-217
Summary....217-218
REFERENCES:....218-223

Chapter 5 : Exploring the Origins and Properties of Oleum....224-240
HISTORY:....224-224
SOURCES OF OLEUM....226-227
APPLICATION OF OLEUM....228240
1. Chlorosulfonic Acid....229-229
2. Alkyl Aryl Sulfonates....229-230
3. Nitrocellulose....230-231
4. TNT....231-231
5. Treflan....232-232
6. DDT....232-233
7. Dyes....233-234
8. Diethyl Sulphate....234-235
9. Sodium Dodecyl Benzene Sulfonate....236-237
10. Sulfonating Agents....237-237
11. Amidosulphuric Acid....238-238
Summary....239-239
REFERENCES:....240-240

Chapter 6: Hydrogen Sulphide: Extraction, Risks & Utilization....241-253
HISTORY....241-241
SOURCES OF HYDROGEN SULPHIDE....243-246
ADVANTAGES OF HYDROGEN SULPHIDE....247-247
1. Sulphur....248-248
2. Thiol....248-248
3. Thioaldehyde....249-249

4. Dimethyl Sulfoxide 249-250
5. Thioketones 250-250
6. Thiophosgene 250-251
7. Thiophene 251-251
Summary 251-252
REFERENCES: 252-252

SECTION 2

Chapter 7: Evolution of the Chemical Industry 254-264
HISTORICAL FOUNDATIONS: 254-254
APPLICATION OF SULPHUR & SULPHUR COMPOUNDS IN CHEMICAL INDUSTRY 256-264
1. Sulphur-Based Solutions for Plant Growth 257-257
2. Synthetic Rubber Vulcanization with Sulphur 257-258
3. Elevating Textile Harnessing Sulphur Compounds for Dyeing 258-258
4. Sulphur Compounds Role in Resin Production 259-259
5. Sulphur Compounds Impact on Plastic Production 259-260
6. Sulphur Compounds Vital Role in Pharmaceuticals 260-261
7. Sulphuric Acid's Role in Fertilizer Production 261-262
8. Sulphuric Acid's Contribution to Detergent Production 262-262
9. Sulphuric Acid Role in Dye Production 262-263
Summary 263-264
REFERENCES: 264-264

Chapter 8: Fueling Progress: Exploring the Petroleum Industry 265-275
HISTORY: 265-265
APPLICATION OF SULPHUR & SULPHUR COMPOUNDS IN PETROLEUM INDUSTRY 267-275
1. Sulphur Compounds as Crude Oil Desulphurization 268-268
2. Sulphur Compounds as Additives in Gasoline and Diesel Fuels 269-269
3. Sulphur and Sulphur Compounds in Crude Oil Refining 269-270

4. Recovery of Sulphur During Petroleum Refining 270-271
5. Desulphurization from Sulphuric Acid 271-171
6. Alkylation from Sulphuric Acid 272-272
7. Acidizing with Sulphuric Acid 272-273
8. Role of Sulphuric Acid in the Refining Process 273-274
Summary 274-275
REFERENCES: 275-275

Chapter 9: The Evolution of the Rubber Industry 276-286
HISTORY: 276-276
APPLICATION OF SULPHUR & SULPHUR COMPOUNDS IN RUBBER INDUSTRIES 278-286
1. Sulphur in Rubber Vulcanization 279-279
2. Sulphur Compounds as Vulcanization Accelerators 279-280
3. Acid-Catalyzed Vulcanization in Rubber Production 280-281
4. Sulphuric Acid in Dehydration Processes 281-282
5. Utilizing Sulphuric Acid for Cleaning in the Rubber Industry 282-283
6. Utilizing Sulphuric Acid for Neutralization 283-283
7. Application of Sulphuric Acid in Production of Rubber Chemicals 284-285
Summary: 285-286
REFERENCES: 286-286

Chapter 10: Tracing the Path of Agriculture: An Evolutionary Journey 287-299
ANCIENT ROOTS: 287-287
APPLICATION OF SULPHUR & SULPHUR COMPOUNDS IN AGRICULTURE INDUSTRY 290-299
1. Soil Amendment by Sulphur 291-291
2. Sulphur-Based Pesticides 292-192
3. Harnessing Sulphur for Fertilizer Production 292-193
4. Enhancing Livestock Feed with Sulphur Supplements 293-294
5. Utilizing Sulphur Compounds for Soil Sterilization 294-295

6. Enhancing Water Quality with Sulphuric Acid 295-296
7. Utilizing Sulphuric Acid in Fertilizer Production 296-297
Summary 298-299
REFERENCES: 299-299

Chapter 11: Evolution of the Pharmaceutical Industry 300-306
HISTORICAL PERSPECTIVE 300-300
APPLICATION OF SULPHUR & SULPHUR COMPOUNDS IN PHARMACEUTICAL INDUSTRY 302-306
1. Drugs from Sulphur 303-303
2. Antioxidant or Reducing Agent from Sulphur Compounds 303-304
3. Pharmaceutical Products from Sulphuric Acid 304-305
Summary 305-305
REFERENCES: 306-306

Chapter 12 : Purifying H2O: Advances in Water Treatment 307-316
HISTORICAL EVOLUTION: 307-311
APPLICATION OF SULPHUR COMPOUNDS & SULPHURIC ACID IN WATER TREATMENT 309-316
1. Removing Sulphur Compounds & Metallic Impurities from Water 310-311
2. Purification of Water by Disinfection 311-312
3. Eliminating Sulphur Compound Impurities for Odorless Water 312-313
4. PH Adjustment of Water Using Sulphuric Acid 313-314
5. Corrosion Control in Water Treatment with Sulphuric Acid 314-314
Summary 314-316
REFERENCES 316-316

Chapter 13: From Papyrus to Print: A Journey Through the Paper Industry 317-325
HISTORY: EVOLUTION OF THE PAPER INDUSTRY 317-317
APPLICATION OF SULPHUR COMPOUND IN PAPER INDUSTRY 319-325
1. Utilizing Sulphur Compounds for Bleaching Purposes 320-320

2. Sizing with Sulphuric Acid....320-321
3. Sulphuric Acid's Role in Paper Industry pH Control....321-322
4. Managing Waste Containing Sulphuric Acid in the Paper Industry....322-323
5. Recovering Chemicals from Sulphuric Acid in the Paper Industry....323-324
Summary....324-325
REFERENCES:....325-325

Chapter 14: Unveiling the Leather Industry: From Ancient Craft to Modern Innovation 326-331
HISTORY OF THE LEATHER INDUSTRY....326-326
APPLICATION OF SULPHUR & SULPHUR COMPOUNDS IN LEATHER INDUSTRY 328-331
1. Sulphur Applications....329-329
2. Sulphur Compound Applications....329-330
3. Sulphuric Acid Applications....330-330
Summary:....330-331
REFERENCES:....331-331

Chapter 15: From Farm to Fork: A Journey Through the Food Industry 332-339
A HISTORICAL PERSPECTIVE....332-332
APPLICATION OF SULPHUR, SULPHUR COMPOUNDS AND SULPHURIC ACID IN FOOD INDUSTRY....335-339
1. Sulphur as a Nutrient....336-336
2. Utilization of Sulphur Compounds....336-337
3. Utilization of Sulphuric Acid....337-337
Summary: Application of Sulphur and Sulphur Compounds....338-339
REFERENCES:....339-339

Chapter 16: Exploring the Depths: An Overview of the Mining Industry 340-350
HISTORICAL EVOLUTION....340-340

APPLICATION OF SULPHUR & SULPHUR COMPOUNDS IN MINING INDUSTRY 342-350

1. Sulphuric Acid in Explosive Production..344-344
2. Sulphuric Acid in Flotation Products...345-345
3. Application of Sulphuric Acid in Ore Processing......................................345-346
4. Application of Sulphuric Acid in Smelting Processes 346-347
5. Application of Sulphuric Acid in Mining 347-348
6. Application of Sulphuric Acid in Acid Mine Drainage (AMD) Treatment 348-349

Summary 349-350

REFERENCES..350-350

About the Authors: 351-351

FOREWORD

Welcome to the fascinating world of sulphuric acid and its derivatives. In this comprehensive volume, "Sulphur's Versatilities: A journey through, SO_2, SO_3, H_2SO_4 and Oleum", the author Mahendra and Aditya Sheth leads us on a journey through the intricate chemistry, industrial processes, and myriad applications of sulphur, sulphuric acid, oleum, sulphur trioxide, sulphur dioxide, and hydrogen sulphide.

It is with great pleasure and a profound sense of pride that I introduce this invaluable resource on sulphuric acid and its derivatives. As the MD & CEO of a company with over a century of experience in sulphuric acid production, I am acutely aware of the pivotal role this chemical plays in our industrial landscape. Moreover, I am delighted to commend the author of this book, who brings a wealth of expertise in manufacturing equipment essential for sulphuric acid production.

For more than a century, our company DMCC, has been at the forefront of sulphuric acid production, pioneering advancements in technology, safety standards, and environmental stewardship. Through decades of dedication and innovation, we have witnessed the transformative impact of sulphuric acid on various industries, from agriculture to pharmaceuticals, and from manufacturing to environmental protection. I recognize the immense value of this book in elucidating the intricate processes and applications associated with sulphur-based inorganic chemicals. The author's expertise in manufacturing equipment for sulphuric acid production lends a unique perspective to the exploration of manufacturing processes, offering insights into the engineering marvels that underpin the production of this essential chemical.

Sulphuric acid, often referred to as the "King of Chemicals," holds a pivotal position in the realm of industrial chemistry. Its versatility and indispensability span a wide range of sectors, from agriculture to pharmaceuticals, from manufacturing to environmental protection. This book endeavours to shed light on the multifaceted nature of sulphuric acid and its derivatives, exploring their origins, production methods, and the vast array of products derived from them.

Within these pages, you will delve into the origins of raw materials, tracing the sources of sulphur and other essential components crucial for the synthesis of these compounds. Understanding the intricacies of raw material sourcing lays a foundation for comprehending the broader industrial landscape and the environmental impacts associated with sulphuric acid production.

The core of this book lies in its exploration of the manufacturing processes involved. From the traditional contact process to modern advancements in catalysis and reactor design, readers will gain insight into the evolution of manufacturing techniques and the technological innovations driving efficiency, safety, and sustainability in production.

Furthermore, this volume provides a comprehensive overview of the downstream applications of sulphuric acid derivatives. From its indispensable role in fertilizer production, where it serves as a cornerstone for global agriculture, to its pivotal role in industrial processes such as petroleum refining, metal processing, and the synthesis of numerous chemicals, the impact of sulphuric acid and its derivatives permeates nearly every aspect of modern life.

Moreover, as we navigate through the pages of this book, we will also explore the critical role of sulphuric acid in environmental stewardship. Its applications in pollution control, wastewater treatment, and the production of renewable energy sources underscore its significance in mitigating environmental challenges and fostering sustainable development.

As you embark on this journey through the world of sulphuric acid and its derivatives, I invite you to immerse yourself in the wealth of knowledge presented within these pages.

In closing, I extend my sincere gratitude to the author for their dedication to elucidating the complexities of sulphuric acid chemistry and production. This book stands as a testament to their expertise and passion for advancing our understanding of one of the most fundamental chemicals in modern industry. Whether you are a seasoned professional in the field of chemistry and chemical engineering, a student eager to expand your understanding of industrial processes, or simply a curious mind seeking to unravel the mysteries of one of the most essential chemicals in our modern world, this book offers a comprehensive and insightful exploration that is sure to captivate and enlighten.

Bimal L. Goculdas
Managing Director and CEO
DMCC Speciality Chemicals Ltd.
(Formerly known as The Dharamsi Morarji Chemical Company Ltd.)

FOREWORD

In the annals of industrial chemistry, certain voices resonate with profound wisdom and experience, shaping the trajectory of entire industries. Mahendra Sheth, the author of "Sulphur's Versatilities: A journey through, SO_2, SO_3, H_2SO_4 and Oleum," is undeniably one of these voices—a stalwart whose insights have illuminated the path for countless manufacturers within the sulphuric acid industry.

I have had the privilege of knowing Mahendra personally for over three decades, our paths intertwining in the intricate tapestry of the sulphuric acid industry since 1989. Over the course of our enduring friendship, I have witnessed firsthand Mahendra's unwavering commitment to his craft, his passion for innovation, and his relentless pursuit of excellence.

It is with great pleasure and anticipation that I introduce this invaluable book penned by Mahendra and Aditya Sheth, a luminary in the sulphuric acid industry whose wealth of knowledge and experience has illuminated countless paths in our field. As someone deeply immersed in the intricacies of chemical engineering, energy, and environmental sustainability, I am particularly enthralled by the insights and wisdom encapsulated within these pages. My own journey in the sulphuric acid industry spans a significant period of more than three decades during which I embarked as a process engineer, honing my skills in the design of sulphuric acid plants following a formative four-year tenure in the petroleum refinery sector. It is within this shared realm of expertise that I have come to recognize the profound impact of Mahendra and Aditya's contributions—a testament to his deep-seated understanding of the industry's nuances and intricacies.

This book is more than just a compendium of knowledge; it is a testament to the author's dedication to serve his fraternity. Drawing from his wealth of experience, Mahendra and Aditya unveils a comprehensive guide that not only elucidates the manufacturing processes of sulphur-based chemicals but also unveils the potential for value addition that lies within these compounds.

In this seminal work, Mahendra and Aditya embark on a captivating journey through the intricate world of sulphur-based chemicals, unravelling their complex nuances and elucidating their diverse applications across industries. From elemental sulphur to sulphuric acid and beyond, Mahendra and Aditya masterfully navigated the intricacies of these essential compounds, offering profound insights and practical guidance for manufacturers, researchers, and industry stakeholders alike.

The preface sets the stage, resonating with poignant anecdotes that underscore the challenges faced by manufacturers amidst fluctuating demands and the spectre of excess inventory. Mahendra and Aditya's narrative voice, infused with a blend of pragmatism and optimism, invites readers into a world where adversity births opportunity, and where every raw material holds the promise of myriad downstream products.

As one delves deeper into the pages of this book, the richness of its content becomes evident. Each section meticulously explores the sources, manufacturing processes, and diverse applications of sulphur-based chemicals, laying bare a treasure trove of insights for manufacturers seeking to diversify their product portfolio and enhance their profitability.

From elemental sulphur to hydrogen sulphide, sulphur dioxide to sulphuric acid, and beyond, Mahendra and Aditya's exposition navigates the intricacies of these compounds with clarity and precision. Moreover, the inclusion of detailed charts and descriptions further enhances the utility of this book, serving as a practical toolkit for manufacturers embarking on the journey of value addition.

Indeed, "Sulphur's Versatilities: A journey through, SO_2, SO_3, H_2SO_4 **and Oleum" transcends the boundaries of a mere handbook; it emerges as a trusted companion for manufacturers, empowering them to make informed decisions and chart a course towards sustainable growth. As the author envisions, this book has the potential to become an indispensable resource, guiding manufacturers through the nuances of technical viability and economic analysis, thereby fostering a culture of innovation and resilience within the industry.**

What distinguishes this book is its profound depth of knowledge and its unwavering commitment to practicality and relevance. Through meticulous research and insightful analysis, Mahendra and Aditya empowers readers to harness the full potential of sulphur-based chemicals, unlocking new avenues of innovation and sustainability in the process.

As we embark on this enlightening journey through the pages of Mahendra and Aditya's book, let us not only glean knowledge but also draw inspiration from his unwavering commitment to excellence and his enduring passion for driving positive change in our industries and communities.

Warm Regards,

Shiv Shukla

Founder & Managing Director

Shiv Sulphuric Solutions, India

FOREWORD

As the world continues to innovate and adapt, the role of chemistry in shaping our future becomes increasingly evident. Among the myriad elements that contribute to this transformative landscape, sulphur stands out as a versatile and indispensable component. From its historical roots to its modern-day applications, sulphur-based chemicals have played a pivotal role in various industries, driving innovation and facilitating progress.

In this insightful volume, "Sulphur's Versatilities: A journey through, SO_2, SO_3, H_2SO_4 and Oleum" readers are invited on a journey of exploration into the fascinating world of sulphur compounds. Authored by experts in the field, this book offers a comprehensive overview of sulphur's diverse applications, production methods and economic significance. From its transformation into Sulphur Dioxide, Sulphur Trioxide, Sulphuric Acid and Oleum to its utilization in Chemical, Petroleum, Rubber, Agriculture, Pharmaceuticals, Water, Paper, Leather, Food, and Mining industries, every aspect of sulphur's versatility is meticulously examined.

With a keen focus on practical insights and real-world applications, "Versatilities of Sulphur" serves as an invaluable resource for professionals, academics, and enthusiasts alike. Whether you are seeking to expand your knowledge of chemical engineering, enhance your understanding of industrial processes, or simply exploring the remarkable properties of sulphur, this book provides a wealth of information to enrich your journey.

As we embark on this exploration of sulphur's boundless potential, let us not only uncover its versatility but also recognize the immense impact it continues to have on our lives and industries. I commend the authors for their dedication and expertise in bringing forth this enlightening volume, and I am confident that "Versatilities of

Sulphur" will serve as a valuable guide in unlocking the transformative power of sulphur-based chemicals."

Sharad Chandra Nigam
Proprietor cum Director (Process)
Samrudhi Consultants, Kanpur, U.P., India
Ex-Corporate Vice President
DMCC Speciality Chemicals Limited, Mumbai, Maharashtra, India

PREFACE

Sulphur's remarkable versatility stems from a combination of its unique chemical properties and its widespread influence across several critical domains. Its ability to form an extensive array of compounds, ranging from simple oxides, Sulphuric Acid and to complex organic molecules, contributes to its multifaceted nature. Sulphur's chemical diversity enables its participation in a multitude of reactions and applications. This book helps anyone in Sulphur based chemical manufacturing to understand how sulphur and its derivatives can enhance your chemical plant's profitability in multiple different ways.

During my career, I've realised that Sulphur and its derivatives like Sulphur Dioxide, Sulphur Trioxide, Sulphuric Acid and Oleum are vital for making many other products. The byproducts made from Sulphur and its derivatives often fetch higher prices than simply manufacturing these derivatives.

Approximately 90% of the sulphur produced worldwide is used in the production of sulfuric acid. Sulphur, the main raw material, is readily available and other necessary elements like oxygen and water are abundant in nature. Manufacturing processes for these base products are well-established, making easy entry into sulphuric acid manufacturing. However, because of the high number of manufacturers, profit margins are often slim, and companies operate on thin margins. Even small price fluctuations can have a big impact.

In my over 40 years working in the Sulphuric Acid Industry, I've often seen companies face challenges when the demand for Sulphuric Acid drops. Sometimes, they end up with too much acid, and must decide whether to shut down their plant or sell their product at a loss. For instance, in a US metal-producing company where I worked, we had to sell excess acid at no profit and even pay for delivery just to keep the smelter running.
When there is excess acid production, companies can utilise Sulphur or its derivatives to turn it into more valuable products, boosting profitability. Most Sulphur-derived

products, except Hydrogen Sulphide, are made through the oxidation of Sulphur. These reactions generate significant amounts of heat, which can be harnessed in various ways such as producing steam, heating water for industrial processes that demand energy input, or serving as a sustainable method for electricity generation.

Lately, I have felt the desire to support the Sulphuric Acid industry, a field I've been closely associated with for the past four decades. I am convinced that by gathering crucial insights on downstream products, companies can optimise surplus production and strategize for more lucrative endeavours. Hence, I've made the decision to embark on writing this book.

The book comprises sixteen chapters organised into two sections. Section 1, spanning six chapters, delves into Sulphur, Sulphur Dioxide, Sulphur Trioxide, Sulphuric Acid, Oleum, and Hydrogen Sulphide. Each chapter offers insights into:

- Sources of raw materials
- Manufacturing processes
- Resulting downstream products

These concepts are conveniently illustrated in charts, depicting raw materials as roots, main products as trunks, and downstream products as branches. Additionally, the book furnishes details on raw materials and wherever feasible a concise manufacturing process descriptions for each downstream product.
The subsequent Section 2, encompassing chapters 7 to 16, explores the applications of sulphur products across various industries.
I aspire for this book to become a widely utilised reference within the Sulphur-based industry, facilitating the production of value-added products and enhancing the industry's overall economic landscape.

By Mahendra Sheth
Location: Mumbai, Maharashtra

INTRODUCTION

Welcome to "The Versatilities of Sulphur," a comprehensive guide designed to deepen your understanding of sulphur's origins and its pivotal role in the production of valuable products. Tailored for both industry professionals and enthusiasts alike, this book serves as an essential resource.

Our aim is to equip manufacturers with the knowledge to select the optimal raw materials and products, enhancing the profitability and diversity of their operations. From the exploration of raw materials to the intricacies of downstream products, we provide detailed insights to enhance your decisions.
By focusing on the diversification of sulphur-based products, we strive to bolster manufacturers' resilience against price volatility and dependency, thereby paving the way for greater profitability and sustainability in their business models.

Below are some simple and straightforward examples showcasing how this book will provide insights and enhance value by either transforming or reacting sulfur-based chemicals to yield more valuable products:

- **Insoluble Sulfur**, commanding a higher price compared to regular sulfur, can be readily produced through transformation into Insoluble Sulphur, extensively employed in the rubber industry for manufacturing items like tires and footwear.
- **Liquid sulfur dioxide (SO_2)**, albeit more expensive, may fetch better price realization due to its preference over its gaseous form, facilitating ease of handling, transport, and storage.
- **Sulfur trioxide** assumes a pivotal role in sulfonation, a process essential for the production of surfactants and detergents.
- **Battery-grade sulfuric acid**, distinguished by its high-quality and purified form, is indispensable for battery manufacturing, including car batteries.
- **Chlorosulfonic acid,** derived from Oleum but pricier than Oleum, and serves as a potent acid utilized in various industrial applications.

These uncomplicated examples underscore the conversion of sulfur-based chemicals into valuable products through specialized processes. Their reactions with other chemicals can yield a myriad of more valuable products, underscoring their economic and functional significance across industries.

This book is an indispensable resource for newcomers and established businesses alike in the sulfur chemicals sector, providing valuable insights into profitable ventures and product expansion opportunities.

Section 1 Gives comprehensive reference based on sulfur and its derivatives, delving into their sources and myriad of byproducts...

In contrast, **Section 2** explores the utilization of sulfur and its derivatives across diverse industrial applications, providing readers with alternative perspectives.

This dual structure enhances accessibility to information, empowering readers to leverage the book in a multitude of ways.
Join us in envisioning a future where this book serves as a vital resource within the Sulphur-based industry, empowering companies to produce value-added products and contribute to the industry's growth and prosperity.

Galiakotwala Engineering Company stands to assist manufacturers in enhancing their product portfolio by conducting technical and economic analysis for Sulphur based byproducts.

We invite you to immerse yourself in the wealth of knowledge presented within these pages and to embark on a journey of discovery and innovation in the world of Sulphur.

Chapter 1: Comprehensive Exploration of Sulphur

HISTORY

The term "Sulphur" finds its roots in the Latin "sulphurium" and the Sanskrit "Sulvere" [Royal Society of Chemistry]. In ancient times, Sulphur was known as "brenne stone" for its combustible properties, from which "brimstone" is derived. Initially believed to contain hydrogen and oxygen, its elemental nature was confirmed in 1809 by the French chemists Louis-Joseph Gay-Lussac and Louis-Jacques Thenard.

OCCURRENCE

Sulphur, ranked as the tenth most abundant element in the universe, has been recognized since antiquity. It constitutes a part of various common minerals, including galena (PbS), gypsum ($CaSO_4.2(H_2O)$), pyrite (FeS_2), sphalerite (ZnS or FeS), cinnabar (HgS), stibnite (Sb_2S_3), epsomite ($MgSO_4.7(H_2O)$), celestite ($SrSO_4$), and barite ($BaSO_4$). Currently, about 25% of sulphur is sourced from petroleum refining, with the remainder being a by-product of extracting other materials from sulphur-containing ores. Most of the world's sulphur is obtained via the Frasch process from underground deposits typically found alongside salt deposits.

India presently lacks mineable elemental sulphur reserves. Sulphur forms direct compounds with almost all elements, except for gold, platinum, and noble gases.

PROPERTIES

Sulphur exists as a yellow crystalline solid and may occur as a pure element or as sulphate or sulphide minerals. Its crystallography is intricate, with sulphur allotropes forming various crystal structures, with rhombic and monoclinic S8 being the most recognized.

Sulphur is a pale yellow, odourless, and brittle material, exhibiting three allotropic forms: **Orthorhombic**, **Monoclinic**, and **Amorphous**. The **orthorhombic** form is the most stable, while **monoclinic** sulphur exists within a temperature range of 96°C and 119°C, converting back to the orthorhombic form upon cooling. **Amorphous** sulphur, obtained by rapid cooling of the crystalline form, is soft and elastic and eventually reverts to the orthorhombic form.

While insoluble in water, sulphur is soluble in carbon disulphide. In all states - gas, liquid, or solid - elemental sulphur exists in more than one allotropic form, presenting a complex array of forms yet to be fully understood.

In 1975, scientists at the University of Pennsylvania synthesised polymeric sulphur nitride, exhibiting metallic properties despite the absence of metal atoms, and featuring unusual optical and electrical characteristics.

High-purity sulphur is commercially available, with purities exceeding 99.999+%. Other physical properties include:

- Atomic weight: 32.066 g/mol
- Specific heat: 0.706 J/g°C
- Density: 2.067 g/cm^3
- Melting Point: 115.2°C
- Boiling Point: 444.6°C

EFFECT OF HEAT ON SULPHUR

As Sulphur undergoes changes with varying temperatures:

- At its boiling point, S_8 transitions to S_6, S_4, and eventually S_1.
- At 2000°C, it transforms to S_1.

SOURCES OF SULPHUR

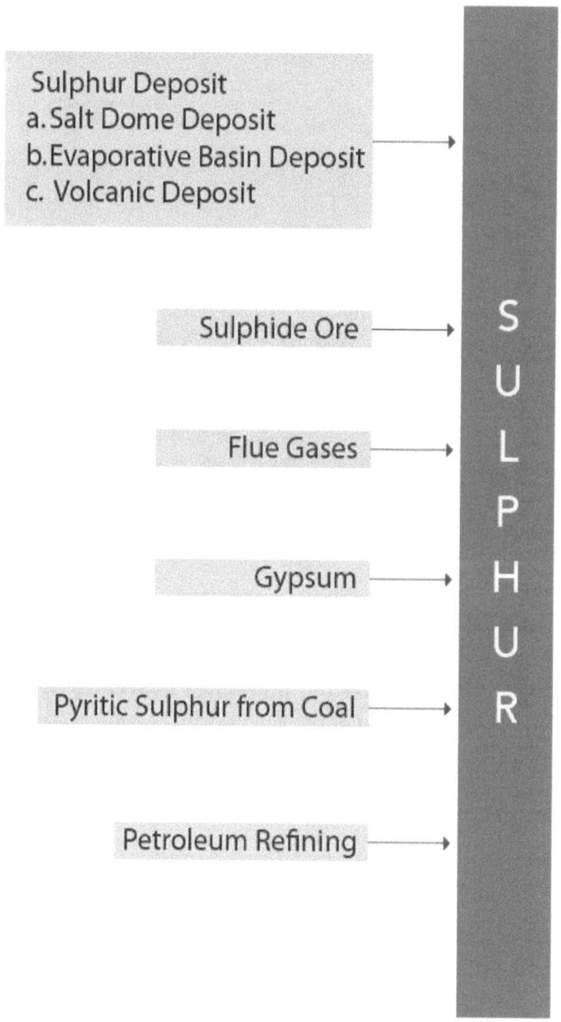

1. Sulphur Deposits

A. Sulphur Extraction from Dome Deposits

Sulphur, a vital industrial commodity, is commonly extracted from dome deposits using the Frasch process, a method pioneered by Herman Frasch in the late 19th century. This process capitalises on sulphur's low melting point to facilitate its extraction.

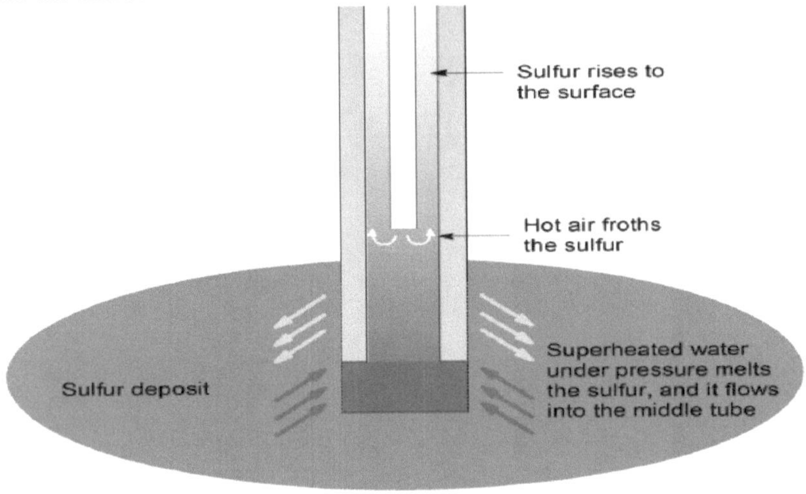

Figure 1: Pictorial representation of the Frasch process.

The Frasch process initiates with the injection of superheated water down the outermost pipe of a three concentric pipe system. Simultaneously, compressed air is pumped down the central tube. This process generates a mixture of elemental sulphur, hot water, and air that ascends through the middle pipe. At a temperature of approximately 170°C under high pressure, the sulphur melts and combines with the superheated water, forming a slurry that is propelled to the surface of the earth.

Drilling three wells into the sulphur deposit constitutes a crucial aspect of the Frasch process. The first well serves to introduce hot water, melting the sulphur and enabling its upward flow through the second well. The third well functions to collect the molten sulphur, which is subsequently pumped to the surface. Here, it undergoes cooling and solidification before being transported to processing facilities for further purification.

The Frasch process boasts remarkable efficiency, enabling the recovery of up to 90% of the sulphur within the deposit. Most of the sulphur obtained through this method is utilised in the industrial production of sulphuric acid, a versatile compound extensively employed in various sectors, particularly the fertiliser industry.

The extraction of sulphur from dome deposits typically involves several stages:

a. **Exploration:** Geologists undertake surveys and analyses to identify prospective dome deposits. This process encompasses the examination of geological data, field surveys, and potentially drilling test wells to evaluate sulphur content and quality.

b. **Mining:** Upon the identification of a viable deposit, mining operations commence. The Frasch process is predominantly employed, involving the following key steps:
   - Drilling: Wells are strategically drilled into the sulphur deposit from the surface.
   - Superheating: Superheated water or steam is injected into the deposit, initiating the melting of sulphur.
   - Pumping: Compressed air is introduced into the deposit to propel the molten sulphur up the well to the surface.
   - Collection: At the surface, the molten sulphur is collected in substantial containers or tanks, where it solidifies upon cooling.

c. **Refining:** The collected sulphur may undergo further refining processes to eliminate impurities and enhance its quality. These processes encompass filtration, centrifugation, or chemical treatments tailored to the intended application of the sulphur.

While dome deposits have historically served as primary sulphur sources, other extraction methods, such as underground mining and extraction from natural gas or oil refining, have gained prominence in recent years, diversifying the sulphur supply chain and ensuring its continued availability for industrial applications.

B. Sulphur Extraction from Evaporative Basin Deposits

Sulphur, a valuable industrial resource, can be obtained from evaporative deposits formed by the evaporation of sulphur-containing fluids like saltwater, brine, or oil. Extracting sulphur from these deposits involves a meticulous process that combines drilling, collection, and refining techniques.

a. **Initial Extraction Process:**

- Drilling: The extraction begins with drilling into the deposit to access the sulphur-rich fluids trapped within.

- Fluid Collection: Once drilled, the sulphur-rich fluids are collected and transported to a processing facility for further refinement.

b. Refinement Techniques:

- Sulphur Recovery Unit (SRU): In industrial settings such as petroleum oil refineries, sulphur compounds present in the by-product gases are removed and converted into elemental sulphur using specialised equipment like the SRU.
- Ludox Process: Another method involves extracting sulphur from brine solutions by introducing a flocculant, causing sulphur to precipitate out of the solution for subsequent collection and processing.

c. Mining Strategies:

- Exploration: Geologists conduct comprehensive surveys and studies to identify potential evaporative basin deposits containing sulphur. This process entails analysing geological data, conducting field surveys, and possibly drilling test wells to assess sulphur content and quality.
- Mining Methods: Once a viable deposit is located, various mining techniques are employed:
    - Open-pit Mining: Suitable for shallow sulphur deposits, this method involves removing overlying soil and rock to expose the sulphur-bearing layers. Heavy machinery like excavators and trucks is then utilised to extract the sulphur.
    - Underground Mining: When sulphur deposits are deeper, underground mining techniques are employed. This entails constructing tunnels or shafts to access the sulphur-bearing layers. Miners extract the sulphur using specialised equipment and transport it to the surface for further processing.

d. Processing and Refining: Once extracted, the sulphur undergoes processing and refining to remove impurities and enhance its quality.

- Crushing and Grinding: The sulphur ore is crushed and ground into a fine powder.
- Flotation Process: The powdered sulphur ore undergoes a flotation process or other separation techniques to isolate sulphur from other minerals or impurities.

It's essential to note that while evaporative deposits may not be as common as dome or volcanic deposits, they still represent a significant source of sulphur. Moreover, the extraction process may vary depending on the type of deposit and its specific location. By employing a combination of exploration, mining, and refining techniques, sulphur extraction from evaporative basin deposits ensures a consistent supply of this critical industrial element.

C. Sulphur Extraction from Volcanic Deposits

Sulphur, a vital industrial resource, can also be obtained from volcanic deposits formed by the precipitation of sulphur compounds from volcanic gas emissions. Extracting sulphur from these deposits involves a combination of mining, collection, processing, and refinement techniques.

Fig.2: Volcanic sulphur deposits (yellow) on the rim of a crater (Photo Credit: MICHAEL MARTEN / SCIENCE PHOTO LIBRARY)

a. Extraction Methods:

- Mining: Sulphur deposits in volcanic areas are typically found in the form of rock or mineral deposits. Mining operations involve extracting these sulphur-rich materials from the earth's crust.
- Natural Seepage: In some cases, sulphur naturally seeps out of the volcano and can be collected by constructing channels or trenches to direct the flow of sulphur into holding ponds.

b. Identification and Exploration: Geologists conduct thorough surveys and studies to identify potential sulphur deposits within volcanic areas. This process includes analysing volcanic gases, studying the composition of volcanic rocks, and conducting field surveys to determine the presence and quality of sulphur deposits.

c. **Mining Techniques:**

- Open-pit Mining: When sulphur deposits are near the surface and accessible, open-pit mining may be employed. This method involves removing overlying materials to expose the sulphur-bearing layers, which are then extracted using heavy machinery and transported for further processing.
- Underground Mining: In cases where sulphur deposits are deeper, underground mining techniques are utilised. Tunnels or shafts are constructed to access the sulphur-bearing layers, and miners extract the sulphur using specialised equipment before transporting it to the surface.

d. **Processing and Refinement:** Once extracted, sulphur undergoes processing and refinement to remove impurities and enhance its quality. This process may involve crushing and grinding the sulphur ore, followed by flotation or other separation techniques to isolate sulphur from other minerals and impurities.

e. **Utilisation:** Sulphur obtained from volcanic deposits finds various industrial applications, including its use in fertilisers, chemical production, rubber manufacturing, petroleum refining, pharmaceuticals, food preservation, and other industrial processes.

While sulphur extraction from volcanic deposits offers significant potential, it is not as common as extraction from dome deposits due to environmental impacts and the potential hazards associated with working in close proximity to active volcanoes. Nevertheless, with careful exploration and extraction techniques, sulphur from volcanic deposits remains an essential resource for various industries.

2. Sulphur Extraction from Sulphide Ore

Recovering sulphur from sulphide ore involves a series of crucial steps aimed at efficiently harnessing this valuable element:

a. **Crushing and Grinding:** The sulphide ore undergoes initial crushing and grinding, breaking it down into a fine powder.

b. **Flotation:** The powdered ore is mixed with water and a flotation agent like xanthate, creating a slurry. This mixture is then agitated while air is bubbled through it. As a result, sulphur-containing minerals rise to the surface as a froth, which is collected and dried to form a concentrate rich in sulphur.

c. **Roasting:** The sulphur-rich concentrate undergoes roasting in a furnace at temperatures typically ranging from 900-1000°C in the presence of air. During this process, sulphur compounds are converted into sulphur dioxide gas, which is then collected for further processing.

d. **Conversion to Sulphuric Acid:** Sulphur dioxide gas obtained from roasting is converted into sulphur trioxide, which, upon reaction with water, yields sulphuric acid. This versatile chemical finds extensive use in various industries, including fertiliser and detergent production.

The recovery of sulphur from sulphide ore stands as a vital industrial process, enabling the production of numerous essential chemicals and materials. However, it's essential to address potential environmental concerns associated with the process, particularly emissions from roasting. Therefore, ongoing efforts are directed towards developing sustainable and eco-friendly techniques for sulphur extraction from sulphide ore.

3. Sulphur Recovery from Flue Gases

Flue gas, the byproduct of burning fossil fuels like coal, oil, natural gas, or wood for heat or power, contains sulphur compounds that contribute to air pollution. Flue gas desulphurization is the process of removing these sulphur compounds from the exhaust emissions of fossil-fuelled power stations. This process, capable of removing up to 95% of sulphur dioxide from flue gas, employs various methods for sulphur recovery:

a. **Wet Scrubbing:** Flue gas is directed through a scrubber tower where a chemical solution, often limestone or lime slurry, is sprayed onto the gas. The sulphur dioxide reacts with the solution, forming calcium sulphite or calcium sulphate, which are then collected and processed to recover the sulphur.

b. **Dry Scrubbing:** Dry sorbent materials like lime, sodium bicarbonate, or trona are injected into the flue gas stream via a dry sorbent injection system. These materials react with sulphur dioxide, forming a solid product that can be collected and further processed for sulphur recovery.

c. **Catalytic Oxidation:** A catalyst is employed to convert sulphur dioxide in the flue gas to sulphur trioxide, which is then absorbed by a sorbent material such as alumina, silica, or activated carbon. The absorbed sulphur trioxide is then released by heating the sorbent material, enabling further processing for sulphur recovery.

d. **Biological Treatment:** Microorganisms are utilised to oxidise sulphur dioxide in the flue gas, resulting in the formation of sulphuric acid. This acid is then collected and processed to recover the sulphur content.

Extracting sulphur from flue gases not only mitigates air pollution but also provides a valuable source of elemental sulphur. However, the method of sulphur extraction from flue gases may vary depending on factors such as location, flue gas composition, and regulatory requirements. As environmental concerns persist, there is a growing emphasis on developing more efficient and sustainable techniques for sulphur recovery from flue gases.

4. Sulphur Recovery from Gypsum

Gypsum, a byproduct generated by industries such as fertiliser production, mining, and power stations, presents both environmental and economic challenges. However, these issues can be mitigated by extracting valuable byproducts like sulphur and calcium carbonate from low-quality gypsum.

The process for sulphur recovery from gypsum involves several key stages:

a. **Reduction of Gypsum to Calcium Sulphide:** Gypsum ($CaSO_4$) is reduced to calcium sulphide (CaS) using activated carbon in a tube furnace operating at 1100°C. This thermal reduction process yields CaS with an efficiency of 96%.

b. **Stripping of Sulphide and Sulphur Production:**
   - The CaS formed is then mixed with water to create a slurry. Carbon dioxide (CO_2) gas is introduced to the slurry, resulting in the stripping of sulphide to form hydrogen sulphide (H_2S) gas and the precipitation of calcium carbonate ($CaCO_3$).
   - The generated H_2S is subsequently reacted with sulphur dioxide (SO_2) to produce elemental sulphur.

c. **Heating to Produce Sulphur Dioxide and Calcium Oxide:** The calcium sulphide (CaS) obtained from the previous step is heated in the presence of oxygen to generate sulphur dioxide (SO_2) gas and calcium oxide (CaO).

- Sulphur Recovery via the Claus Process: Sulphur dioxide (SO_2) produced in the earlier stage is captured, cooled, and processed using the Claus process to obtain solid sulphur.

While this method presents a viable means of sulphur extraction, it is not extensively utilised in industry due to the higher cost compared to obtaining sulphur from sulphide ores. However, it remains an environmentally conscious approach to managing gypsum waste and recovering valuable sulphur resources.

5. Extraction of Sulphur from Pyritic Sulphur in Coal

Fossil fuels like coal harbour sulphur in two main inorganic forms: pyritic sulphur (FeS_2) and sulphates (Na_2SO_4, $CaSO_4$, $FeSO_4$), alongside organic forms such as sulphides, mercaptans, and bisulphides. Coal typically contains sulphur ranging from 0.2 to 5 percent by dry weight. When subjected to high-temperature gasification, most of the sulphur content in coal transforms into hydrogen sulphide (H_2S) and carbonyl sulphide (COS) due to the oxygen-deprived environment. Acid gas removal systems then extract 95% to over 99% of H_2S and COS from the fuel gas, converting it into saleable sulphur or sulphuric acid byproducts.

a. Extraction Process Overview

- Coal Preparation: Coal undergoes crushing and grinding to increase its surface area, aiding subsequent processing steps.

b. Desulphurization Methods

- Physical Separation: Techniques like gravity separation, magnetic separation, or froth flotation exploit the differing physical properties of coal and pyrite for separation.
- Coal Washing: This common method involves mixing coal with water and applying processes like density separation, centrifugation, or cyclones to remove impurities, including pyritic sulphur.
- Chemical Treatment: Leaching or oxidation processes utilise chemicals to dissolve or react with sulphur minerals selectively while preserving the coal.

c. Sulphur Recovery Techniques

- Roasting or Oxidation: Separated pyrite minerals are subjected to roasting or oxidation to convert sulphur into sulphur dioxide (SO_2) gas, which can be processed further to produce sulphuric acid or elemental sulphur.

- Chemical Processing: Various chemical processes transform pyritic sulphur into valuable sulphur compounds. For instance, pyritic sulphur can yield iron sulphide (FeS), crucial in sulphuric acid production or other sulphur-derived compounds.

While this approach offers a means of sulphur extraction, it's not extensively adopted in industry due to the cost-effectiveness of obtaining sulphur from sulphide ores. Nonetheless, it remains an environmentally mindful strategy for managing coal-derived sulphur and recovering valuable sulphur resources.

6. Sulphur Extraction from Hydrogen Sulphide

Sulphur extraction from hydrogen sulphide (H_2S) employs the Claus process, a method renowned for its efficiency. In this process, H_2S undergoes a series of reactions catalysed by an oxidising agent like Bauxite ($Al_2O_3H_2O$), in the presence of excess air, to yield sulphur dioxide (SO_2) and water vapour. The generated SO_2 then reacts again with additional H_2S, facilitated by the catalyst, to produce elemental sulphur and water. The elemental sulphur obtained can exist in liquid or solid forms, finding extensive utility across diverse industrial sectors, notably in sulphuric acid and fertiliser production.

Another notable technique, the "Crown-Eckroth process," employs specialised catalysts to convert hydrogen sulphide directly into elemental sulphur.

The overall reactions can be summarised as follows:

a. Claus Process:

Reaction 1: $SO_2+\frac{3}{2}O_2 \rightarrow SO_2+H_2O$

Reaction 2: $SO_2+2H_2S \rightarrow 3S+2H_2O$

Reaction 3: $3H_2S+\frac{3}{2}O_2 \rightarrow 3S+3H_2O$

These processes are pivotal for harnessing sulphur resources efficiently, contributing to various industrial endeavours while ensuring optimal utilisation of hydrogen sulphide byproducts.

7. Sulphur Extraction from Petroleum Refinery

Sulphur constitutes a significant by-product in oil refining and gas processing. Most crude oil grades inherently contain sulphur, necessitating its removal during refining to meet stringent sulphur content regulations in refined products.

The extraction process primarily involves hydrotreating, wherein sulphur-containing fractions are subjected to high-pressure hydrogen gas treatment with a catalyst, resulting in the generation of hydrogen sulphide (H_2S) gas. This H_2S gas is then converted into elemental sulphur in a sulphur recovery unit. Much of the sulphur extracted from refineries is subsequently channelled into the petrochemicals market for the production of sulphuric acid.

The extraction process unfolds through several stages:

a. **Desalting:** Initial treatment of crude oil eliminates any salt impurities present.

b. **Distillation:** Crude oil is subjected to heating, leading to its fractionation into various fractions, among which are sulphur-containing fractions.

c. **Hydrotreating:** Sulphur-containing fractions undergo treatment with hydrogen gas at high pressure and temperature, facilitated by a catalyst. This process effectively removes sulphur compounds like hydrogen sulphide (H_2S) and mercaptans, known for their adverse environmental and health impacts.

d. **Claus Process:** Remaining sulphur compounds undergo conversion into elemental sulphur via the Claus process. This involves the reaction of sulphur compounds with oxygen to yield sulphur dioxide (SO_2), which further reacts with hydrogen sulphide (H_2S) to yield elemental sulphur.

e. **Purification:** The elemental sulphur undergoes purification to eliminate impurities and moisture, ensuring its quality before transport to its final destination.

f. In summary, sulphur extraction from petroleum refining not only mitigates harmful sulphur compound emissions into the environment but also yields a valuable commodity with myriad industrial applications.

Summary

a. **Frasch Process:** Widely regarded as the most economical means of sulphur production, the Frasch process involves extracting sulphur from underground deposits by heating it with superheated water. Praised for its cost-effectiveness and environmental friendliness, this method generates minimal harmful emissions and demands less energy compared to other techniques.

b. **Claus Process:** Employed to produce sulphur from by-product gases in oil refineries and natural gas processing plants, the Claus process stands out for its energy efficiency and cost-effectiveness. Although it does emit some gases, it remains a favoured method for sulphur production.

c. **H_2S to Sulphur Conversion:** This method focuses on converting hydrogen sulphide (H_2S) into elemental sulphur. Renowned for its energy efficiency and economical nature, it does result in emissions, albeit to a lesser extent compared to other processes.

d. **Sulphur Recovery Units:** Widely utilised in the oil and gas sector, sulphur recovery units are instrumental in extracting sulphur from natural gas and other hydrocarbon streams. While considered energy-efficient and cost-effective, they do emit some gases during operation.

e. **Sulphur Burning:** This method involves the combustion of elemental sulphur or sulphide ores to produce sulphur. Regrettably, it is deemed the least cost-effective and environmentally hazardous of all methods due to its significant emissions and high energy consumption.

f. In summary, while each method has its advantages and drawbacks, ongoing efforts focus on minimising environmental impacts and maximising efficiency in sulphur production.

APPLICATION OF SULPHUR

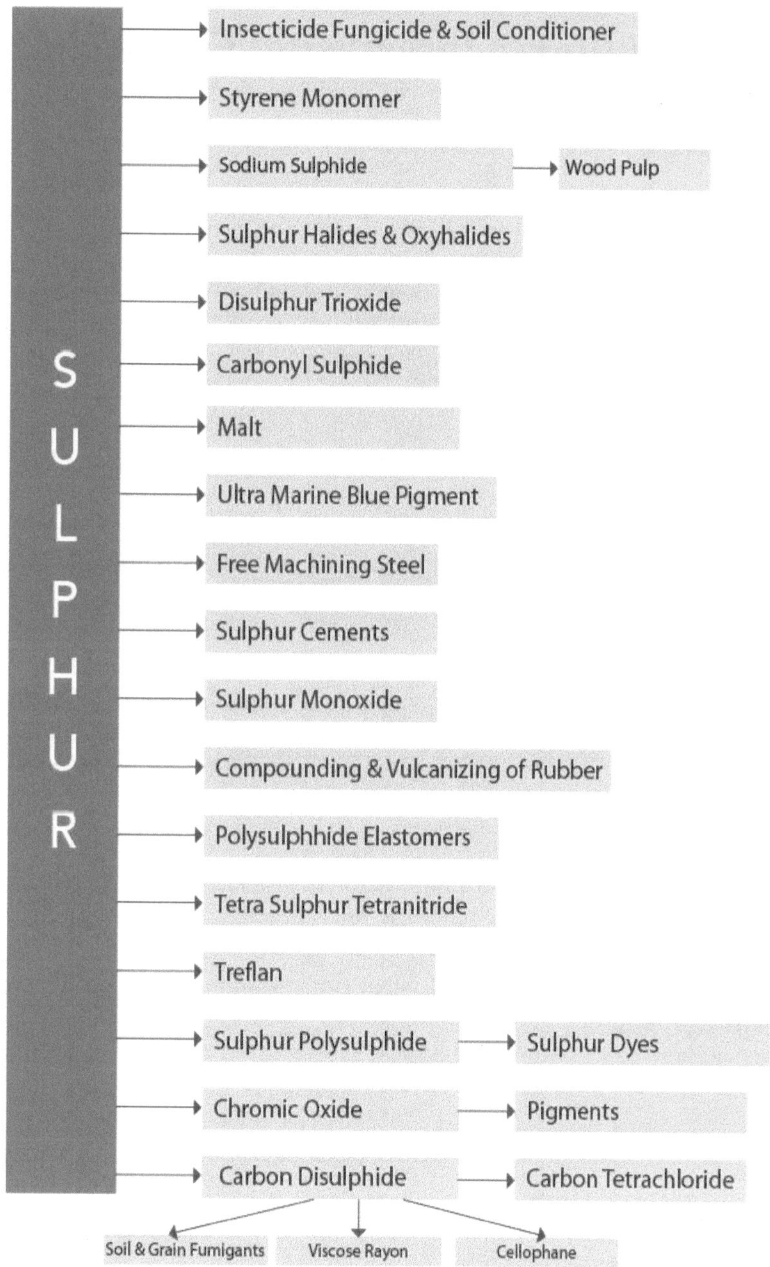

Sulphur serves as a fundamental resource for numerous chemical industries, primarily utilised in the production of sulphuric acid. This acid, in turn, forms the basis for manufacturing chemical fertilisers, textiles, dyestuffs, as well as facilitating processes such as pickling and galvanising of steel, refining petroleum, constructing storage batteries, producing explosives, and synthesising other acids.

1. Agriculture: Insecticide, Fungicide, and Soil Conditioner

In the realm of agriculture, sulphur finds multifaceted utility as an insecticide, fungicide, and soil conditioner:

a. **Insecticide:** Employed to combat a broad spectrum of pests, including mites, thrips, and specific beetle species, sulphur disrupts insect metabolism, impeding growth and reproduction. It operates primarily as a contact insecticide, necessitating direct interaction with targeted insects for efficacy.

b. **Fungicide:** Against fungal diseases like powdery mildew, black spot, and rust, sulphur proves effective by interfering with fungal metabolism, impeding growth and proliferation. Similar to its role as an insecticide, sulphur functions as a contact fungicide, requiring direct contact with the fungus for optimal performance.

c. **Soil Conditioner:** Sulphur contributes to soil improvement by enhancing both its physical and chemical properties. It aids in soil acidification, particularly beneficial for acid-loving plants such as blueberries and azaleas. Moreover, sulphur facilitates the breakdown of heavy clay soils, enhancing permeability and drainage.

Sulphur also serves as a vital nutrient for plants and is incorporated into fertilisers. Deficiency in sulphur can manifest in stunted growth, leaf discoloration, and diminished crop yields.

It's imperative to exercise caution when using sulphur-based products in agriculture, as overuse or improper application may lead to environmental harm and health risks. Always adhere to recommended dosage rates and safety measures outlined on product labels.

2. Styrene Monomer Production

While sulphur is not directly involved in the production of styrene, a crucial component in various plastics and synthetic materials, it can indirectly impact the process as a catalyst modifier or impurity in feedstock.

a. **Traditional Production Method :** The conventional approach to producing styrene entails the dehydrogenation of ethylbenzene, a process involving the removal of a hydrogen atom from the benzene ring. Though sulphur itself isn't directly utilised in this process, it can influence it in several ways.

b. **Sulphur as a Catalyst Modifier :** Sulphur-containing compounds can be integrated into catalytic systems as promoters or modifiers during the dehydrogenation of ethylbenzene. Compounds like alkyl disulphides or thiophenes have been investigated for their potential to enhance catalytic performance and stability, thereby improving overall styrene yield.

c. **Feedstock Impurities :** Furthermore, sulphur may exist as impurities within the feedstock used for styrene production. Ethylbenzene, the primary precursor, may contain trace amounts of sulphur compounds based on its source. These impurities can influence reaction kinetics and the efficacy of catalysts employed in the dehydrogenation process.

d. **Evolution of Production Methods :** Over time, styrene production methods have evolved, with alternatives like oxidative dehydrogenation being explored for enhanced efficiency and reduced environmental impact. These newer processes may involve different catalyst systems and reaction conditions, potentially altering the role of sulphur or sulphur-containing compounds in the production process.

e. **Conclusion :** In essence, while sulphur itself doesn't directly participate in styrene production, its presence can exert an indirect influence as a catalyst modifier or feedstock impurity. The specific impact of sulphur in styrene production hinges on the catalyst system and process utilised, underscoring the dynamic nature of industrial chemical synthesis.

3. Utilising Sulphur in Sodium Sulphide Production

Sulphur plays a pivotal role in the creation of sodium sulphide (Na_2S) through a chemical process known as the "sulphide process." In this method, sulphur reacts with sodium hydroxide (NaOH) to yield sodium sulphide and water. The overarching reaction is succinctly outlined as follows:

$S + 2NaOH \rightarrow Na_2S + H_2O$

This reaction typically occurs within a reactor maintained at elevated temperatures (approximately 150-200°C) and pressures (around 10-20 atmospheres) to enhance both the reaction kinetics and yield.

Sodium sulphide emerges as a yellowish to reddish-hued salt, possessing high solubility in water. It finds widespread utility across diverse industrial domains, including:

a. **Textile Industry**: Sodium sulphide serves as a fundamental component in the manufacture of rayon and cellophane, contributing to their production processes.

b. **Dye and Pigment Production**: Its role extends to the creation of vibrant dyes and pigments, enriching various applications in the colouring industry.

c. **Leather and Rubber Processing**: Sodium sulphide finds application in leather and rubber processing, aiding in the treatment and modification of these materials.

d. **Desulphurizing Agent**: Within the oil and gas sector, sodium sulphide serves as a potent desulphurizing agent, facilitating the removal of sulphur compounds from petroleum products.

e. **Chemical Synthesis**: It participates in the synthesis of both inorganic and organic compounds, contributing to diverse chemical manufacturing processes.

f. **Pulp and Paper Industry**: Sodium sulphide features prominently in the Kraft process utilised for pulping wood chips. It assists in fibre softening and lignin removal, crucial steps in paper production.

The multifaceted applications of sodium sulphide underscore its significance across various industrial sectors, from textiles to chemicals and beyond. Its versatility and efficacy make it an indispensable component in numerous manufacturing processes, highlighting the essential role of sulphur in facilitating industrial advancement.

10a. Enhancing Wood Pulp Production with Sodium Sulphide

Sulphur assumes a crucial role in the production of wood pulp, particularly in the renowned Kraft process. Widely acknowledged for its remarkable efficiency and effectiveness, the Kraft process, also termed the sulphate process, stands as the predominant method for pulping wood chips.

Within the Kraft process, wood chips undergo cooking in a solution comprising sodium hydroxide (NaOH) and sodium sulphide (Na_2S), meticulously designed to soften the fibres and eliminate lignin. This cooking process unfolds within a digester under conditions of elevated temperature and pressure.

The overarching chemical transformation can be succinctly encapsulated as follows:

$$C_6H_{10}O_5 + NaOH + Na_2S \rightarrow C_6H_{10}O_5 + Na_2SO_4 + H_2S$$

During this process, the lignin—a complex polymer responsible for binding cellulose fibres within wood—is effectively degraded by the combined action of NaOH and Na_2S. This breakdown of lignin facilitates the separation of fibres, paving the way for the production of pulp suitable for various applications, including paper, cardboard, and cellulose-based products.

Furthermore, the Kraft process yields sulphur dioxide (SO_2) as a by-product, which is diligently recovered and repurposed in the synthesis of sulphuric acid—a ubiquitous industrial chemical.

While the Kraft process boasts efficiency, its environmental footprint raises concerns due to pollutant emissions and substantial waste generation. Consequently, alternative pulping methods such as mechanical pulping, chemi-mechanical pulping, and Organosolv pulping have gained traction, offering more environmentally sustainable alternatives to traditional processes.

In essence, sulphur's indispensable role in the Kraft process underscores its significance in enhancing wood pulp production while necessitating a balanced approach to mitigate environmental impacts.

4. Halides and Oxyhalides Production

Sulphur serves a pivotal function in the synthesis of halides and oxyhalides, compounds crucial in various industrial applications. Halides consist of a halogen element (chlorine, bromine, iodine) combined with a metal, while oxyhalides encompass a blend of a halogen, oxygen, and a metal. The process of "halogenation" leverages sulphur's versatility to generate these compounds.

Sulphur undergoes transformation into sulphur dichloride (SCl_2) through reaction with chlorine gas (Cl_2) under catalytic conditions. This compound finds utility in the production of pesticides, fungicides, and serves as an intermediate in diverse chemical processes.

Additionally, sulphur monochloride (S_2Cl_2) can be derived from sulphur through reaction with chlorine gas, facilitated by a catalyst such as iron(III) chloride ($FeCl_3$). The balanced chemical equation illustrates this reaction:

$S_8 + 4Cl_2 + FeCl_3 \rightarrow 4S_2Cl_2 + FeCl_2$

Here, eight sulphur molecules (S_8) react with four chlorine molecules (Cl_2) in the presence of iron(III) chloride ($FeCl_3$) catalyst, yielding four molecules of sulphur monochloride (S_2Cl_2) and iron(II) chloride ($FeCl_2$) as a byproduct.

Furthermore, sulphur's interaction with chlorine gas and oxygen yields sulphuryl chloride (SO_2Cl_2), which finds application in pesticide production, fumigation, and serves as an intermediate in various chemical processes.

While the halogenation process offers invaluable synthetic pathways, it demands meticulous handling due to the reactive nature of chlorine gas, necessitating high temperatures and pressures. Diligent adherence to safety protocols is paramount to mitigate potential hazards associated with this process.

5. Disulphur Trioxide Production

Disulphur trioxide (S_2O_3) is a chemical compound synthesised through the oxidation of sulphur and sulphur dioxide (SO_2) with oxygen. Typically, this process is catalysed by substances like vanadium pentoxide (V_2O_5) and conducted at high temperatures ranging from 200 to 500 degrees Celsius.

The reaction is succinctly depicted as:

$SO_2 + S + \frac{1}{2} O_2 \rightarrow S_2O_3$

This production process is exothermic, meaning it releases heat. The thermal energy generated can be harnessed to facilitate additional chemical reactions or processes, thereby enhancing overall efficiency.

Disulphur trioxide finds wide-ranging applications across various industrial sectors. It serves as a crucial reagent in the synthesis of numerous chemicals, including sulphuric acid. Furthermore, it acts as a pivotal feedstock for the production of other chemicals such as sulphur trioxide (SO_3) and sulphuric acid. Additionally, it functions as an intermediate in the manufacture of fertilisers, dyes, pigments, and an array of other chemical compounds.

It's imperative to highlight that Disulphur trioxide is highly toxic and corrosive. Therefore, stringent safety measures and proper handling protocols are essential when working with this substance to ensure the well-being of personnel and the environment.

6. Carbonyl Sulphide Production

Carbonyl sulphide (COS) stands as a pivotal chemical compound with diverse industrial applications. It is synthesised through the reaction between carbon disulphide (CS_2) and oxygen, facilitated by a catalyst. The predominant method for COS production is the contact process, wherein CS_2 undergoes oxidation in the presence of a copper-chromium catalyst at elevated temperatures.

Raw materials for COS generation encompass carbon disulphide and oxygen. Under catalyst-mediated conditions, these compounds react to yield COS and carbon dioxide (CO_2). The reaction unfolds at high temperatures, typically ranging from 150 to 200 degrees Celsius, and under substantial pressures, typically between 50 and 100 atmospheres. Subsequently, the produced COS undergoes purification and isolation from other reaction byproducts.

COS finds multifaceted utility across industrial domains. It serves as a crucial reagent in organic synthesis and acts as a precursor for various chemicals. Additionally, it serves as a fundamental feedstock for the synthesis of carbon dioxide and sulphur dioxide.

Furthermore, COS exhibits notable reactivity with sulphur, engendering the formation of sulphur dioxide (SO_2) and carbon disulphide (CS_2) through the reaction:
$COS + 2S \rightarrow SO_2 + CS_2$

This exothermic reaction releases heat and occurs readily under catalyst-assisted conditions or at elevated temperatures. It is frequently employed in CS_2 production, a significant industrial chemical employed in the fabrication of synthetic fibres, rubber, and assorted products.

7. Impact on Malt Production

Sulphur-containing compounds exert a profound influence on the aroma, taste, and flavour of food and beverage items. Notably, dimethyl sulphide (DMS) stands as a significant contributor to the flavour profile of lager-type beers. While beneficial at concentrations between 30 and 100 μg/l in certain lagers, DMS originates from precursors like S-methylmethionine (SMM) and dimethyl sulphoxide (DMSO) found in malt.

Sulphur finds application in malt production primarily as a fungicide to curb the proliferation of unwanted microorganisms during the malting process. Typically administered as a gas, sulphur exposure precedes the germination and drying stages of barley, ensuring consistent malt quality and freedom from contaminants. Furthermore, sulphur dioxide (SO_2) serves as a preservative in beer brewing, thwarting microbial growth that could compromise beer quality.

Here's a breakdown of sulphur's involvement in malt production and usage:

a. **Malting Process:** Malting encompasses the germination and drying of cereal grains, primarily barley, to yield malt. Sulphur dioxide (SO_2) plays several vital roles throughout this process:

b. **Grain Preservation:** SO_2 acts as a fumigant, safeguarding stored grains against mould, fungi, and pests, thereby preserving barley quality pre-malting.

c. **Germination Control:** SO_2 regulates the germination phase, moderating enzyme activity, curbing excessive growth, and ensuring malt uniformity.

d. **Kilning:** Following germination, malted barley undergoes kilning to halt germination and cultivate desired flavours, hues, and aromas. SO_2 usage in kilning deters undesirable compound formation and preserves malt characteristics.

e. **Enzyme Inhibition:** During mashing, SO_2 inhibits specific enzymes that may jeopardise beer stability and flavour, contributing to consistent brewing outcomes.

f. **Antioxidant and Preservative:** With antioxidant properties, SO_2 preserves malt freshness during storage, staving off oxidation and extending shelf life.

g. **Flavour Enhancement:** Trace sulphur compounds derived from SO_2 contribute nuanced flavours to select beer styles, such as traditional German lagers, imparting subtle sulphur or "doughy" undertones.

h. In essence, sulphur's multifaceted role in malt production underscores its significance in shaping the quality and characteristics of brewed beverages.

8. Ultramarine Blue Pigment Production

Sulphur serves as a cornerstone in the creation of ultramarine blue pigment, a hue derived from the natural mineral lapis lazuli. The process of crafting ultramarine blue pigment entails meticulous steps, including sulphur's pivotal involvement, resulting in a vibrant blue suitable for a spectrum of applications across industries.

a. **Raw Materials**: The foundation of ultramarine blue pigment lies in a blend of kaolin clay, sodium carbonate (soda ash), sulphur, and silica (typically sourced from quartz). Sulphur stands out as a key player, furnishing essential sulphur atoms pivotal to the pigment's blue hue.

b. **Mixing and Heating**: The amalgamation of raw materials, including sulphur, occurs in carefully measured proportions, followed by exposure to intense heat within a furnace. Temperatures soar to a range of 700 to 800 degrees Celsius, laying the groundwork for chemical transformations.

c. **Reaction and Formation**: Within the crucible of heat, the "sulphur fusion process" unfolds. Sulphur engages in a chemical tango with other constituents, birthing sodium polysulphides. These intermediates then engage with the surrounding matrix, culminating in the birth of ultramarine blue pigment—a dazzling testament to sulphur's transformative prowess.

d. **Washing and Drying**: With the reaction concluded, the resulting blue marvel undergoes thorough washing to expunge impurities and residual materials. Post-cleansing, the pigment embarks on a drying journey, ultimately manifesting as the coveted ultramarine blue pigment ready for deployment in an array of artistic and industrial pursuits.

e. In essence, sulphur's indispensable contribution to ultramarine blue pigment synthesis underscores its pivotal role in bestowing upon us a timeless, captivating shade of blue.

9. Free Machining Steel

Sulphur serves as a key additive in steel production, particularly in the creation of free machining steel. This specialised alloy involves the deliberate inclusion of sulphur in modest proportions to enhance the material's machinability, referring to its ease of cutting, shaping, and refining.

Incorporating sulphur into steel results in the formation of minute inclusions within the material. These inclusions function as effective chip breakers during machining processes, preventing the formation of lengthy and cumbersome chips. By interrupting the chip flow, sulphur significantly improves the workability of the steel, thereby diminishing the risk of tool breakage and enhancing machining efficiency.

Typically introduced into steel in the form of a sulphur-bearing alloy, such as leaded steel, the sulphur content in free machining steel typically ranges between 0.15% and 0.35%. This carefully balanced addition ensures optimal machinability while preserving the steel's structural integrity and performance across various applications.

10. Sulphur Cement

Sulphur cement (SC) has emerged as an innovative solution for managing the surplus elemental sulphur generated by petroleum refineries. It stands as a sustainable and cost-effective construction material renowned for its robust chemical resistance.
The production of this thermoplastic material involves heating a blend of mineral aggregates and sulphur within a temperature range of 130−149 °C while continuously mixing. As the liquefied sulphur cools and solidifies, it effectively bonds the aggregates, resulting in the formation of sulphur concrete.

SC exhibits impressive compression strength, manageable shrinkage, and remarkable durability. A notable advantage lies in its water-free mixing and placement process, making it particularly suitable for arid and semi-arid regions where water resources are scarce. Furthermore, its sustainability credentials are bolstered by its cost-effectiveness and resilience against harsh environmental conditions, coupled with its high strength.

Sulphur cement, also known as sulphur concrete or sulphur polymer cement, is crafted by blending ground gypsum with small quantities of sulphur. Acting as a hardening accelerator, sulphur expedites the setting process compared to traditional Portland cement.

The manufacturing process entails heating gypsum to temperatures typically around 150-160°C in the presence of sulphur, inducing a chemical reaction that yields a durable solid material. This versatile cement finds applications across various sectors, including pre-fabrication, infrastructure construction, and as a binding agent in the oil and gas industry.

Central to the production of sulphur cement is elemental sulphur, augmented by additional components like fillers, aggregates, or plasticizers to tailor specific properties. The process unfolds through melting and mixing the sulphur, incorporating additives or fillers, followed by moulding into desired shapes and undergoing a curing phase to solidify and fortify the material.

11. Sulphur Monoxide Production

a. **Sulphur Mining**

- Elemental sulphur is extracted from sulphur deposits through mining operations.
- Sulphur deposits are often found in underground formations, salt domes, or extracted as a by-product of natural gas and oil refining.

b. **Sulphur Melting**

- Mined sulphur is melted to produce molten sulphur, typically by heating it to temperatures above its melting point of 115.21 °C.
- Molten sulphur is then stored or transported to the next processing facility.

c. **Sulphur Burning:** Molten sulphur is burned in the presence of oxygen to convert it into sulphur dioxide (SO_2) through the combustion reaction:

   a. $S + O_2 \rightarrow SO_2$

d. **Conversion to Sulphur Trioxide:** Sulphur dioxide is further oxidised to form sulphur trioxide (SO_3) using a catalyst such as vanadium pentoxide (V_2O_5) or platinum: $2SO_2 + O_2 \leftrightharpoons 2SO_3$

e. **Reduction to Sulphur Monoxide:** Sulphur trioxide is then reduced to form sulphur monoxide (SO) through suitable reduction reactions, such as:

   a. $SO_3 + H_2 \rightarrow SO + H_2O$

f. **Purification and Collection:**

- The produced (SO) gas is purified to remove any impurities or unreacted gases.
- Sulphur monoxide is then collected and stored for various industrial applications.

g. Applications of Sulphur Monoxide

- Chemical Synthesis: (SO) is used as a reagent in various chemical synthesis processes, including the production of sulphuric acid (H_2SO_4), sulfites, and other sulphur-containing compounds.
- Industrial Processes: It finds application in metallurgy, pharmaceuticals, and agricultural chemicals.
- Research and Development: Sulphur monoxide is also used in research laboratories for studying sulphur chemistry and developing new chemical processes.

h. Environmental Considerations:

- Emission Control: Effluent treatment measures should be in place to minimise the release of sulphur dioxide (SO_2) and other pollutants into the atmosphere, ensuring environmental compliance and reducing ecological impact.
- Waste Management: Proper disposal methods for sulphur waste and by-products must be implemented to prevent environmental contamination and ensure sustainability.
- Producing sulphur monoxide from sulphur deposits involves a combination of mining, processing, and chemical reactions, with careful consideration given to environmental and safety aspects throughout the production process.

12. Vulcanizing of Rubber

Sulphur plays a crucial role as a vulcanizing agent in both the compounding and vulcanization processes of rubber. Vulcanization enhances rubber properties by interlinking polymer chains with the aid of a vulcanizing agent like sulphur, boosting strength, durability, and resistance to heat and chemicals.

During compounding, sulphur is incorporated into rubber alongside fillers, pigments, and antioxidants in the form of a sulphur-based vulcanizing agent. This mixture undergoes heating and agitation to ensure even distribution of sulphur throughout the rubber matrix.

Vulcanization bestows rubber with its characteristic elasticity, achieved by blending latex with sulphur and subjecting it to heat through either pressure or free vulcanization methods. Pressure vulcanization, conducted under hydraulic press compression at 150°C, is common for articles moulded in hydraulic presses. Conversely, free vulcanization, utilised for continuous extruded products, employs steam or hot air when pressure vulcanization isn't feasible.

In the vulcanization process, rubber and the vulcanizing agent, heated to approximately 150-170°C, prompt sulphur atoms to establish cross-links with rubber molecules, forming a network of interconnected polymer chains and enhancing rubber properties.

The widespread use of sulphur in rubber production, notably in tires, hoses, belts, and seals, is attributable to its ability to bolster rubber's durability and resistance to wear and tear while elevating its overall performance.

13. Poly-Sulphide Elastomer

Sulphur plays a pivotal role in the production of polysulphide elastomers, also referred to as "thiokol" or "thiothene" elastomers. These materials are crafted by merging polysulphide polymers with a curing agent—typically a metal oxide or polyamine—in the presence of sulphur. The resultant substance boasts excellent chemical resistance, commendable compression set, and flexibility at low temperatures, rendering it apt for various industrial and aerospace uses, including seals, gaskets, and O-rings.

Here's a comprehensive look at sulphur's role in the manufacturing process of polysulphide elastomers:

a. **Polymerization**: The creation of polysulphide elastomers involves polymerization reactions. Usually, di-functional compounds like dichlorides or diols serve as the starting materials, featuring sulphur-containing groups at each end. These functional groups offer sites for cross-linking with sulphur atoms during polymerization.

b. **Cross-Linking with Sulphur**: Once the di-functional compounds are blended with curing agents and additives, sulphur is introduced into the mix. Sulphur atoms engage in a reaction with the sulphur-containing functional groups of the di-functional compounds via vulcanization or cross-linking. This process forms a three-dimensional network structure within the polymer, imbuing the material with elastomeric properties.

c. **Curing Process**: Post the sulphur and ingredient amalgamation, the mixture typically undergoes a curing process. This may entail heating the mixture to a specific temperature range, often with accelerators or activators added to facilitate the cross-linking reaction. The curing process fosters the bonding of polymer chains, culminating in the desired polysulphide elastomer.

d. **Application and Properties**: Once cured, the polysulphide elastomer exhibits desirable traits such as flexibility, chemical resistance, and resilience. It can be shaped, moulded, or extruded into diverse forms to meet specific application requisites. Polysulphide elastomers find widespread use in sealants, gaskets, coatings, adhesives, and other products where resistance to fuels, oils, chemicals, and weathering is imperative.

14. Tetra Sulphur Tetranitride

Tetrasulphur tetranitride (S_4N_4) is a chemical compound composed of four sulphur atoms and four nitrogen atoms. Its synthesis typically involves subjecting sulphur and nitrogen to high temperatures, often in a vacuum or under high pressure conditions. The resultant compound manifests as a yellow or orange solid, exhibiting high reactivity, and serving as a valuable reagent in organic synthesis. Moreover, its potential applications extend to the realm of materials science, as it demonstrates superconductivity at low temperatures. Nonetheless, S_4N_4 is exceedingly sensitive to air and moisture, necessitating meticulous handling precautions.

15. Treflan

Treflan, a herbicide utilised to manage weed growth in diverse agricultural settings, contains trifluralin as its active ingredient—a pre-emergence herbicide. Although sulphur is not directly engaged in Treflan's production, it plays a crucial role in trifluralin's synthesis.

Trifluralin is derived from various starting materials, one of which is thiophosphoryl chloride ($PSCl_3$). This compound, housing both phosphorus and sulphur, originates from phosphorus trichloride (PCl_3) reacting with elemental sulphur (S) as follows:

$PCl_3 + S \rightarrow PSCl_3$

Following this, thiophosphoryl chloride undergoes further transformation with 2,6-dinitrochlorobenzene in the presence of a base, yielding trifluralin:

$PSCl_3$ + 2,6-dinitrochlorobenzene + base → trifluralin

The precise Treflan production methodologies may vary based on the manufacturer and chosen synthesis route. However, the indirect involvement of sulphur in synthesising one of the crucial trifluralin precursors underscores its significance in the herbicide's overall production process.

16. Sodium Polysulphide

Sodium polysulphide (Na_2Sx) is a compound crucial for many industrial processes, and sulphur plays a pivotal role in its production. Formed by the reaction of elemental sulphur with sodium hydroxide (NaOH), sodium polysulphide finds wide application across various industries.

Here's an overview of its production process:

a. **Preparation:** Sodium polysulphide is typically prepared by dissolving sulphur in an aqueous solution of sodium hydroxide. The concentration and temperature of the solution can vary based on the desired properties of the sodium polysulphide.

b. **Mixing and Reaction:** Elemental sulphur is introduced into the sodium hydroxide solution, and thorough mixing ensures complete dissolution of the sulphur. This process triggers the formation of sodium polysulphide (Na_2Sx), where 'x' denotes the number of sulphur atoms in the polysulphide chain.

c. **Filtration and Purification:** Following the reaction, the sodium polysulphide solution may undergo filtration or purification steps to eliminate impurities or solid residues.

d. **Concentration and Storage:** The resultant sodium polysulphide solution may undergo concentration by water removal through evaporation or other methods. The concentrated solution is then stored or subjected to further processing based on specific application requirements.

e. Sodium polysulphide serves diverse roles across industries, including mining, water treatment, textile processing, and chemical manufacturing. It functions as a reducing agent, a sulphur source, and a flotation agent in mining operations, while also serving as an intermediate in the production of various chemicals. Its versatile nature makes it indispensable in industrial processes worldwide

16a. Sulphur Dye

Sulphur dyes constitute a class of dyes synthesised through the reaction of a diazo compound with a disulphide compound in the presence of sulphur. The resulting dye molecules incorporate sulphur atoms, imbuing them with vibrant colours. Renowned for their affordability, sulphur dyes find extensive use in dyeing cotton, wool, and other natural fibres, as well as synthetic fibres like polyester, nylon, and acrylic.

These dyes fall into two primary categories:

a. **Direct dyes:** These dyes are directly applied to fibres, subsequently fixed using reducing agents such as hydrosulfite or sodium dithionite.

b. **VAT dyes:** Initially applied to fibres in an oxidised state, these dyes undergo reduction to achieve their colour form, typically utilising sodium hydrosulfite or similar reducing agents.

Notably, sulphur dyes boast commendable light fastness and washing resistance. However, their susceptibility to chlorine bleach renders them unsuitable for fabrics subjected to regular bleaching.

17. Chromic Oxide

Chromic oxide (Cr_2O_3) comprises chromium and oxygen and is typically synthesised by subjecting chromium(III) oxide (Cr_2O_3) to high temperatures alongside sulphur. In this process, sulphur interacts with chromium(III) oxide, yielding chromium(VI) oxide, more commonly referred to as chromic oxide. This reaction unfolds within a furnace environment, typically maintained at temperatures ranging between 800-1000°C.

Resulting as a dark green powder insoluble in water, chromic oxide finds multifaceted applications. It serves as a pigment in ceramics, glass, and enamels, enhancing their coloration. Additionally, it functions as a polishing compound for both metals and glass surfaces. Moreover, chromic oxide acts as a catalyst in the production of various chemicals and features prominently in certain abrasive formulations.

Beyond industrial applications, chromic oxide finds utility in laboratory settings as an oxidising agent, prized for its stability and minimal toxicity. Furthermore, it serves as a vital reference material for calibrating spectrophotometers and other analytical instruments.

17a. Pigments

Sulphur plays a vital role in the production of various pigments, especially synthetic ones, utilised across industries.

Here are some examples:

a. **Sulphur Yellows**: These vibrant yellow pigments are created by combining arylide yellow (diarylide yellow) with sulphur in the presence of a catalyst. Widely employed in paints, inks, and plastics, they impart vivid hues.

b. **Sulphur Browns**: Produced by reacting a disulphide compound with a diazo compound in the presence of sulphur, these brown pigments find applications in paints, inks, and plastics, offering earthy tones.

c. **Sulphur Black**: Derived from burning elemental sulphur, this pigment yields a distinctive black powder. Valued in rubber production, printing inks, and paints, it undergoes a chemical process involving multiple steps. Additionally, sulphur black serves as a dye in textile industries.

d. **Sulphur Green**: Through a reduction process, sulphur transforms into sodium sulphide, yielding sodium polysulphides upon reaction with sodium carbonate. The resulting chromophores form a green ultramarine pigment, enriching colour palettes.

The production process typically involves several steps:

a. **Sulphur Dissolution:** Finely ground sulphur is blended with a sodium hydroxide solution, generating sodium polysulphides.

b. **Sulfonation:** The sodium polysulphide solution reacts with sulfonating agents, yielding intermediate sulphur compounds.

c. **Oxidation:** Intermediates undergo oxidation, commonly facilitated by agents like hydrogen peroxide or air.

d. **Neutralisation:** The resultant mixture is neutralised, often with mineral acids, achieving a pH near 7.

e. **Filtration and Drying:** Filtration eliminates impurities, followed by drying to obtain the final sulphur black dye.

While this overview simplifies the process, it's imperative to prioritise safety and adhere to proper protocols when handling chemicals like sulphur. Consulting chemical engineers and adhering to industry regulations is essential for commercial production. Sulphur pigments exhibit commendable light fastness and heat resistance but are susceptible to acids, alkalis, solvents, and sunlight-induced fading, rendering them unsuitable for outdoor applications.

18. Carbon Disulphide

The CS_2 manufacturing unit, established in 2010 at the Jhagadia plant in Gujarat, India, played a pivotal role in augmenting the production capabilities of various products at United Phosphorus Limited (UPL). CS_2 serves as a crucial raw material for manufacturing Mancozeb, WDG, and Antracol. In this process, hydrogen sulphide (H_2S) is produced as a by-product, which is either converted back to sulphur or utilised in the synthesis of sodium hydrosulphide (NaHS).

Carbon disulphide (CS_2) is a colourless, flammable liquid comprising carbon and sulphur atoms. Its production typically involves the reaction of carbon with sulphur at elevated temperatures, often catalysed by a catalyst. This reaction is known as the carbonylation of sulphur.

Another method for industrially producing carbon disulphide is the Friedel-Crafts process. This process entails the reaction between carbon monoxide and sulphur under high pressure and temperature, facilitated by a catalyst like aluminium chloride. The reaction yields carbon disulphide along with sulphur dioxide as a by-product.

Additionally, the Birkeland-Eyde process, primarily utilised for nitrogen compound synthesis like nitric acid, can be adapted for carbon disulphide synthesis from sulphur.

The process involves several steps:

a. **Sulphur Preparation:** Solid sulphur (S_8) is melted and purified to eliminate impurities. This purification process usually involves melting the sulphur and filtering it.

b. **Arc Furnace:** An electric arc furnace provides the requisite energy for the reaction. This furnace comprises two electrodes, typically constructed of carbon or graphite, connected to a low-voltage and high-current power supply.

c. **Reaction Chamber:** The arc furnace is enclosed within a reaction chamber designed to contain reactants and products. The chamber is sealed and equipped with appropriate gas inlet and outlet ports.

d. **Arc Ignition:** A high current at low voltage is applied across the electrodes to initiate the electric arc, generating temperatures of approximately 2000 ± 500 degrees Celsius.

e. **Sulphur Vaporisation:** The intense heat from the electric arc causes the sulphur to vaporise, transitioning from a solid to a gaseous state.

f. **Reaction with Carbon:** The sulphur vapour reacts with carbon from the electrodes to form carbon disulphide (CS_2) according to the chemical equation: $2S + C \rightarrow CS_2$.

g. **Collection and Separation:** The gaseous mixture containing carbon disulphide is collected and cooled to condense it back into liquid form. The condensed liquid is then separated from any unreacted sulphur and other by-products using various separation techniques.

Given the energy-intensive nature of the process and the toxicity of carbon disulphide, rigorous safety precautions are essential. Industrial-scale production of carbon disulphide necessitates strict adherence to safety protocols to safeguard workers and prevent environmental hazards.

Carbon disulphide serves diverse industrial purposes, including the manufacture of pesticides, rubber, rayon, and cellophane. Additionally, it finds applications in various organic chemical syntheses and serves as a solvent for cellulose pulp in rayon and cellophane production. Despite its utility, carbon disulphide requires careful handling due to its toxic and flammable nature.

18a. Carbon Disulphide use in Soil & Grain Fumigants

Carbon disulphide (CS_2) serves as a prevalent soil and grain fumigant, deployed to combat a range of pests including nematodes, fungi, insects, and mites. Employed as a soil fumigant, it is administered prior to planting to deter pests residing within the soil. Its inherent volatility facilitates deep penetration, ensuring thorough pest control. Similarly, in grain fumigation, CS_2 is applied to stored grain to eradicate insects and mites, safeguarding the grain from infestation.

The efficacy of carbon disulphide as a fumigant stems from its potent toxicity towards pests coupled with its rapid evaporation due to its low boiling point. This property enables it to swiftly permeate soil layers or grain structures, effectively eradicating pests. However, its toxicity poses significant risks to human health and animal well-being, prompting stringent regulation of its use. Concerns over its adverse environmental impact and high toxicity have initiated the gradual phasing out of carbon disulphide in soil and grain fumigation practices.

In response to these concerns, efforts are underway to develop alternative fumigants with lower toxicity profiles. Promising alternatives such as methyl bromide, chloropicrin, and 1,3-dichloropropene are being explored, offering effective pest control while mitigating the risks associated with carbon disulphide usage.

18b. Carbon Disulphide in Production of Viscose Rayon

Carbon disulphide (CS_2) plays a pivotal role in the production of viscose rayon, a prominent form of artificial silk. In the viscose process, cellulose, typically derived from wood pulp, undergoes dissolution in a solution of sodium hydroxide (NaOH) and carbon disulphide (CS_2), yielding a thick, syrupy liquid known as viscose. This viscose is then extruded through spinnerets—small metallic cups with multiple apertures—into a coagulating bath containing sulphuric acid. The acid bath induces solidification of the cellulose, resulting in the formation of fibres. Subsequently, these fibres are subjected to washing, stretching, and drying processes to yield viscose rayon.

The inclusion of carbon disulphide in the viscose process is indispensable, as it facilitates the dissolution and subsequent solidification of cellulose fibres, enabling the formation of thread-spinning fibres. However, it is imperative to acknowledge that carbon disulphide poses significant health and safety risks due to its highly toxic and flammable nature. Moreover, the disposal of waste generated during viscose rayon production can potentially lead to environmental contamination.

In response to these concerns, alternative methods for rayon production are actively being explored. For instance, the lyocell process offers a promising alternative by utilising a non-toxic solvent, N-Methylmorpholine N-oxide (NMMO), thereby mitigating health, safety, and environmental risks associated with carbon disulphide usage.

18c. Carbon Disulphide in Production of Cellophane

Carbon disulphide (CS_2) serves a vital role in the production of cellophane, a transparent film derived from cellulose. The process involves dissolving cellulose in a solution of sodium hydroxide (NaOH) and carbon disulphide (CS_2). This amalgamation yields a viscous solution, which is then extruded through a narrow aperture into a coagulating bath, typically comprising sulphuric acid. Here, the cellulose fibres congeal to form a film, which is subsequently stretched, dried, and wound onto rolls, resulting in cellophane.

The inclusion of carbon disulphide in cellophane production is indispensable, facilitating the dissolution and solidification of cellulose fibres to form a film that can be processed further. However, it's imperative to recognize the significant health and safety risks posed by carbon disulphide, given its highly toxic and flammable nature. Moreover, the disposal of waste generated during cellophane production can contribute to environmental pollution.

In response to these concerns, alternative production methods are under development, including the utilisation of non-toxic solvents like N-Methylmorpholine N-oxide (NMMO). Cellophane, renowned for its moisture-proof and gas-impermeable properties, finds extensive application in packaging various items such as food products, fruits, toffees, incense sticks, tea leaves, and grains, ensuring their freshness over extended periods.

18d. Carbon Disulphide in Production of Carbon Tetrachloride

Carbon tetrachloride (CCl_4) can be synthesised through the chlorination of carbon disulphide (CS_2) with chlorine gas (Cl_2) as the primary reactant. The chemical equation for this process is: $CS_2 + 3Cl_2 \rightarrow CCl_4 + S_2Cl_2$.

This reaction occurs under elevated temperatures and pressures in the presence of a catalyst, typically aluminium chloride ($AlCl_3$).

Carbon disulphide, a highly flammable liquid, serves as a precursor in the production of various chemicals, including pesticides. Conversely, carbon tetrachloride is a colourless liquid with a sweet, ether-like aroma. While it was once extensively used as a cleaning solvent and refrigerant, its usage has significantly decreased due to environmental and health concerns.

SUMMARY:

Sulphur, a highly versatile element, serves as a fundamental building block for various industrial applications through its derivatives. Let's explore the most prevalent sulphur derivatives:

a. **Sulphuric Acid (H_2SO_4):** As one of the foremost chemicals in global production and consumption, sulphuric acid finds extensive utilisation across multiple sectors. It contributes to the manufacture of fertilisers, dyes, detergents, and an array of chemical products. Additionally, it plays essential roles in petroleum refining, pigment production, and pharmaceutical manufacturing.

b. **Sulphur Dioxide (SO_2):** Functioning as a pivotal intermediate, sulphur dioxide plays a vital role in sulphuric acid synthesis. It is instrumental in the production of sulfites and bisulfites, serving as preservatives and reducing agents in diverse industrial processes.

c. **Sulphates:** Widely employed in various applications including fertilisers, detergents, and paper production, sulphates showcase versatile properties catering to diverse industrial needs.

d. **Sulphides:** With applications spanning lubricants, fungicides, and pigments, sulphides contribute significantly to industrial processes. Moreover, sulphur compounds serve as crucial intermediates in the synthesis of pesticides, dyes, and rubber accelerators.

e. **Elemental Sulphur:** Elemental sulphur finds application in industrial sectors such as fertiliser production and sulphuric acid manufacturing, reflecting its broad utility and importance.

f. **Conclusion**: The choice of sulphur derivative depends on specific application requirements and the availability of raw materials. Notably, sulphuric acid emerges as the most prevalent and indispensable chemical globally, underscoring its critical role in various industrial processes.

REFERENCES:

1. https://www.rsc.org/periodic-table/element/16/sulphur
2. https://pubchem.ncbi.nlm.nih.gov/element/sulphur
3. Sulphur: History, Technology, Applications & Industry
4. By Gerald Kutney ChemTec Publishing 2007
5. SULPHUR AND PYRITES Indian Minerals Yearbook 2020
6. (Part- III: MINERAL REVIEWS) 59th Edition, September 2021
7. Petrucci, Ralph H. General Chemistry, Principles & Modern Applications. Macmillan Publishing Company, Ninth Edition. Page 930-937.
8. Karchmer, J.H. The Analytical Chemistry of sulphur and its Compounds. New York: John Wiley & Sons, Inc., 1970.
9. Grant M. Young, George E. Williams, in Encyclopedia of Geology, Vol.5, Second Edition, 2021
10. https://www.sciencephoto.com/media/168265/view/volcanic-sulphur-deposits
11. Electrochemistry in mineral and metal processing 8 (Proceedings of the 8th International Symposium, From the 217th Meeting of The Electrochemical Society, Vancouver, Canada, April 26-28, 2010, ECS Transactions, Vol. 28, No. 6), F.M. Doyle, R. Woods, and G.H. Kelsall (editors), ECS, Pennington, NJ 2010.
12. Industrial Electrochemistry (2nd edition), D. Pletcher and F.C. Walsh, Kluwer, Dordreck, 1993.
13. https://www.wwdmag.com/what-is-articles/article/10940461/what-is-flue-gas-desulphurization

14. GELDENHUYS AJ, MAREE JP, DE BEER M and HLABELA P (2001) An integrated limestone/lime process for partial sulphate removal. Proc. Environmentally Responsible Mining Conference, South Africa. September. CSIR, Pretoria, South Africa.
15. National Energy Technology Laboratory https://netl.doe.gov/research/coal/energysystems/gasification/gasifipedia/sox-emissions
16. https://www.mckinseyenergyinsights.com
17. https://environment.govt.nz/facts-and-science/air/air-pollutants/sulphur-dioxide-and-effects-on-health
18. Recovery of sulphur dioxide from flue gas for sulphuric acid production Kamila Jóźwiak, Krzysztof Czajka, INTERNATIONAL SCIENTIFIC JOURNAL SCIENCE. BUSINESS. SOCIETY" WEB ISSN 2534-8485; PRINT ISSN 2367-8380
19. PRODUCTION OF CARBON DIsulphide FROM MOLTEN sulphur AND NATURAL GAS By Pururaj Singh Solanki, Ashutosh Nehete and Amogh Gawaskar
20. Preparation and characterization of ultramarine blue pigments from fly ash by using the X-ray photoelectron spectroscopy (XPS) for the determination of chemical states of sulphur in chromophores, Anna Škvarlová, Mária Kaňuchová, Microporous and Mesoporous Materials, Volume 284, August 2019, Pages 283-288
21. Effects of sulphur on yield and malting quality of barley, F.J. Zhao, S. Fortune, V.L. Barbosa, Journal of Cereal Science, Volume 43, Issue 3, May 2006, Pages 369-377
22. Integrated mixing machine for sulphur concrete production, Maisa El Gamal, Khaled El-Sawy, Abdel-Mohsen O. Mohamed, Case Studies in Construction Materials, Volume 14, June 2021

Chapter 2: Sulphur Dioxide: A Historical and Chemical Exploration

HISTORICAL INSIGHTS

- Ancient Rome: SO_2 found utility as a preservative, bleaching agent for wool, and as a fumigant.
- 18th Century: Swedish chemist Carl Wilhelm Scheele successfully isolated SO_2 in 1772.
- Early Scientific Exploration: English physicist and chemist Joseph Priestley pioneered the detailed study of SO_2, developing methods for gas collection over water.
- 19th Century: SO_2's industrial significance surged, especially in sulphuric acid production, pivotal for fertilizers, detergents, and dyes.
- 20th Century: SO_2 diversified its applications, serving as a refrigerant and gaining prominence in paper manufacturing and food preservation.
- Mid-20th Century: Growing environmental concerns spurred regulatory efforts to curtail SO_2 emissions from coal-fired power plants and industries.
- 21st Century: The gradual decline in SO_2 usage accompanies the adoption of eco-friendly alternatives and cleaner energy sources, though niche applications persist in products like wine and dried fruit.

CHEMICAL CHARACTERISTICS

- Physical Appearance: Colorless gas with a choking odor
- Boiling Point: -10°C
- Density: 2.619 g/L (heavier than air)
- Toxicity: Highly toxic upon inhalation; potential irritant to eyes and mucous membranes.

SUMMARY OF PROPERTIES

- Density: 2.619 g/L
- Boiling Point: -10°C
- Melting Point: -75.5°C

This summary encapsulates the historical journey, chemical attributes, and key properties of sulphur dioxide, delineating its varied roles across centuries.

SOURCES OF SULPHUR DIOXIDE

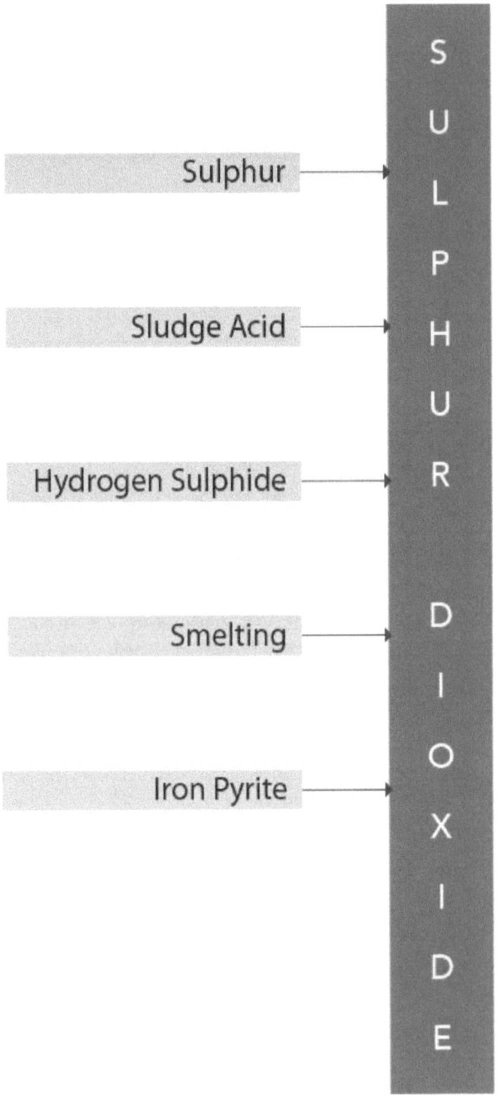

1. Production of Sulphur Dioxide from Sulphur

Sulphur dioxide (SO_2) is generated from sulphur through combustion or thermal oxidation processes. This transformation involves heating elemental sulphur to elevated temperatures in the presence of oxygen, leading to the formation of sulphur dioxide gas.

Reaction Equation: $S + O_2 \rightarrow SO_2$

(Exothermic reaction) -70.9 Kcal/mole

The specific method for producing sulphur dioxide from sulphur may vary, but one common approach involves burning sulphur in the presence of oxygen. The balanced chemical equation for this reaction is: $S (s) + O2 (g) \rightarrow SO_2 (g)$

In this reaction, solid sulphur (S) combusts in the presence of gaseous oxygen (O_2) to yield gaseous sulphur dioxide (SO_2). This reaction is exothermic, releasing heat energy.

Another prevalent method involves burning liquid sulphur (S) with oxygen (O_2) in a furnace, as represented by the equation: $S(l) + O_2 \rightarrow SO_2$

This exothermic reaction liberates a substantial amount of heat. The produced sulphur dioxide is commonly utilized in the manufacturing of sulphuric acid, a key industrial chemical.

2. Production from Sludge Acid

The Wellman–Lord process offers a regenerable solution for extracting sulphur dioxide from flue gas. In this method, sulphur dioxide present in flue gas undergoes absorption into a sodium sulphite solution dissolved in water, forming sodium bisulphite. Upon reducing the temperature, the bisulphite transforms into sodium pyrosulphite, which then precipitates.

Upon heating, these reactions reverse, leading to the conversion of sodium pyrosulphite into a concentrated stream of sulphur dioxide and sodium sulphite. The sulphur dioxide produced can be utilized for subsequent reactions, such as the manufacture of sulphuric acid, while the sulphite is reintegrated into the process.

In some desulphurization processes, flue gas derived from sludge can be purified by scrubbing it with a recirculating limestone slurry. This method effectively eliminates nearly 95% of the sulphur dioxide from the flue gas.

3. Production from Hydrogen Sulphide

Sulphur dioxide (SO_2) can be generated from hydrogen sulfide (H_2S) through an oxidation process. This method involves the reaction of H_2S with oxygen, facilitated by a catalyst such as a metal oxide. As a result of this reaction, SO_2 gas and water are produced as byproducts.

In the widely-utilized multistage Claus sulphur-recovery process, a segment of the H_2S present in the feed gas undergoes oxidation to form sulphur dioxide (SO_2) and water within a reaction furnace, using either air or enriched oxygen.

The produced SO_2 finds application in various industrial sectors, including the manufacturing of sulphuric acid and the formulation of preservatives for food and beverages.

4. Production from Smelting of Ores

Sulphur dioxide (SO_2) is a byproduct of the smelting process used for sulfide ores like zinc, copper, nickel, and lead. During smelting, the ores are heated to high temperatures, leading to the release of SO_2. This byproduct is often captured and transformed into sulphuric acid, a crucial chemical used across various industries, including fertilizer production, detergent manufacturing, and dye synthesis.

However, the emission of SO_2 during smelting poses significant environmental concerns, contributing to issues like acid rain and air pollution. To mitigate these impacts, strict regulations govern SO_2 emissions, prompting the implementation of advanced abatement technologies. Among these technologies, flue-gas desulphurization (FGD) systems stand out, as they effectively remove SO_2 from flue gasses before their release into the atmosphere. These measures help minimize the environmental footprint of ore smelting operations while ensuring compliance with environmental standards.

5. Production from Iron Pyrite

Sulphur dioxide (SO_2) is generated from iron pyrite (FeS_2), commonly known as "fool's gold", through a process involving roasting or burning the pyrite. When iron pyrite is exposed to heat, it undergoes a chemical reaction that releases SO_2 gas as a byproduct. This SO_2 can then find application in various industrial sectors, including the production of sulphuric acid and food and beverage preservatives.

The chemical equation representing the reaction of iron pyrite with oxygen to produce sulphur dioxide is:

$4\ FeS_2\ (s) + 11\ O_2\ (g) \rightarrow 2\ Fe_2O_3\ (s) + 8\ SO_2\ (g)$

This reaction illustrates how iron pyrite reacts with oxygen to yield iron oxide (Fe_2O_3) and sulphur dioxide (SO_2). The resulting sulphur dioxide can be harnessed for various industrial processes, highlighting the versatility of iron pyrite in sulphur dioxide production.

SUMMARY

Sulphur dioxide (SO_2) is a versatile compound used in various industrial applications, and its production methods vary depending on the source and intended use.

Some of the most common methods for producing sulphur dioxide include:

a. **Contact Process:** This method involves the combustion of sulphur with dry air to produce sulphur dioxide. It is the most widely used industrial method for SO_2 production.
b. **Lead Chamber Process:** An older method that involves the reaction of sulphur with nitrogen oxides in the presence of a catalyst to produce sulphur dioxide.
c. **Smelting of Sulfide Ores:** Sulphur dioxide is generated as a by-product during the smelting of sulfide ores like zinc, copper, nickel, and lead.
d. **Acid Sulphite Oxidation of Sludge:** This method entails the acidification of sludge to release sulphur dioxide.
e. **Desulphurisation of Sludge or Flue Gases:** Processes aimed at the recovery of sulphur dioxide from sludge or flue gases, often using scrubbing techniques.
f. **Burning of Iron Pyrite:** Iron pyrite, also known as "fool's gold," can be roasted or burned to release sulphur dioxide.

It's important to note that sulphur dioxide emissions contribute to acid rain and air pollution. Therefore, preventive measures, such as employing emission control technologies and regulatory frameworks, are crucial to limit its release into the environment and mitigate its harmful effects.

APPLICATION OF SULPHUR DIOXIDE SHEET 1

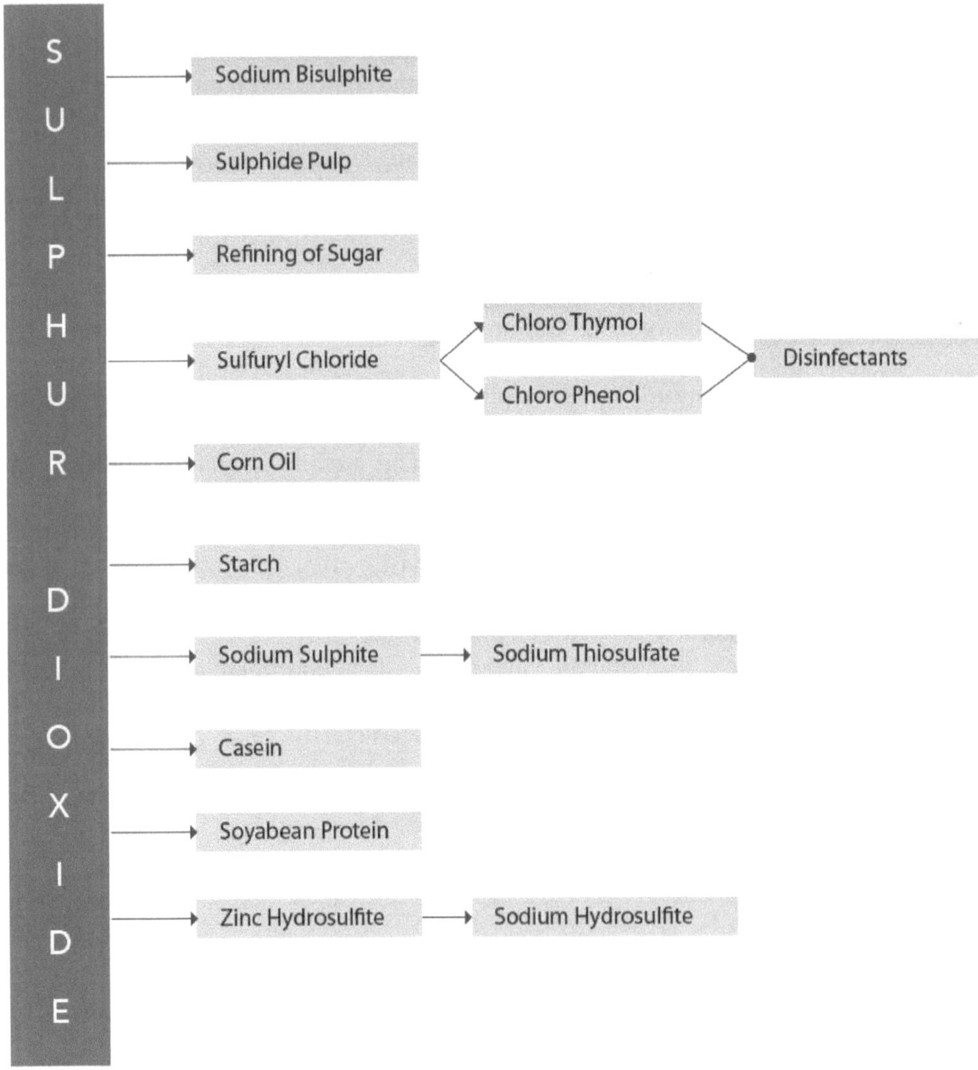

1. Sodium Bisulphite

Sodium bisulphite ($NaHSO_3$) is a versatile compound used as a reducing agent and preservative in various industrial applications. It can be produced from sulphur dioxide (SO_2) through several methods involving reactions with sodium hydroxide (NaOH) or sodium carbonate (Na_2CO_3).

Here are the main processes:

a. **Reaction with Sodium Carbonate**

- Sulphur dioxide (SO_2) is bubbled into a solution containing sodium carbonate (Na_2CO_3) and water (H_2O).
- The chemical reaction produces sodium bisulphite ($NaHSO_3$) and carbon dioxide (CO_2).
- Equation: $SO_2 + Na_2CO_3 + H_2O \rightarrow 2NaHSO_3 + CO_2$

b. **Reaction with Sodium Hydroxide**

- Sulphur dioxide (SO_2) is bubbled into a solution of sodium hydroxide (NaOH).
- Initially, sodium sulphite (Na_2SO_3) and water (H_2O) are formed.
- Then, the excess sulphur dioxide reacts with sodium sulphite to produce sodium bisulphite ($NaHSO_3$).

  Equations:
- $SO_2 + 2NaOH \rightarrow Na_2SO_3 + H_2O$
- $Na_2SO_3 + H_2O + SO_2 \rightarrow 2NaHSO_3$

c. **The resulting sodium bisulphite solution is used in various industries**

- Food and Beverage Industry: Acts as a preservative to prevent spoilage.
- Paper Industry: Used in bleaching pulp.

These production processes are essential for ensuring the availability of sodium bisulphite for industrial applications where its reducing and preserving properties are invaluable.

2. Sulphite Pulp

Sulphite pulping is a method commonly used in the papermaking industry to produce high-quality pulp. This process involves the reaction of sulphur dioxide (SO_2) and/or bisulphite ions with lignin, a component of wood, resulting in water-soluble sulfonated lignin.

Here's a detailed overview of the production process:

a. **Cooking**:
   - Wood chips are immersed in a sulphite liquor containing sulphur dioxide (SO2).
   - The mixture is heated under acidic or neutral conditions to facilitate the extraction of lignin, hemicellulose, and other impurities from the wood fibres.

b. **Bleaching:**
   - The cooked pulp undergoes a bleaching process where additional sulphur dioxide (SO_2) is used. This bleaching step further removes residual lignin and other impurities, resulting in a bright white pulp suitable for papermaking.
   - Washing: The bleached pulp is thoroughly washed to remove any residual sulphur dioxide (SO_2) and other chemicals used in the process.
   - Screening:The washed pulp is screened to remove any remaining large impurities and to achieve the desired pulp quality. The sulphite pulping process offers several advantages:
   - Cost-Effectiveness: It is relatively low-cost compared to other pulping methods.
   - Efficiency: The process efficiently extracts lignin and other impurities from the wood fibres.
   - Quality: It produces bright white pulp, suitable for high-quality paper products.
   - Environmental Impact: The use of sulphur dioxide (SO_2) helps to conserve energy and resources and reduces the environmental impact of pulping operations compared to alternative methods.

- Overall, sulphite pulping remains a popular choice in the papermaking industry due to its efficiency, cost-effectiveness, and ability to produce high-quality pulp while minimizing environmental impact.

3. Sugar Refining Enhanced by Sulphur Dioxide

Refining sugar is a meticulous process aimed at purifying raw sugar to yield pristine sugar crystals suitable for consumption. This refining journey involves various meticulous steps, including washing, filtration, crystallization, and centrifugation.

Sulphur dioxide (SO_2) plays a pivotal role in refining sugar, offering multifaceted benefits:

a. Bleaching Enhancement:

- Sulphur dioxide serves as a potent bleaching agent in sugar refining endeavors.
- It effectively eliminates impurities and colorants from the sugar, resulting in a brighter and whiter sugar product.
- The SO_2 actively interacts with specific compounds present in the sugar, such as reducing sugars and organic pigments, thereby breaking them down and significantly enhancing the overall appearance of the sugar.

b. Antimicrobial Protection and Preservation:

- Leveraging its antimicrobial properties, sulphur dioxide functions as a natural preservative during sugar refining.
- It actively inhibits the growth of microorganisms within the sugar, effectively thwarting spoilage and extending the shelf life of the refined sugar product.
    - pH Regulation:
- In certain refining scenarios, sulphur dioxide is judiciously employed to fine-tune the pH level throughout the sugar refining process.
- Its addition aids in reducing the alkalinity of the sugar solution, thereby fostering a more conducive environment for subsequent processing stages.
    - Oxidation Mitigation:
- Sugar refining operations encompass various stages that may inadvertently trigger oxidation, potentially jeopardizing the integrity of the sugar.

- Sulphur dioxide acts as a robust antioxidant, forming a protective shield around the sugar and safeguarding it against oxidative degradation, thus ensuring its sustained quality and stability.

In essence, the strategic integration of sulphur dioxide into the sugar refining process not only elevates the visual appeal of the sugar but also augments its longevity and overall quality, thereby enriching the consumer experience.

4. Sulphuryl Chloride Production and Its Derivatives

Sulphuryl chloride (SO_2Cl_2) serves as a vital precursor in various chemical synthesis processes. It is commonly synthesized from sulphur dioxide (SO_2) and chlorine gas (Cl_2) in the presence of a catalyst, as depicted in the balanced chemical equation:

$$2SO_2 + 2\,Cl_2 \xrightarrow{catalyst} 2SO_2Cl_2$$

The inclusion of a catalyst enhances the reaction rate, with activated carbon or other catalysts commonly employed to facilitate the reaction between SO_2 and Cl_2.

4a. Production of Chloro Thymol from Sulphuryl Chloride

Chlorothymol, a valuable compound with applications as a fungicide and bactericide, can be synthesized from sulphuryl chloride through a controlled chemical reaction. This process requires expertise and adherence to safety protocols due to the reactive nature of the compounds involved.

Synthesis Process:

a. **Chlorothymol production involves the following steps:**

   Reaction with Base: Sulphuryl chloride reacts with a base like sodium hydroxide to yield sodium chlorosulfonate and sodium hydroxide:

   $SO_2Cl_2 + 2NaOH \rightarrow NaHSO_3Cl + NaCl + H_2O$

- Hydrolysis: Sodium chlorosulfonate undergoes hydrolysis to produce chlorothymol and sodium sulphate:

  $NaHSO_3Cl + NaOH \rightarrow C_6H_7ClO + Na_2SO_4 + H_2O$

- Controlled Conditions: Optimal reaction conditions, including temperature and duration, are critical to ensure high yields of chlorothymol and minimize by-product formation. This process is typically conducted on an industrial scale under controlled conditions.

- Characteristics and Applications: Chlorothymol, a white crystalline solid with a thyme-like aroma, finds diverse applications in the chemical industry
- Fungicide and Bactericide: Effective in controlling mold and bacterial growth in food, feed, seed, cosmetics, and personal care products.
- Preservative: Used in paints and varnishes to inhibit fungal proliferation and extend product shelf life.

The production of chlorothymol from sulphuryl chloride involves a series of controlled chemical reactions, yielding a valuable compound with diverse industrial applications. Adherence to safety protocols and controlled conditions are imperative to ensure efficient synthesis and handling of chlorothymol.

4b. Production of Chlorophenol from Sulphuryl Chloride

Chlorophenol, a versatile compound utilized in various industries, can be synthesized from sulphuryl chloride through a controlled chemical process. This method requires precision and adherence to safety standards due to the reactive nature of the involved compounds.

a. **Synthesis Process:** Chlorophenol production involves the following steps:

- Reaction with Base: Sulphuryl chloride reacts with a base such as sodium hydroxide to form sodium chlorosulfonate and sodium hydroxide:
  $SO_2Cl_2 + 2NaOH \rightarrow NaHSO_3Cl + NaCl + H_2O$
- Hydrolysis: Sodium chlorosulfonate undergoes hydrolysis to yield chlorophenol and sodium sulphate:
  $NaHSO_3Cl + NaOH \rightarrow C_6H_5ClO + Na_2SO_4 + H_2O$

b. **Characteristics and Applications:** Chlorophenols, characterized by a phenol ring structure with chlorine atoms, find diverse applications:

- Wood Preservatives: Effective in protecting wood from decay and termite infestation.
- Herbicides: Used to control weed growth in agricultural and forestry applications.
- Disinfectants: Employed in sanitation and water treatment processes for microbial control.

- Chemical Intermediates: Serve as intermediates in the synthesis of various chemicals.

c. **Safety Considerations:** While chlorophenols have widespread utility, some variants may pose health risks if mishandled:
- Toxicity: Certain chlorophenols may exhibit toxicity and require careful handling to mitigate health hazards.
- Professional Expertise: Production processes should be overseen by trained professionals equipped with appropriate safety gear and facilities.

The production of chlorophenol from sulphuryl chloride involves controlled chemical reactions, yielding a valuable compound with diverse industrial applications. Adherence to safety protocols and controlled conditions are paramount to ensure efficient synthesis and safe handling of chlorophenol.

4c. Production of Disinfectants from Chloro Thymol & Chlorophenol

Both chloro thymol and chloro phenol serve as key components in disinfectant formulations. By reacting with a strong base like sodium hydroxide, they yield the corresponding sodium salts, which serve as effective disinfectants. The production process involves purification, concentration, and formulation steps, tailored to achieve desired disinfectant properties.

In conclusion, sulphuryl chloride serves as a versatile precursor in the synthesis of chloro thymol, chloro phenol, and subsequently, disinfectants. Adhering to stringent safety protocols and precise reaction control ensures optimal yield and quality of the derived compounds, thus underlining the importance of professional expertise and careful handling in industrial-scale production.

5. Corn Oil Production

Corn oil, a valued edible oil extracted from corn kernels, undergoes a meticulous production process, with sulphur dioxide (SO_2) contributing to certain crucial stages:

a. **Steeping Corn:**
- Initial Soaking (Steeping): Corn kernels are immersed in water at approximately 50°C (122°F) for 30-40 hours. This process, known as steeping, serves to soften the kernels and enhance their permeability.

- Sulphur Dioxide Addition: Sulphur dioxide is introduced at very low levels (around 0.1%) during steeping. Its key functions include:
- Controlling undesirable bacterial growth.
- Facilitating swelling and softening of the corn kernel.
- Increasing permeability of corn hulls.

b. **Corn Oil Extraction:**

- Germ Separation: The germs within the corn kernel, containing approximately 85% of the oil, are separated post-steeping.
- Washing and Oil Extraction: Germs undergo repeated washing to eliminate residual starch. Mechanical and solvent processes are then employed to extract oil from the germs.
- Refining and Filtration: Extracted oil undergoes refining and filtration processes to yield finished corn oil.

c. **High Oil Corn:**

- Oil-Rich Varieties: Corn varieties with an oil content exceeding 6% are categorized as high oil corn.
- Oil Distribution: The germ region of corn seeds contains the bulk of the oil (80-84%), followed by the aleurone layer (12%) and endosperm (5%).

d. **Health Benefits of Corn Oil:**

- Culinary Versatility: Corn oil is favored for cooking methods like frying and sautéing.
- Rich in Unsaturated Fats: It is abundant in both polyunsaturated (PUFA) and monounsaturated fatty acids (MUFA), contributing to its nutritional value.
- High Oil Content: Among grain crops, corn boasts one of the highest oil contents, rivaling oats in this regard.

In summary, sulphur dioxide plays a vital role in corn oil production, aiding in the preparatory steeping process by facilitating kernel softening and hull permeability. Its controlled addition ensures the extraction of high-quality oil, contributing to the nutritional and culinary value of corn oil as a preferred cooking medium.

6. Production of Starch

Sulphur dioxide (SO_2) finds application in the production of starch, particularly in processes such as corn starch production. Here's a detailed look at its role and impact:

a. Sulphur Dioxide in Steeping Process:

- Steeping Enhancement: Sulphur dioxide is introduced at minimal concentrations (0.1%) during the steeping phase of starch production.
- Bacterial Control: Its presence helps control undesirable bacterial growth, ensuring a sanitary environment for starch production.
- Kernel Permeability: Sulphur dioxide aids in increasing the permeability of corn hulls, facilitating the release of starch during subsequent processing stages.
- Softening Effect: By promoting swelling and softening of dense corn kernels, sulphur dioxide assists in the extraction of starch.

b. Sulphur Dioxide in Corn Starch Production:

- Bleaching Agent: In the production of corn starch, sulphur dioxide serves as a bleaching agent. It targets residual proteins and impurities, enhancing starch purity and translucency.
- Maillard Reaction Inhibition: Sulphur dioxide mitigates the formation of undesired Maillard reactions, which can lead to discoloration and off-flavors in the final starch product.
- Regulatory Compliance: Its usage in starch production is strictly regulated by food safety authorities to ensure consumer safety. Stringent guidelines are enforced to guarantee the final starch product meets safety standards for human consumption.

In the starch production process, sulphur dioxide plays a multifaceted role, contributing to bacterial control, kernel permeability enhancement, and starch quality improvement. Its judicious application, coupled with regulatory oversight, ensures the production of high-quality starch products that meet stringent safety and quality standards for consumers.

7. Sodium sulphite from Sulphur Dioxide

Sodium sulphite (Na_2SO_3) is a key chemical compound produced from sulphur dioxide (SO_2) through a series of chemical reactions.

Let's explore this process in detail:

a. **Sulphur Dioxide Oxidation:**

- Initial Reaction: Sulphur dioxide undergoes oxidation to form sulphur trioxide (SO_3) in the presence of oxygen (O_2):

  $SO_2 + \frac{1}{2} O_2 \rightarrow SO_3$

b. **Sodium sulphite Synthesis:**

- Reaction with Sodium Carbonate: The produced sulphur trioxide then reacts with sodium carbonate (Na_2CO_3) to yield sodium sulphite (Na_2SO_3) along with the liberation of carbon dioxide (CO_2):

  $SO_3 + Na_2CO_3 \rightarrow Na_2SO_3 + CO_2$
- Reaction Medium: This reaction can occur either in the gas phase or in an aqueous solution, depending on the production setup.
- Industrial Process: Typically, the reaction is carried out continuously in industrial plants to ensure efficient and continuous production of sodium sulphite.

c. **Applications of Sodium sulphite:**

- Textile Bleaching Agent: Sodium sulphite finds extensive use as a bleaching agent in the textile industry, where it helps remove colorants and impurities from fabrics.
- Chemical Reducing Agent: It serves as a vital reducing agent in the synthesis of various chemicals, facilitating the conversion of other compounds in industrial processes.
- Food Preservative: Sodium sulphite is employed as a food preservative, where its antioxidant properties help extend the shelf life of perishable food products.

7a. Production of Thiosulphate from Sodium sulphite

The synthesis of thiosulphate ($S_2O_3^{2-}$) from sodium sulphite (Na_2SO_3) involves chemical reactions that are crucial in various industrial applications:

a. **Reaction Mechanism:** Sodium sulphite reacts with either sulphur or sulphur dioxide to produce sodium thiosulphate ($Na_2S_2O_3$):

   $3\ Na_2SO_3 + 2S \rightarrow 2\ Na_2S_2O_3$

   $4\ Na_2SO_3 + 2SO_2 + O_2 \rightarrow 3\ Na_2S_2O_3$

b. **Industrial Variation:** Industrial processes may utilize different reaction conditions and catalysts to optimize reaction efficiency and product purity.

c. **Applications:** Sodium thiosulphate serves various functions, including photographic fixing, water treatment, and medical applications.

Through these reactions, sodium sulphite serves as a precursor for the production of sodium thiosulphate, contributing to a diverse range of industrial processes and applications.

8. Utilization in Casein Production

Casein, a prominent protein found in milk, contributes significantly to the nutritional profile and functional properties of dairy products. With its gel-forming ability under acidic conditions, casein plays a crucial role in various food and industrial applications.

a. **Casein in Food and Industry:**

- Nutritional Component : Casein constitutes approximately 80% of the total protein content in cow's milk, making it a valuable nutrient source in dairy products.
- Functional Properties : Its unique gelation properties render casein suitable for diverse applications, including cheese production, dairy-based desserts, and nutritional supplements.
- Industrial Applications : Beyond food, casein finds use in adhesives, coatings, and plastics due to its capacity to form robust gels when combined with other compounds.

b. **Role of Sulphur Dioxide in Casein Production:**

Sulphur dioxide (SO_2) serves as a vital component in the production of dairy products, particularly in cheese-making, where it fulfills several essential functions:

- Preservative Action: Sulphur dioxide acts as a preservative, helping to inhibit the growth of bacteria and microorganisms in cheese. By controlling microbial proliferation, it safeguards the cheese from spoilage and off-flavors.
- Regulatory Compliance: The utilization of sulphur dioxide in cheese production is subject to stringent regulations enforced by food safety agencies. These guidelines ensure that the concentration of sulphur dioxide in the final product remains within safe limits for consumption.
- Allergen Considerations: It's important to recognize that some individuals may exhibit sensitivity or allergic reactions to sulphur dioxide. Hence, its use in

food production necessitates careful consideration and adherence to labeling requirements to alert consumers of potential allergens.

9. Soybean Protein Preservation

Soy protein is a valuable component of many food products due to its nutritional benefits and versatile functional properties. It serves as a crucial ingredient in various food items, including plant-based meat substitutes, protein bars, and dairy alternatives.

a. **Sulphur Dioxide in Soybean Protein Preservation:**

- Preservative Function: Sulphur dioxide (SO_2) is employed sporadically as a preservative in soy protein products. Its primary role is to preserve the color, flavor, and overall quality of the product by inhibiting microbial growth.
- Spoilage Prevention: By deterring the proliferation of bacteria and fungi, sulphur dioxide helps extend the shelf life of soy protein products, thereby reducing the risk of spoilage and enhancing product stability.
- Health Considerations: While sulphur dioxide aids in product preservation, excessive usage can pose health risks to consumers. Some individuals may experience adverse effects such as headaches, breathing difficulties, and allergic reactions upon consumption of soy products containing high levels of sulphur dioxide.
- Worker Safety: Furthermore, sulphur dioxide is recognized as an irritant and may cause respiratory issues among workers involved in its handling and processing. Thus, stringent safety measures and protective equipment are essential to mitigate occupational hazards.
- Exploration of Alternatives: In response to health concerns, some companies are exploring alternative preservation methods or employing reduced levels of sulphur dioxide in soy protein processing. These strategies aim to minimize health risks while maintaining product quality and safety.

10. Zinc Hydrosulfite Production

Zinc hydrosulphite, a vital compound in various industries, serves as both a reducing agent and a bleaching agent. Its versatility finds applications in textile, paper, and photographic sectors, contributing to the production of high-quality materials.

Production Process:

a. **Reactant Role of Sulphur Dioxide:** Sulphur dioxide (SO_2) plays a pivotal role as a reactant in the synthesis of zinc hydrosulphite. The production process involves the reaction of sulphur dioxide with zinc oxide in an aqueous medium. This reaction forms zinc hydrosulphite and water. Maintaining precise pH levels in the solution is crucial for achieving optimal product yield.
b. **Manufacturing Method:** A commonly employed method entails absorbing sulphur dioxide gas into a slurry of zinc dust and water at controlled temperatures, typically around 40-50°C. This method ensures efficient absorption and reaction kinetics.
c. **Chemical Reaction:** The chemical equation representing the reaction of sulphur dioxide in zinc hydrosulphite production is:
$Zn + 2SO_2 + 4H_2O \rightarrow Zn(S_2O_4)(H_2O)_4$

This reaction yields zinc hydrosulphite as the primary product, accompanied by the generation of hydrogen gas as a byproduct.

10a. Utilization in Sodium Hydrosulphite Production

a. **Role as a Starting Material:** Zinc hydrosulphite serves as a key starting material in the synthesis of sodium hydrosulphite, another essential industrial compound. Sodium hydrosulphite is renowned for its applications as a potent reducing and bleaching agent across diverse industrial processes.

b. **Production Process:** The production of sodium hydrosulphite involves the reaction of zinc hydrosulphite with sodium carbonate. This reaction yields sodium hydrosulphite and zinc carbonate. Similar to zinc hydrosulphite production, maintaining precise pH levels is imperative to ensure optimal reaction kinetics and product purity.
c. **Industrial Significance:** Utilizing zinc hydrosulphite as a starting material enables the production of sodium hydrosulphite with consistent purity and quality, rendering it suitable for various industrial applications demanding high performance and reliability.

Sulphur dioxide plays a crucial role in the production of zinc hydrosulphite, facilitating its synthesis through a carefully controlled chemical reaction. The resulting zinc hydrosulphite serves as a fundamental building block for the production of sodium hydrosulphite, further extending its utility across a broad spectrum of industrial processes.

APPLICATION OF SULPHUR DIOXIDE SHEET 2

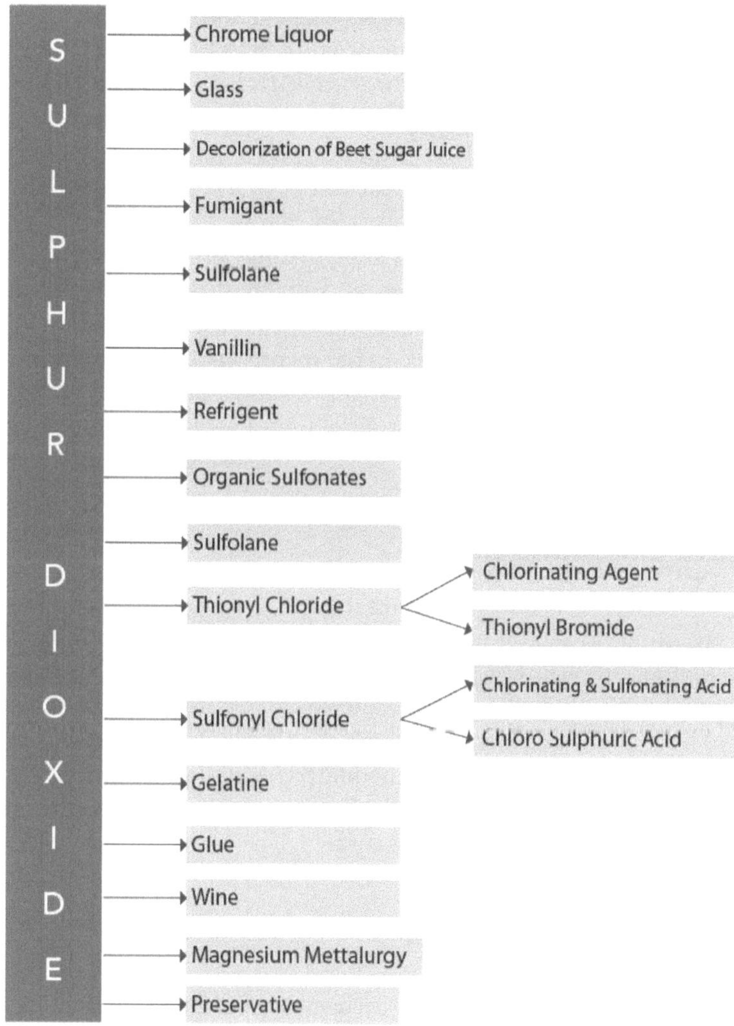

11. Chrome Liquor Production

Chrome liquor, a pivotal component in leather tanning, comprises a solution of sodium dichromate utilized to impart specific properties to leather. Its role extends to enhancing durability, color fixation, and resistance to heat and moisture, contributing to the production of high-quality leather goods.

a. **Incorporation of Sulphur Dioxide:**

- pH Adjustment and Oxidation: Sulphur dioxide (SO_2) finds application in chrome liquor production primarily for pH adjustment and oxidation purposes. It aids in the conversion of chromium(III) ions to chromium(VI) ions, a crucial step in enhancing the solubility and reactivity of chromium in the tanning process.
- Reaction Mechanism: The reaction between chromium(III) ions and sulphur dioxide leads to the formation of chromium(VI) ions. These ions exhibit greater solubility in water, facilitating their availability for subsequent utilization in the tanning process. Sulphur dioxide's role in oxidation enhances the efficiency and effectiveness of the chrome liquor solution.
- Enhanced Process Efficiency: The incorporation of sulphur dioxide in chrome liquor production contributes to the optimization of the tanning process. By promoting the conversion of chromium ions to a more soluble form, sulphur dioxide enhances process efficiency, ensuring consistent leather quality and meeting industry standards.

b. **Chrome Liquor Synthesis Process:**

- Chemical Reaction: Chrome liquor is typically synthesized through the reaction of sodium dichromate ($Na_2Cr_2O_7$) or chromium trioxide (CrO_3) with sodium sulphate (Na_2SO_4) or sulphuric acid (H_2SO_4). This reaction yields basic chromium sulphate, a crucial precursor in chrome liquor production.
- Dissolution in Water: Following the formation of basic chromium sulphate, it is dissolved in water to generate the chrome liquor solution. This solution serves as a vital ingredient in leather tanning processes, imparting desirable properties to the leather substrate.

12. Glass Production

Glass, a ubiquitous material with diverse applications ranging from architectural structures to optical devices, undergoes meticulous manufacturing processes to achieve desired properties. Sulphur dioxide (SO_2) assumes a crucial role in glass production, particularly in refining, decolorization, and oxidation state control.

a. **Refining Agent:**
Impurity Removal: In the glass melting process, raw materials may contain impurities like iron oxide, which can compromise the clarity and quality of the final product. Sulphur dioxide acts as a refining agent, reacting with impurities to form sulphates that are easily extractable from the molten glass, thereby enhancing clarity and purity.

b. **Decolorization:**
Color-Causing Impurities Reduction: Certain glass formulations, especially those intended for colorless or low-iron glass, require the removal of color-causing impurities like iron ions. Sulphur dioxide serves as a reducing agent, facilitating the reduction or removal of these impurities to achieve a clearer and more transparent glass product.

c. **Oxidation State Control:**
Maintaining Optimal Conditions: The oxidation state of glass melt constituents significantly influences the glass's properties and appearance. Sulphur dioxide aids in controlling the oxidation state within the glass furnace, ensuring the maintenance of optimal conditions for desired glass characteristics, including color and transparency.

d. **Batch Adjustment:**
Customized Glass Properties: Sulphur dioxide is occasionally incorporated into glass batch compositions to fine-tune the chemical composition and tailor specific properties of the final glass product. Its addition can influence parameters such as viscosity, refractive index, and other physical attributes, allowing for the production of glass with customized characteristics.

13. Decolorization of Beet Sugar Juice

Beet sugar juice extraction from sugar beets initiates a series of purification steps to eliminate impurities and color compounds, ensuring the production of high-quality sugar products. Sulphur dioxide (SO_2) serves as a crucial decolorizing agent in this process, enhancing the clarity and purity of beet sugar juice.

a. **Decolorization Process:**

- Impurity Removal: The decolorization step targets the removal of natural pigments like anthocyanins and other colored substances from the beet sugar juice, thereby enhancing its visual appeal and quality.

b. Sulphur Dioxide Addition:

- Introduction of Sulphur Dioxide: Sulphur dioxide, either in gaseous form or as a sodium metabisulphite ($Na_2S_2O_5$) solution, is introduced into the beet sugar juice. Acting as a potent reducing agent, sulphur dioxide interacts with the color-causing compounds present in the juice.

c. Reaction with Color Compounds:

- Chemical Reduction: Sulphur dioxide initiates a chemical reduction process upon contact with the colored compounds. This reaction effectively breaks down the chromophores responsible for the juice's color, resulting in a lighter and more translucent juice.

d. pH Adjustment:

- Acidic Environment Creation: Concurrently, sulphur dioxide addition lowers the pH of the juice, inducing an acidic environment. This acidic pH facilitates the decolorization process, enhancing the stability and efficacy of sulphur dioxide.

e. Filtration and Separation:

- Solid Impurity Removal: Following sulphur dioxide treatment, the decolorized beet sugar juice undergoes filtration or centrifugation to separate solid impurities and residual color precipitates. This step ensures the production of a clarified juice, ready for subsequent sugar crystallization.

14. Fumigant

Fumigants play a vital role in agricultural and industrial settings by controlling pests and pathogens. Sulphur dioxide (SO_2) is a key ingredient in the production of fumigants, contributing to their effectiveness in safeguarding stored food products and controlling plant diseases.

a. Role of Sulphur Dioxide in Fumigants

- Pest and Pathogen Control: Sulphur dioxide acts as a potent fumigant capable of penetrating deeply into stored grain, food products, and soils. Its ability to reach inaccessible areas ensures effective eradication of pests and pathogens, thereby preserving the safety and quality of stored commodities.

b. Applications of Fumigants

- Stored Grain Preservation: Fumigants containing sulphur dioxide are extensively used to control pests in stored grain and other food products. By targeting pests at various life stages, sulphur dioxide-based fumigants prevent infestations and maintain the integrity of stored food supplies.

- Soil Treatment: In agricultural practices, fumigants are applied to soil to control plant diseases and nematodes. Sulphur dioxide's broad spectrum of activity enables it to combat diverse pathogens present in the soil, thereby promoting crop health and productivity.

c. Environmental and Health Considerations

- Regulatory Measures: While sulphur dioxide-based fumigants are effective, their use can pose environmental and health risks. Many countries have established regulations to govern the application of sulphur dioxide as a fumigant, aiming to mitigate adverse impacts on human health and the environment.
- Alternative Solutions: In response to environmental concerns, research efforts are directed towards developing alternative fumigation methods that are safer and more sustainable. These initiatives focus on reducing reliance on sulphur dioxide and exploring eco-friendly alternatives to meet pest and pathogen control needs.

15. Sulfolane

a. Production Process:

- Conversion of Sulphur Dioxide: Sulphur dioxide reacts with water to form sulphurous acid.
- Oxidation to sulphite: Sulphurous acid is oxidized with oxygen to produce sulphite.
- Conversion to Sulfolane: sulphite reacts with ethylene oxide to yield sulfolane.
- Purification: Sulfolane undergoes distillation for purification, ensuring high quality and purity.

b. Alternative Production Method:

Sulfolane can also be synthesized by reacting butadiene with sulphur dioxide and hydrogen sulfide. This process involves oxidation, condensation, and hydrogenation reactions, tailored to meet specific purity requirements and industrial standards.

c. Applications of Sulfolane:

- Solvent: Sulfolane serves as a high-boiling solvent in various industrial processes due to its exceptional solvency power. Its high boiling point and low vapor pressure make it suitable for applications requiring elevated temperatures and minimal evaporation.

- Gas Treating: In gas sweetening processes, sulfolane effectively removes undesirable sulphur compounds like hydrogen sulfide from natural gas streams, ensuring compliance with environmental regulations.
- Extraction and Purification: Sulfolane is utilized in extraction and purification processes to separate aromatic hydrocarbons, extract compounds from aqueous solutions, and purify chemical intermediates and final products.
- Polymer Processing: In the polymer industry, sulfolane acts as a solvent and plasticizer, facilitating processing, blending, and film formation of polymers.
- Pharmaceuticals and Fine Chemicals: Sulfolane serves as a reaction medium and solvent in the synthesis of pharmaceuticals and fine chemicals, offering unique properties for specific chemical reactions and separations.

16. Vanillin: A Synthetic Approach

a. **Synthesis from Sulphur Dioxide:**

- Conversion of Sulphur Dioxide: Sulphur dioxide reacts with water to form sulphurous acid.
- Oxidation to sulphite: Sulphurous acid undergoes oxidation with oxygen to yield sulphite.
- Conversion to Vanillin: sulphite is subjected to a series of chemical reactions, including oxidation and reaction with guaiacol, ultimately yielding vanillin.
- Purification: The synthesized vanillin undergoes purification processes involving distillation and crystallization to ensure high quality and purity.

b. **Commercial Production Methods:**

- Natural Extraction: Vanillin is traditionally extracted from vanilla beans through solvent extraction or steam distillation methods. However, this approach is costly and environmentally challenging.
- Chemical Synthesis: Synthetic vanillin is produced on a large scale using chemical synthesis methods. One common approach involves the conversion of guaiacol, derived from lignin or other sources, into vanillin through oxidation and methylation reactions.
- Biotechnological Production: Vanillin can also be synthesized via biotechnological processes using microbial fermentation. Microorganisms such as yeasts or bacteria are employed to convert specific precursors into vanillin.

c. **Applications and Significance:** Vanillin is a key flavor and fragrance compound widely used in the food, cosmetic, and pharmaceutical industries.

Its synthetic production offers a cost-effective and sustainable alternative to natural extraction methods from vanilla beans. Synthetic vanillin ensures consistent quality and availability while minimizing environmental impact.

17. Refrigerants

a. **Dimethyl Sulfoxide (DMSO):** DMSO is synthesized from sulphur dioxide by reacting it with dimethyl sulfide. The reaction yields DMSO and residual sulphur dioxide, which are then separated by distillation. DMSO finds application as a refrigerant in specialized scenarios, such as cooling high-temperature superconductors.

b. **Sulfolane:** Sulfolane, another refrigerant derived from sulphur dioxide, undergoes synthesis via a series of chemical reactions, including its reaction with ethylene oxide. This process results in the production of sulfolane, a colorless and flammable liquid characterized by a low boiling point. Due to these properties, sulfolane is suitable for refrigeration purposes in select industrial settings.

c. **Considerations:**

- Limited Usage: While sulphur dioxide can be utilized to produce refrigerants like DMSO and sulfolane, its application in this capacity remains relatively uncommon. Most refrigerants in widespread use are derived from hydrocarbons or synthetic compounds with reduced environmental impact.
- Environmental Impact: The choice of refrigerant is crucial due to environmental considerations. Manufacturers and industries prioritize refrigerants with minimal ozone depletion potential (ODP) and low global warming potential (GWP) to mitigate their environmental impact.
- Specialized Applications: Despite the limited use of sulphur dioxide-based refrigerants, their application in specialized contexts underscores their importance in certain industrial cooling requirements.

18. Organic Sulphonates

- Step 1: Conversion to Sulphurous Acid
  Sulphur dioxide undergoes a reaction with water to form sulphurous acid.
- Step 2: Oxidation to sulphite
  Sulphurous acid is oxidized with oxygen to yield sulphite.
- Step 3: Conversion to Sulfonic Acid

sulphite reacts with olefins like ethylene or propylene to form sulfonic acid.

- Step 4: Derivatization to Organic Sulphonates
  The sulfonic acid undergoes derivatization by reacting with appropriate alcohols or phenols to produce various organic sulfonates, such as alkyl benzene sulfonates or alkyl sulphates.
- Step 5: Purification
  The resulting organic sulfonates are purified through processes like distillation or crystallization to obtain the desired purity.

Applications

Organic sulfonates serve as crucial surfactants across a spectrum of industrial and consumer applications:

a. **Detergents:** They are key components in detergent formulations, aiding in the removal of dirt and stains from fabrics and surfaces.
b. **Textiles:** Organic sulfonates find usage in textile processing, facilitating dyeing and finishing processes.
c. **Personal Care Products:** They contribute to the foaming and cleansing properties of personal care items such as shampoos, soaps, and body washes.
d. **Oilfield Chemicals:** In the oil and gas industry, organic sulfonates are employed as surfactants and emulsifiers in drilling fluids and enhanced oil recovery operations.

19. Sulfones

Sulfones are synthesized from sulphur dioxide through a reaction with organic compounds containing multiple carbonyl groups, such as dialkyl ketones or aryl ketones. This reaction typically occurs in the presence of a strong base like sodium hydroxide.

a. **Mechanism:**

- Preparation of Sulfonyl Intermediate: Sulphur dioxide reacts with the strong base to generate a sulfonyl intermediate.
- Addition to Carbonyl Group: The sulfonyl intermediate then undergoes nucleophilic addition to the carbonyl group (C=O) present in the organic compound.
- Formation of Sulfone: The resulting intermediate undergoes further reactions, leading to the formation of the sulfone product.

b. **Variations and Optimization**

- The specific conditions of the reaction, such as temperature, pressure, and concentration of reactants, are optimized to achieve the desired yield of sulfone.
- Catalysts or additional reagents may be employed to enhance the reaction rate or control the selectivity of the process.
- Different organic compounds and sulfones may require variations in reaction conditions and reagents for optimal synthesis.

c. **Applications of Sulfones:** Sulfones find extensive use in various industrial and pharmaceutical applications:

- Solvents: Sulfones serve as high-boiling point solvents in organic synthesis and chemical processes.
- Pharmaceuticals: They are key structural components in several pharmaceutical compounds, exhibiting diverse pharmacological properties.
- Polymers: Sulfones are utilized in the synthesis of specialty polymers, including high-performance plastics and engineering materials.
- Electronics: Sulfones find application in electronic devices as dielectric materials and heat-resistant components.

20. Thionyl Chloride

Thionyl chloride ($SOCl_2$) is synthesized from sulphur dioxide and chlorine gas through the following reaction:

$SO_2 + Cl_2 \rightarrow SOCl_2 + Cl_2O$

The chlorine gas oxidizes sulphur dioxide to produce thionyl chloride along with dichlorine oxide. Thionyl chloride is then separated from the reaction mixture.

Uses of Thionyl Chloride:

Thionyl chloride finds various applications in chemical synthesis and industry:

- Ester Synthesis: Thionyl chloride is utilized in the synthesis of esters from alcohols and carboxylic acids through esterification reactions.
- Carboxylic Anhydride Production: It is employed in the production of carboxylic anhydrides from carboxylic acids.
- Sulfonyl Chloride Synthesis: Thionyl chloride is used to synthesize sulfonyl chlorides from sulfonic acids.
- Dechlorination: It serves as a dechlorinating agent, converting organochlorine compounds into more reactive species.
- Battery Electrolyte: Thionyl chloride is employed as an electrolyte in certain primary batteries, like lithium thionyl chloride batteries.

Due to its highly reactive and toxic nature, thionyl chloride must be handled with extreme caution:

- It should be used in a well-ventilated area, and protective gear like gloves and goggles must be worn.
- Proper storage and handling procedures should be followed to prevent accidental exposure and reactions.

20a. Production of a Chlorinating agent from Thionyl Chloride

The production of a chlorinating agent from thionyl chloride involves a sequence of chemical reactions. Thionyl chloride ($SOCl_2$) is a versatile compound extensively utilized as a chlorinating reagent across diverse chemical processes. Here's a breakdown of the process by which a chlorinating agent can be obtained from thionyl chloride:

Reaction Pathway:

a. **Thionyl Chloride ($SOCl_2$) Hydrolysis**: Thionyl chloride reacts with water (H_2O) to produce hydrogen chloride (HCl) and sulphur dioxide (SO_2).
$SOCl_2 + H_2O \rightarrow 2HCl + SO_2$

b. **Sulphur Dioxide (SO_2) Oxidation:** The sulphur dioxide generated from the hydrolysis reaction undergoes further oxidation to form sulphur trioxide (SO_3) with oxygen (O_2) acting as the oxidizing agent. $2SO_2 + O_2 \rightarrow 2SO_3$

c. **Sulphur Trioxide (SO_3) Hydration:** sulphur trioxide reacts with water to yield sulphuric acid (H_2SO_4). $SO_3 + H_2O \rightarrow H_2SO_4$

d. **Sulphuric Acid (H_2SO_4) Neutralization:** sulphuric acid can be neutralized using a suitable base, such as sodium hydroxide (NaOH), resulting in the production of sodium sulphate (Na_2SO_4). $H_2SO_4 + 2NaOH \rightarrow Na_2SO_4 + 2H_2O$

Chlorinating Agent: The resulting sodium sulphate (Na_2SO_4) from the neutralization step does not function as a chlorinating agent. However, the hydrogen chloride (HCl) formed during the hydrolysis step serves as a potent chlorinating agent. Hydrogen chloride exhibits high reactivity and readily reacts with various organic and inorganic substrates, enabling the introduction of chlorine atoms.

20b. Production of Thionyl Bromide from Thionyl Chloride Thionyl chloride ($SOCl_2$) undergoes a bromination reaction with bromine (Br_2), resulting in the formation of thionyl bromide ($SOBr_2$) and hydrogen chloride gas (HCl). Chemical Equation: $SOCl_2 + Br2 \rightarrow SOBr_2 + Cl_2$

Detailed Process:

a. **Preparation of Thionyl Chloride ($SOCl_2$):** Thionyl chloride is commonly synthesized by reacting sulphur dioxide (SO_2) with phosphorus pentachloride (PCl_5).

$SO_2 + PCl_5 \rightarrow SOCl_2 + POCl_3$

b. **Bromination Reaction:** Thionyl chloride ($SOCl_2$) reacts with bromine (Br_2) to produce thionyl bromide ($SOBr_2$).
   $SOCl_2 + Br_2 \rightarrow SOBr_2 + Cl_2$
   Careful control of temperature and pressure optimizes thionyl bromide yield while minimizing by-product formation.
c. **Separation and Purification:** Post-reaction, separation techniques like distillation or extraction isolate thionyl bromide from reaction by-products and unreacted materials.
d. **Characterization and Quality Control:** Thionyl bromide undergoes rigorous analysis to confirm its identity and purity, meeting predefined standards for various applications.

Environmental Impact: Effluent treatment and proper waste disposal mitigate environmental impact, ensuring regulatory compliance and minimizing hazards posed by by-products.

21. Sulphonyl Chloride

Sulphonyl chloride (SO_2Cl_2) is synthesized from sulphur dioxide (SO_2) and phosphorus pentachloride (PCl_5) through the following reaction:
$SO_2 + PCl_5 \rightarrow SO_2Cl_2 + POCl_3$

Phosphorus pentachloride reacts with sulphur dioxide to produce sulphonyl chloride and phosphorus oxychloride. It's crucial to note that this reaction is highly exothermic and produces toxic fumes, necessitating proper safety precautions.

Applications of Sulphonyl Chloride:
Sulphonyl chloride serves as a precursor for various chlorinating and sulfonating agents:

a. **N-Chloramines Production:** Reacting sulphonyl chloride with primary amines yields N-chloramines, which can be hydrolyzed to sulfonamides.
   $SO_2Cl_2 + RNH_2 \rightarrow RSO_2NH_2 + HCl$
   $RSO_2NH_2 + H_2O \rightarrow RSO_2NH_3 + HOCl$
   $RSO_2NH_3 + OH \rightarrow RSO_2NH_2 + H_2O$
b. **Alkyl sulphites Formation:** Interaction of sulphonyl chloride with alcohols produces alkyl sulphites.
   SO_2Cl_2 + ROH rightarrow $ROSO_2Cl$ + HC
c. **Sulfonic Acid Derivatives:** Sulphonyl chloride reacts with various nucleophiles (e.g., water, alcohols, amines) to yield sulfonic acids and sulfonates, crucial intermediates in dye and pharmaceutical production.

21a. Production of Chlorinating Sulphonating Agent from Sulphonyl Chloride

The reaction involves the chlorination of sulphonyl chloride (RSO_2Cl) with chlorine gas (Cl_2) to produce a chlorinating sulphonating agent, typically represented as (RSO_2Cl)n, where (R) represents an organic group.

Chemical Equation: $RSO_2Cl + Cl_2 \rightarrow (RSO_2Cl)n$

Detailed Process:

a. **Synthesis of Sulphonyl Chloride (RSO_2Cl):** Sulphonyl chloride is synthesized by the reaction of sulphur dioxide (SO_2) with a chlorinating agent such as phosphorus pentachloride (PCl_5) or thionyl chloride ($SOCl_2$).
$SO_2 + 2Cl_2 \rightarrow SO_2Cl_2 \rightarrow SOCl_2 + Cl_2$

b. **Chlorination Reaction:** Sulphonyl chloride (RSO_2Cl) undergoes chlorination by reacting with excess chlorine gas (Cl_2).
$RSO_2Cl + Cl_2 \rightarrow (RSO_2Cl)n$
Precise control of reaction conditions ensures the desired degree of chlorination.

c. **Isolation and Purification:** Following the reaction, the chlorinating sulphonating agent is isolated and purified using techniques such as distillation or crystallization to eliminate impurities.

d. **Characterization and Quality Control:** The chlorinating sulphonating agent undergoes comprehensive characterization to verify its chemical structure and purity. Quality control measures are implemented to ensure compliance with specified standards for various applications.

Environmental Impact: To mitigate environmental impact and adhere to regulations concerning hazardous substances, effective effluent treatment and waste management procedures are imperative for the handling and disposal of by-products.

21b. Production of Chloro Sulphuric Acid from Sulphonyl Chloride

Chlorosulphuric acid (HSO_3Cl) is produced from sulphonyl chloride (SO_2Cl_2) by reacting it with sulphuric acid (H_2SO_4) in the following reaction:

$SO_2Cl_2 + H_2SO_4$ rightarrow $HSO_3Cl + HCl$

Sulphonyl chloride reacts with sulphuric acid to yield chlorosulphuric acid and hydrogen chloride.

Applications of Chlorosulphuric Acid: Chlorosulphuric acid finds applications as a sulfonating agent in organic synthesis, particularly in detergent and dye manufacturing. It serves as a dehydrating agent and a source of sulphur trioxide.

Safety Considerations:

Given its high reactivity and corrosiveness, chlorosulphuric acid must be handled with care. Proper safety measures, including the use of appropriate protective equipment and ventilation, are essential during its handling and storage.

22. Gelatine

Sulphur dioxide (SO_2) plays a crucial role in the production of gelatine, primarily serving as a preservative and processing aid.

Here's a detailed overview of its utilization:

a. **Preservative Function:** Sulphur dioxide serves as a potent preservative in gelatine production. It effectively inhibits the growth of bacteria and other microorganisms that could lead to spoilage and compromise the quality of the gelatine. Moreover, sulphur dioxide helps to maintain the color, flavor, and aroma of the final product, ensuring its overall quality and shelf stability. However, it's essential to note that some individuals may be sensitive to sulphites and may experience allergic reactions upon consuming products containing SO_2.

b. **Regulatory Considerations:** In the production of gelatine used for capsules or tablet coatings, regulatory guidelines often stipulate specific limits on sulphur dioxide content. Typically, gelatine may contain no more than 0.15 percent of sulphur dioxide, ensuring compliance with safety standards. Additionally, the formulation may include suitable concentrations of sodium lauryl sulphate and other antimicrobial agents to further enhance product safety and stability.

c. **Production Process:**

- Pretreatment of Raw Materials: Raw materials, such as animal hides, bones, or skin, undergo pretreatment processes to remove impurities and extract collagen—the primary protein source for gelatine. Sulphur dioxide may be employed during this stage to mitigate microbial contamination and inhibit the growth of bacteria, yeasts, and molds.
- Gelatine Extraction: Following pretreatment, the collagen-containing raw materials are subjected to an extraction process to obtain gelatine. Sulphur dioxide can be introduced during extraction to prevent microbial proliferation and safeguard the integrity of the collagen.
- Preservation: Gelatine is susceptible to microbial spoilage and degradation, particularly during storage and transportation. Sulphur dioxide acts as an antimicrobial agent, extending the shelf life of gelatine by impeding the growth of harmful microorganisms and maintaining its freshness and quality over time.
- Bleaching and Clarification: Gelatine may undergo additional processing steps, such as bleaching and clarification, to enhance its color, clarity, and overall appearance. Sulphur dioxide may be utilized as a bleaching agent in this context, facilitating the removal of any residual impurities and refining the visual characteristics of the gelatine.

23. Glue

Sulphur dioxide (SO_2) finds application in the production of certain types of glue, particularly animal glue. Here's how it contributes to the manufacturing process:

a. **Preservative Function:** SO_2 serves as a preservative in glue production by effectively inhibiting the growth of bacteria and other microorganisms. These contaminants can lead to spoilage and degradation of the glue over time, affecting its adhesive properties and overall quality. By incorporating sulphur dioxide, manufacturers can extend the shelf life of the glue and ensure its stability during storage and use.
b. **Color and Consistency Maintenance:** In addition to its preservative role, sulphur dioxide helps maintain the color and consistency of the glue. It prevents discoloration and deterioration that may occur due to microbial activity or exposure to environmental factors. By preserving the visual and textural attributes of the glue, SO_2 contributes to the overall aesthetics and performance of the final product.

Production Process Overview:

a. **Raw Material Preparation:** The production process begins with the preparation of raw materials, which may include collagen, starch, synthetic polymers, or other adhesive compounds. These materials undergo processing to achieve the desired consistency and properties required for the specific type of glue being manufactured.
b. **Mixing and Formulation:** Prepared raw materials are mixed with water or suitable solvents, along with additives such as plasticizers, cross-linking agents, and preservatives. Sulphur dioxide is added at this stage to impart its preservative properties and ensure the long-term stability of the glue formulation.
c. **Heating and Reaction:** The mixture undergoes heating to initiate various chemical reactions essential for adhesive bond formation. Depending on the type of glue being produced, these reactions may include polymerization, cross-linking, or drying processes. Sulphur dioxide aids in maintaining the integrity of the adhesive compounds throughout these reactions.
d. **Cooling, Packaging, and Quality Control:** Once the desired chemical reactions occur, the glue is cooled, packaged, and subjected to rigorous quality control procedures. Quality checks ensure that the adhesive properties, strength, and durability of the glue meet the specified standards before it is released for commercial distribution and use.

e. By incorporating sulphur dioxide into the glue production process, manufacturers can produce high-quality adhesives with enhanced stability, longevity, and performance characteristics.

24. Wine

Sulphur dioxide (SO_2) plays a pivotal role in the production of wine, serving as a preservative and antioxidant to ensure wine quality and stability. Here's a detailed look at its use in winemaking:

a. **Protection of Grapes:** During grape harvesting, sulphur dioxide may be applied to protect the grapes from oxidation and microbial spoilage. It can be utilized in various forms, including solutions or gaseous state, in storage containers or during transportation. By preserving the freshness and integrity of the grapes, sulphur dioxide contributes to maintaining the quality of the wine.

b. **Control of Fermentation:** Sulphur dioxide is added at different stages of the winemaking process to control fermentation. As an antimicrobial agent, it helps inhibit the growth of unwanted yeasts and bacteria, ensuring a clean and controlled fermentation process. By regulating microbial activity, sulphur dioxide assists winemakers in achieving desired flavor profiles and preventing off-flavors.

c. **Antioxidant Protection:** One of the crucial roles of sulphur dioxide in winemaking is its function as an antioxidant. It prevents the oxidation of wine components, such as phenolic compounds and organic acids, which can lead to undesirable changes in flavor, aroma, and color. By preserving the wine's freshness and vitality, sulphur dioxide enhances its overall sensory characteristics.

d. **Preservation and Stability:** Sulphur dioxide acts as a preservative, extending the shelf life of wine by inhibiting the growth of spoilage microorganisms. It maintains the stability of the wine and prevents unwanted fermentation or refermentation in the bottle, safeguarding its quality over time. This preservation effect ensures that consumers enjoy wine at its peak condition.

e. **Sanitization of Equipment:** Winemaking equipment, including barrels, tanks, and bottling lines, requires thorough sanitation to prevent contamination and maintain hygiene standards. Sulphur dioxide is employed for sanitization purposes, effectively controlling microbial populations and reducing the risk of spoilage or off-flavors caused by microbial contamination.

25. Magnesium Metallurgy

Sulphur dioxide (SO_2) finds application in magnesium metallurgy as a reducing agent during the production of magnesium metal from magnesium compounds.

Here's a detailed overview of its role in the process:

a. **Reducing Agent in Magnesium Production:** In magnesium metallurgy, sulphur dioxide acts as a crucial reducing agent in the conversion of magnesium oxide (MgO) to magnesium metal. Typically, this process involves reacting magnesium oxide with a suitable reducing agent, such as carbon or aluminum, in the presence of SO_2. The sulphur dioxide serves as a sulphur source and aids in the reduction of magnesium oxide to produce magnesium metal.

b. **Control of Reaction Temperature:** Sulphur dioxide also plays a role in regulating the reaction temperature during the magnesium production process. By carefully adjusting the concentration of SO_2, manufacturers can optimize the reduction process, ensuring efficient conversion of magnesium oxide to magnesium metal while controlling the temperature to prevent unwanted side reactions or overheating.

c. **Efficiency Enhancement:** The presence of sulphur dioxide can enhance the efficiency of the reduction process by facilitating the reduction of magnesium oxide and promoting the formation of magnesium metal. Through its reducing properties, SO_2 contributes to the overall efficacy of the metallurgical process, enabling the production of high-quality magnesium metal with improved yield and purity.

d. **Emission Control and Environmental Considerations:** While sulphur dioxide is instrumental in magnesium metallurgy, it's essential to address its environmental impact. The emission of sulphur dioxide gas during the production process can contribute to air pollution and greenhouse gas emissions. Therefore, manufacturers employ measures to minimize emissions and mitigate environmental impacts, such as utilizing advanced emission control technologies and adhering to regulatory standards.

The primary production of magnesium involves extracting magnesium from mineral sources like magnesite ($MgCO_3$) or dolomite ($MgCO_3 \cdot CaCO_3$). While sulphur dioxide is not directly used in the primary production process, it may be generated as an impurity or byproduct during secondary processes or specific production methods. Efforts are made to control and minimize the presence of sulphur dioxide and other impurities in the final magnesium product to ensure environmental sustainability and product quality.

By leveraging sulphur dioxide as a reducing agent, magnesium metallurgy achieves efficient conversion of magnesium compounds into high-quality magnesium metal, contributing to various industrial applications while addressing environmental considerations.

26. Preservative

Sulphur dioxide (SO_2) serves as a pivotal ingredient in the production of preservatives, safeguarding the quality and extending the shelf life of various food, beverage, pharmaceutical, cosmetic, and industrial products. Here's a detailed exploration of its multifaceted role:

a. **Food Preservation:** In the food industry, sulphur dioxide acts as a potent preservative, effectively inhibiting the proliferation of bacteria, yeasts, and molds in perishable food items. It finds extensive application in preserving dried fruits, fruit juices, wine, and other susceptible food products, thereby extending their shelf life while maintaining their nutritional integrity and sensory attributes.

b. **Beverage Preservation:** Sulphur dioxide plays a vital role in preserving the freshness and quality of beverages like wine, beer, and cider. As an antioxidant, it shields these beverages from oxidation, which can lead to flavor degradation and discoloration. By maintaining the stability of flavor compounds and color pigments, SO_2 ensures that beverages retain their desirable characteristics over time.

c. **Pharmaceutical Preservation:** The pharmaceutical industry relies on sulphur dioxide for the preservation of pharmaceutical formulations. By inhibiting microbial growth, SO_2 safeguards the integrity and efficacy of medications, ensuring that they remain potent and stable throughout their shelf life. This application is crucial for maintaining the safety and efficacy of pharmaceutical products.

d. **Cosmetic and Personal Care Products:** Sulphur dioxide finds utility as a preservative in cosmetic and personal care items such as creams, lotions, and shampoos. By impeding the growth of bacteria and fungi, it preserves the quality and safety of these products, preventing spoilage and ensuring consumer satisfaction. Sulphur dioxide's antimicrobial properties contribute to the longevity and efficacy of cosmetic formulations.

e. **Industrial Preservatives:** In industrial settings, sulphur dioxide is incorporated into preservatives for cutting fluids, lubricants, and other industrial products. By inhibiting microbial proliferation, it extends the usability and shelf life of these substances, enhancing operational efficiency and reducing the risk of contamination-related issues in industrial processes.

While sulphur dioxide serves as a highly effective preservative across diverse applications, it's essential to adhere to regulatory guidelines and monitor its concentration to mitigate the risk of adverse reactions in sensitive individuals. By leveraging sulphur dioxide as a preservative, manufacturers uphold product quality, safety, and longevity across a spectrum of consumer goods and industrial applications.

27. Sulphur Trioxide & Oleum Production

Sulphur dioxide (SO_2) plays a pivotal role as a feedstock in the production of sulphur trioxide (SO_3) and oleum (fuming sulphuric acid) through the intricate chemical process known as the contact process.

Let's delve into the comprehensive utilization of sulphur dioxide in this critical industrial process:

a. **Conversion to Sulphur Dioxide:** The production process initiates with the conversion of elemental sulphur or sulphur-containing compounds, such as hydrogen sulfide (H_2S), to sulphur dioxide (SO_2). This transformation typically occurs through combustion or roasting processes, where sulphur or its compounds react with oxygen to yield sulphur dioxide as the primary product.
b. **Catalytic Conversion:** Subsequently, sulphur dioxide undergoes catalytic conversion to sulphur trioxide (SO_3) in a meticulously controlled environment. This conversion transpires in the presence of a catalyst, most commonly vanadium pentoxide (V_2O_5), at elevated temperatures ranging between 400-500 degrees Celsius. The catalyst facilitates the reaction between sulphur dioxide and oxygen to form sulphur trioxide, a critical precursor in various industrial processes.
c. **Absorption and Formation of Oleum:** The freshly produced sulphur trioxide is swiftly absorbed into a concentrated sulphuric acid (H_2SO_4) solution to engender oleum, also recognized as fuming sulphuric acid. This absorption process leads to the formation of oleum, which may contain varying concentrations of dissolved sulphur trioxide, contingent on the desired product specifications and applications. Oleum possesses distinct chemical properties and finds widespread utility in diverse industrial sectors.
d. **Dilution and Stabilization:** Finally, the oleum is meticulously diluted with water to attain the desired concentration of sulphur trioxide while ensuring optimal stability and safety. Dilution procedures are judiciously controlled to prevent excessive heat generation and to stabilize the resultant sulphur trioxide product, thereby facilitating its seamless integration into downstream industrial processes.

Through this meticulously orchestrated series of steps, sulphur dioxide serves as a vital precursor in the production of sulphur trioxide and oleum, pivotal substances that underpin numerous industrial applications. However, it's imperative for producers to uphold stringent environmental standards and employ sustainable practices to mitigate the emission of sulphur dioxide and its potential impact on air quality and environmental health.

Summary

Sulphur dioxide (SO_2) boasts a plethora of applications across diverse industries, each harnessing its unique properties to serve specific purposes:

a. **Industrial Applications:**
- Production of Sulphuric Acid: SO_2 serves as a precursor in the production of sulphuric acid, a vital industrial chemical utilized in fertilizers, detergents, dyes, and various chemical processes.
- Food Preservation: Widely employed as a preservative in the food industry, SO_2 prevents microbial growth, preserves color in fruits and vegetables, and acts as an antioxidant, commonly used in dried fruits, fruit juices, and wine production.
- Bleaching Agent: SO_2 finds utility as a bleaching agent in paper and textile manufacturing, aiding in the removal of color impurities from these materials.

b. **Environmental Applications**:
- Air Pollution Control: Despite being a byproduct of fossil fuel combustion, especially coal and oil, SO_2 is mitigated through technologies like flue gas desulphurization (FGD) to curtail its environmental impact, reducing air pollution.
- Preservation in Winemaking and Brewing: In winemaking and brewing, SO_2 is pivotal for preventing microbial contamination and oxidation, ensuring the integrity and longevity of wine and beer.

c. **Laboratory Applications**:
- Analytical Chemistry: SO_2 serves as a reducing agent and a source of sulphite ions in analytical chemistry, facilitating various titration methods, particularly for determining oxidizing agents.
- Synthesis of Organic Compounds: In organic synthesis, SO_2 acts as a versatile reagent, enabling the introduction of the sulfonate functional group ($-SO_3H$) into organic molecules, expanding the repertoire of organic synthesis methodologies.

While SO2's multifaceted utility is undeniable, its emissions necessitate stringent regulation due to its adverse effects on human health and the environment. Through meticulous regulation and technological innovation, efforts are underway to mitigate

SO_2's negative impact, ensuring its continued utilization while minimizing environmental harm.

REFERENCES:

1. https://www.encyclopedia.com/science-and-technology/chemistry/compounds-and-elements/sulphur-dioxide
2. The National Institute for Occupational Safety and Health (NIOSH)
3. S.C. Gad, in Encyclopedia of Toxicology (Third Edition), 2014
4. Ray Tricker, Samantha Tricker, in Environmental Requirements for Electromechanical and Electronic Equipment, 1999
5. https://pubchem.ncbi.nlm.nih.gov
6. *Haynes, W.M. (ed.). CRC Handbook of Chemistry and Physics. 95th Edition. CRC Press LLC, Boca Raton: FL 2014-2015, p. 4-92*
7. https://environment.govt.nz/facts-and-science/air/air-pollutants/sulphur-dioxide-and-effects-on-health/
8. PROCESS FOR PRODUCTION OF SULPUR, DIOXIDE Albert C. Mohr, Berkeley, Calif., assignor to Stauffer Chemical Company, a corporation of California, 1937
9. Kohl, Arthur L.; Nielsen, Richard B. (1997). "sulphur Dioxide Removal". Gas Purification. Gulf Professional Publishing. pp. 554–555. ISBN 978-0-88415-220-0.
10. Lee, C. C. (2005). Environmental Engineering Dictionary. Rowman & Littlefield. p. 884. ISBN 978-0-86587-848-8.
11. Kamila Jóźwiak, Krzysztof Czajka, Recovery of sulphur dioxide from flue gas for Sulphuric Acid Production, INTERNATIONAL SCIENTIFIC JOURNAL SCIENCE. BUSINESS. SOCIETY" YEAR VI, ISSUE 2, P.P. 40-41 (2021)
12. Processes in Petroleum Industries, Emission Inventory GuideBook, December 2006
13. A. C. ROBERTS, D. J. McWEENY, The uses of sulphur dioxide in the food industry,

Journal of Food Technology, September 1972

14. L. Usseglio-Tomasset, Properties and use of sulphur dioxide, Journal of Food Additives and Contaminants, Vol 9, Pages 399-404 , 2009
15. James Kisielewski, PROCESS FOR PRODUCTION OF SODIUM BIsulphite, US 2010/0008.845 A1, 2010
16. Paper: Pulping and Bleaching, H.L. Hintz, in Encyclopedia of Materials: Science and Technology, 2001
17. Standard Fabrication Practices for Cane Sugar Mills, Edward Delden, Chapter: Sulphitation Process
18. Patnaik, P. (2002). *Handbook of Inorganic Chemicals*. McGraw-Hill

19. Committee on Food Chemicals Codex, Institute of Medicine, National Academy of Science, Food Chemicals Codex: Fourth Edition, 1996.
20. T. R. Parsons, B. E. Baker, Studies on protein hydrolysis, Journal of Food and Agriculture, 1956
21. Ge, Shaohui; Wu, Zhijie; Zhang, Minghui; Li, Wei; Tao, Keyi.*Industrial & Engineering Chemistry Research*,**2006***45(7)*, 2229-2234,
22. Sharipov, A. Kh.*Russian Journal of Applied Chemistry 2003,76(1)*, 108-113.
23. P.B. DeGroot, T.H. Norris. Isotopic exchange reactions in liquid sulphur dioxide—VII. *Journal of Inorganic and Nuclear Chemistry* **1974**, *36* (5) , 1123-1129
24. The 20th International Scientific and Research Conference "Metallurgy: Technologies, Innovation, Quality. Metallurgy-2017" 15–16 November 2017, Novokuznetsk, Russia

Chapter 3: The Dynamic World of Sulphur Trioxide

HISTORY:

The synthesis of sulphur trioxide (SO_3) dates back to 1866 when the German chemist Carl Friedrich Claus first successfully produced it. This significant achievement marked the discovery of a compound that would play a crucial role in various industrial processes. Initially, sulphur trioxide was observed as a colorless pale yellow gas, known for its distinctive properties. Over time, its production in large quantities became associated with the burning of sulphur-containing fuels like coal and oil.

MAIN APPLICATION:

The primary application of sulphur trioxide lies in its pivotal role in the production of sulphuric acid (H_2SO_4). As one of the most essential and widely used chemicals across industries, sulphuric acid finds its way into the manufacture of fertilizers, detergents, dyes, and various other industrial processes, including petroleum refining and metal processing.

PROPERTIES:

Sulphur trioxide boasts several distinctive properties that make it a versatile compound in industrial applications:

Physical Properties:

- **Appearance**: Under normal conditions, sulphur trioxide exists as a colorless to pale yellow liquid, occasionally solidifying into crystalline form at lower temperatures.
- **Odor**: It emits a strong, pungent odor.
- **Density**: With a density of around 1.92 grams per cubic centimeter (g/cm^3), it exhibits considerable mass relative to volume.
- **Melting Point**: Sulphur trioxide displays a relatively low melting point of 16.8 degrees Celsius (62.2 degrees Fahrenheit).
- **Boiling Point**: Its boiling point is recorded at 44.8 degrees Celsius (112.6 degrees Fahrenheit).

- **Solubility**: Highly soluble in water, sulphur trioxide readily combines with water molecules to form sulphuric acid (H_2SO_4).

Chemical Properties:

- **Reactivity**: Demonstrating high reactivity, sulphur trioxide rapidly reacts with water vapor to yield sulphuric acid, releasing substantial heat in the process.
- **Acidic Nature**: As the anhydride of sulphuric acid, it exhibits strong acidic properties, reacting vigorously with water to form highly corrosive sulphuric acid.
- **Dehydrating Agent**: Known for its potent dehydrating properties, it effectively removes water molecules from various substances.
- **Oxidizing Agent**: Sulphur trioxide acts as a powerful oxidizing agent, instigating combustion or other chemical reactions upon contact with organic and inorganic compounds.
- **Reactivity with Bases**: It forms salts known as sulphates upon reacting with bases.
- **Stability**: Despite its reactivity, sulphur trioxide demonstrates relative instability, tending to polymerize into larger sulphur compounds under specific conditions.

INDUSTRIAL USES:

- Production of Sulphuric Acid: Serving as a critical intermediate compound, sulphur trioxide plays a vital role in the manufacturing process of sulphuric acid, catering to diverse industrial applications.
- Chemical Synthesis: Its involvement in chemical synthesis extends to the production of detergents, pharmaceuticals, dyes, and pigments.
- Catalysts and Desiccants: Widely employed as both a catalyst and desiccant, sulphur trioxide facilitates essential chemical reactions and acts as a drying agent in various industrial processes.
- Petroleum Refining: Within the petroleum industry, it contributes to alkylation reactions and the production of gasoline additives, further augmenting its industrial utility.

This comprehensive understanding of sulphur trioxide underscores its profound significance across various industrial sectors, emphasizing its pivotal role in chemical synthesis, manufacturing processes, and industrial applications.

SOURCES OF SULPHUR TRIOXIDE

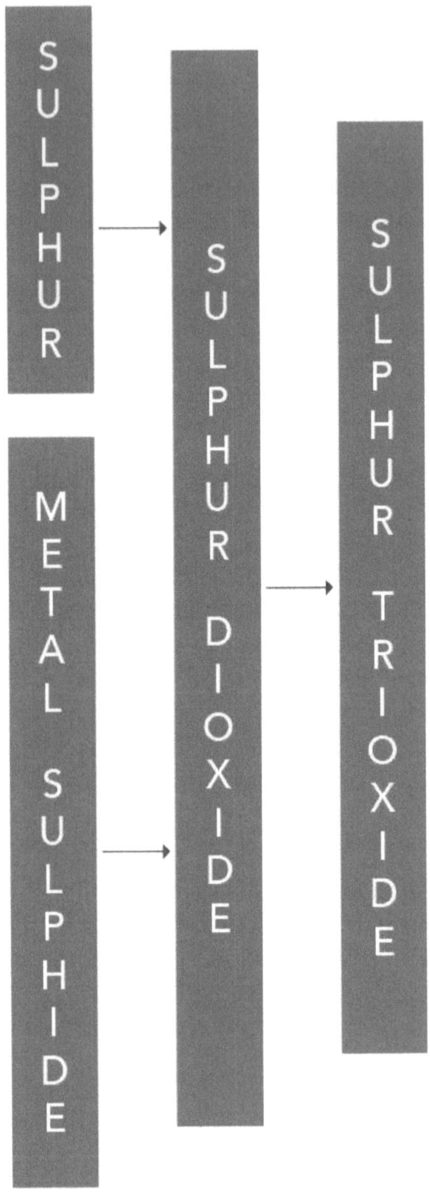

Sulphur trioxide (SO_3) can be generated through various methods, each serving distinct industrial purposes:

a. **Contact Process**: Widely employed for sulphuric acid manufacturing, the Contact Process involves oxidizing sulphur dioxide (SO_2) with air in the presence of a vanadium(V_2O_5) pentoxide catalyst. This catalytic oxidation yields sulphur trioxide as a key intermediate.
b. **Sulphur Combustion**: Combusting elemental sulphur with surplus oxygen offers another avenue for producing sulphur trioxide. This method is particularly valuable in contexts where elemental sulphur is readily available or serves as a byproduct of other processes.
c. **Roasting Sulfide Ores:** The roasting of sulfide ores, such as iron pyrite (FeS_2), initiates the release of sulphur dioxide (SO_2). Subsequently, sulphur dioxide can undergo further oxidation to yield sulphur trioxide, offering a route for its production from naturally occurring sulfide minerals.
d. **Reactions with Metal Oxides:** Sulphur trioxide synthesis can also proceed through reactions between sulphur dioxide and metal oxides, including calcium oxide (CaO), magnesium oxide (MgO), or aluminum oxide (Al2O3). These reactions afford a method for producing sulphur trioxide using metal oxides as reactants.

These methods represent the primary avenues for sulphur trioxide production, with the selection of a specific method contingent upon factors such as resource availability, cost-effectiveness, and environmental considerations. Each method contributes to the diverse industrial applications of sulphur trioxide, ranging from sulphuric acid production to the synthesis of various chemical compounds.

APPLICATION OF SULPHUR TRIOXIDE-SHEET 1

1. Thionyl Chloride

Thionyl chloride ($SOCl_2$) is a widely used reagent in organic synthesis and industrial processes. It is commonly produced by the reaction of sulfur trioxide (SO_3) with sulfur dichloride (SCl_2).

Here's a simplified overview of the production process:

a. **Preparation of Sulfur Dichloride (SCl_2):** Sulfur dichloride is typically prepared by reacting sulfur with chlorine gas. The reaction is exothermic and requires careful control to prevent runaway reactions. $S+Cl_2 \rightarrow SCl_2$

b. **Preparation of Thionyl Chloride ($SOCl_2$):** Once sulfur dichloride is obtained, it can react with sulfur trioxide to produce thionyl chloride. $SO_3+SCl_2 \rightarrow SOCl_2+SO_2$

- This reaction typically occurs at elevated temperatures and may be carried out in the presence of a catalyst or under carefully controlled conditions.

c. **Separation and Purification:** After the reaction, the mixture containing thionyl chloride, sulfur dioxide, and unreacted sulfur dichloride needs to be separated and purified. Distillation is often used to separate thionyl chloride from other components due to its relatively low boiling point.

d. **Storage and Distribution:** Thionyl chloride is usually stored and transported under anhydrous conditions to prevent its decomposition. It's often kept in sealed containers or under an inert gas atmosphere to minimize contact with moisture, which can cause it to hydrolyze and release toxic gases like hydrogen chloride.

e. **Industrial Applications:** Thionyl chloride finds numerous applications in the chemical industry, including:

- Conversion of carboxylic acids to acid chlorides.
- Synthesis of various organochlorine compounds.
- Preparation of lithium batteries.
- Production of pesticides and pharmaceuticals.

It's essential to note that the production and handling of thionyl chloride require proper safety precautions due to its toxicity and reactivity. Workers handling thionyl chloride must be adequately trained, and appropriate protective equipment and procedures should be in place to prevent accidents and exposure.

2. Poly Sulphuryl Chloride

Poly sulphuryl chloride (SxOyClz) is a compound derived from sulphur trioxide (SO_3) and is utilized as a potent oxidizing and chlorinating agent.

Here's an overview of its production process, properties, and applications:

a. **Properties and Distinction from Thionyl Chloride:** Poly sulphuryl chloride, with the chemical formula SO_2Cl_2, exists as a colorless liquid with a pungent odor. Unlike thionyl chloride ($SOCl_2$), which is a source of chloride ions, poly sulphuryl chloride serves as a source of chlorine atoms.

b. **Production Process:** Poly sulphuryl chloride is synthesized through reactions involving sulphur trioxide and either sulphur dichloride (S_2Cl_2) or thionyl chloride ($SOCl_2$).
The reactions are as follows:

- Reaction with Sulphur Dichloride (S_2Cl_2): $6SO_3 + 2S_2Cl_2 \rightarrow S_4O_{10} + 2SO_2 + 2Cl_2$
- Reaction with Thionyl Chloride ($SOCl_2$): $4SO_3 + 2SOCl_2 \rightarrow S_4O_{10} + 2SO_2 + 2Cl_2$
- In both cases, poly sulphuryl chloride (S_4O_{10}) is formed along with sulphur dioxide (SO_2) and chlorine gas (Cl_2) as byproducts.

Poly sulphuryl chloride finds several applications in various industries:

a. **Oxidizing Agent:** It is utilized as a potent oxidizing agent in chemical reactions requiring strong oxidation, such as the conversion of alcohols to carbonyl derivatives.

b. **Chlorinating Agent:** As a strong chlorinating agent, it facilitates the introduction of chlorine atoms into organic molecules, aiding in the synthesis of chlorinated compounds like solvents, pesticides, and pharmaceutical intermediates.

c. **Polymerization Reactions:** It participates in polymerization reactions and contributes to the production of specific polymers and copolymers.

d. **Desiccant:** Poly sulphuryl chloride serves as a desiccant to remove moisture from materials and environments.

e. **Laboratory Reagent:** Widely used in laboratories, it acts as a versatile reagent in various synthetic processes, including oxidation and chlorination reactions.

3. Trimethylamine Sulphur Trioxide

Sulphur trioxide trimethylamine, also known as trimethylamine sulfonate or TMS, serves as a crucial reagent in organic synthesis, offering diverse applications in various industries. Here's an overview of its production process and uses:

Production Process: Sulphur trioxide trimethylamine is synthesized through a reaction between trimethylamine ($N(CH_3)_3$) and sulphur trioxide (SO_3) in an organic solvent like dichloromethane or chloroform. The resulting product is a stable crystalline solid that can be isolated for further applications.

a. **Sulfonating Agent:** TMS is extensively employed as a sulfonating agent in organic synthesis. It facilitates the introduction of sulfonic acid groups (-SO_3H) into various organic compounds, enhancing their properties such as water solubility, acidity, or reactivity. Sulfonated compounds produced using TMS find applications across industries including pharmaceuticals, dyes, detergents, and specialty chemicals.
b. **Sulfonamide Synthesis:** In medicinal chemistry, TMS plays a vital role in the synthesis of sulfonamides, which are essential structural motifs found in antibiotics, diuretics, and other pharmaceuticals. Through a condensation reaction with amines, TMS enables the formation of sulfonamide derivatives, contributing to the development of various therapeutic agents.
c. **Polymerization Catalyst:** TMS exhibits catalytic properties in certain polymerization reactions. For instance, it has been utilized in the production of styrene sulfonate polymers, which serve as ion-exchange resins, flocculants, and viscosity modifiers in various industrial applications.

In conclusion, sulphur trioxide trimethylamine is a versatile compound with significant applications in organic synthesis, pharmaceuticals, and polymer chemistry. Its production process and uses underscore its importance as a valuable reagent in various industrial and scientific endeavors.

4. Sulphur Trioxide Dioxane

Sulphur trioxide-dioxane, also known as sulphuryl dioxane, is a valuable compound synthesized from sulphur trioxide (SO_3) and dioxane ($C_4H_8O_2$). This coordination complex finds diverse applications in chemical synthesis and catalysis.

Here's a detailed overview:

Sulphur trioxide-dioxane is produced through a reaction between sulphur trioxide and dioxane, forming a stable, water-soluble complex. The reaction occurs in a polar solvent such as ethanol, and the coordination complex is obtained as follows:
$SO_3 + C_4H_8O_2 \rightarrow SO_3(C_4H_8O_2)$

Applications:

a. **Chemical Reagent:** The sulphur trioxide-dioxane complex serves as a versatile chemical reagent in various reactions. It participates in chemical syntheses and transformations, facilitating the production of a wide range of compounds.
b. **Catalyst:** Sulphur trioxide-dioxane acts as a catalyst in the synthesis of fine chemicals and pharmaceuticals. Its presence enhances reaction rates and yields, making it valuable in industrial processes.
c. **Solvent System:** The complex's water-soluble nature allows it to function as a solvent system in certain reactions, providing an environment conducive to specific chemical transformations.

5. Sodium Dodecylbenzenesulfonate (SDBS) and its Role in Detergent Manufacturing

Sodium dodecylbenzenesulfonate (SDBS) is a vital component in the production of detergents and cleaning products, offering effective surface-active properties. Here's a comprehensive overview of its production process, applications, and its role in detergent manufacturing:

a. **Synthesis of Dodecylbenzenesulfonic Acid (DBSA):**

- Dodecylbenzene, typically derived from petroleum, undergoes a reaction with sulphur trioxide (SO_3) to yield dodecylbenzenesulfonic acid (DBSA).
- This reaction often employs a catalyst, such as sulphuric acid, to facilitate the formation of DBSA.

b. **Neutralization to Form SDBS:**

- The resulting DBSA is neutralized by the addition of sodium hydroxide (NaOH), leading to the formation of sodium dodecylbenzenesulfonate (SDBS).
- SDBS emerges as a sodium salt of the sulfonic acid, exhibiting desirable surfactant properties.

Uses of Sodium Dodecylbenzenesulfonate (SDBS):

a. **Detergents:**

- SDBS serves as a fundamental ingredient in various detergents, including laundry detergents, dishwashing detergents, and household cleaners.
- Its surfactant nature enables efficient cleaning by reducing the surface tension of water, facilitating the removal of dirt, oils, and grease from surfaces.

b. **Emulsifier:**

- Employed as an emulsifying agent, SDBS facilitates the dispersion of immiscible substances, such as oil and water, forming stable emulsions.
- This property finds applications in formulating products like paints, coatings, and personal care items.

c. **Foaming Agent:** In formulations like shampoos, bubble baths, and shaving creams, SDBS functions as a foaming agent, enhancing the formation of stable foam and providing a luxurious lather.

Industrial Applications: Widely utilized in industrial processes such as textile processing, paper manufacturing, and leather tanning, SDBS aids in impurity removal, enhances dyeing processes, and improves final product quality.

5a. Production of Detergents from SDBS:

a. Detergents are formulated by combining SDBS with other surfactants, builders (e.g., sodium carbonate or sodium silicate), and additional ingredients like fragrances and preservatives.
b. The specific composition of the detergent depends on its intended application and desired performance characteristics.

6. Sulfonating Agents

Sulfonating agents, pivotal compounds in various industries, are synthesized from sulphur trioxide (SO_3) via sulfonation. This chemical process involves the reaction between an organic compound and sulphur trioxide to yield a sulfonic acid, typically represented by the equation:

$R\text{-}H + SO_3 \rightarrow R\text{-}SO_3H$

In the equation, (R) signifies the organic compound subjected to sulfonation, which can range from aromatic to aliphatic compounds. The resultant sulfonic acid can subsequently undergo neutralization with an appropriate base, such as sodium hydroxide, to generate the corresponding sodium salt.

These sulfonating agents find extensive utilization across diverse applications, serving as surfactants in cleaning and personal care products, and playing vital roles in the production of dyes, pigments, and various chemicals. Sulfonation stands as a ubiquitous reaction in the chemical industry, with sulfonating agents manufactured on a large scale to fulfill the requirements of myriad applications.

Following the synthesis of sulfonating agents, the detergent formulation process ensues. This involves blending the sulfonating agent with other essential components, such as surfactants, builders like sodium carbonate or sodium silicate, fragrances, and preservatives. The mixture undergoes further processing, such as spray drying, granulation, or extrusion, to attain its final form, whether as a liquid, powder, or bar. The finished product is then packaged and distributed for use in both household and industrial applications.

7. Ethylene Sulphate

Ethylene sulphate, a significant chemical compound, is primarily synthesized through the reaction of sulphur trioxide (SO_3) with ethylene oxide (EO). This process is meticulously conducted under controlled conditions within a specialized reactor.

Here's an overview of the production process:

a. **Preparation of Reactants:** Ethylene oxide (C_2H_4O) is typically prepared by directly oxidizing ethylene (C_2H_4) with oxygen (O_2), often in the presence of a silver catalyst. Meanwhile, sulphur trioxide (SO_3) can be obtained through the oxidation of sulphur dioxide (SO_2) with oxygen (O_2), or via alternative methods like the contact process.
b. **Reaction:** The prepared ethylene oxide and sulphur trioxide are introduced into a reactor equipped with mechanisms for effective mixing and temperature control. Within this environment, the two compounds undergo a chemical reaction to yield ethylene sulphate ($C_2H_4O_4S$):
    a. $C_2H_4O + SO_3 \rightarrow C_2H_4O_4S$
c. **Purification:** Following the reaction, the resulting mixture undergoes purification to eliminate impurities, unreacted starting materials, and any unwanted byproducts. Various purification techniques, such as distillation, filtration, and crystallization, are employed based on specific requirements and the properties of the compounds involved.

Once ethylene sulphate is obtained, it finds application in various sectors, including:

a. **Surfactant Production:** Ethylene sulphate serves as a pivotal starting material for the synthesis of anionic surfactants, crucial components in detergents, personal care products, and industrial cleaners.

b. **Textile Industry:** It is utilized as a reactive dyeing agent in the textile sector, enhancing color fastness and dye fixation on textile materials.
c. **Chemical Synthesis:** Ethylene sulphate acts as a versatile building block for synthesizing diverse organic compounds, spanning pharmaceuticals, agrochemicals, and specialty chemicals.
d. **Electroplating:** In the electroplating industry, ethylene sulphate may be incorporated into plating baths as an additive, enhancing the properties of the electroplated materials.

It's imperative to emphasize the necessity for meticulous handling and adherence to safety protocols during both the production and utilization of ethylene sulphate, as both ethylene oxide and sulphur trioxide pose potential hazards.

8. Methylene Sulphate

Methylene sulphate, also known as dimethyl sulphate (DMS), is an organic compound with the formula $(CH_3O)_2SO_2$. It is a colorless liquid with a potent odor. Given its toxicity and potential health hazards, strict adherence to safety protocols and regulations is imperative when handling and working with dimethyl sulphate.

Production Methods:

The production of methylene sulphate typically involves the reaction of sulphur trioxide (SO_3) with methanol (CH_3OH) or dimethyl ether (CH_3OCH_3). This reaction can be achieved through various methods:

a. **Direct Sulfation:** In this method, sulphur trioxide reacts directly with methanol or dimethyl ether in the presence of a catalyst, such as sulphuric acid or an acidic ion-exchange resin.
b. **Indirect Sulfation:** Here, methanol or dimethyl ether undergoes an initial reaction with sulphuric acid to form methyl hydrogen sulphate. Subsequently, the methyl hydrogen sulphate is further reacted with sulphur trioxide to yield methylene sulphate.

Both approaches necessitate meticulous control of reaction conditions, including temperature, pressure, and reagent proportions, to ensure optimal conversion and product yield. Moreover, these reactions demand a well-equipped facility with stringent safety measures to manage the hazardous nature of dimethyl sulphate.

Applications: Methylene sulphate finds utility in diverse chemical processes. It serves as a methylating agent, facilitating the transfer of methyl groups to other compounds. This characteristic renders it valuable in the synthesis of pharmaceuticals, dyes, and agricultural chemicals. Additionally, methylene sulphate can act as a solvent in various organic reactions.

9. Ethionic Acid

Ethionic acid, also known as 2-hydroxyethanesulfonic acid, is a compound synthesized from sulphur trioxide (SO_3) through a multi-step process. Here's an outline of its production and applications:

a. **Production Process:**

- Production of Sulphur Trioxide: Sulphur trioxide is typically produced by the reaction of sulphur dioxide (SO_2) with oxygen (O_2) using the contact process.
- Sulfonation: Ethylene oxide (C_2H_4O) undergoes sulfonation with sulphur trioxide to yield ethionic acid. The reaction can be represented as:
  $C_2H_4O + SO_3 + H_2O \rightarrow C_2H_6O_5S$
- This reaction is conducted under controlled conditions to optimize yield and purity.
- Purification: The reaction mixture is purified to eliminate impurities and byproducts, employing techniques such as distillation, filtration, or crystallization.

b. **Applications:**

- Surfactant: Ethionic acid serves as a surfactant in detergents and cleaning agents due to its sulfonic acid group, which enhances water solubility and surface-active properties.
- Textile Industry: It aids in dye dispersion and color retention in textile dyeing and finishing processes.
- Personal Care Products: Included in shampoos, soaps, and cosmetics for its mildness and compatibility with skin, acting as a pH adjuster and stabilizer.
- Pharmaceuticals: Serves as an intermediate in pharmaceutical synthesis.

9a. Sodium Ethylene Sulfonate from Ethionic Acid

Sodium ethylene sulfonate (SES) is an important chemical compound used in various industrial applications, including as a monomer in the production of polymers, as a surfactant, and as a corrosion inhibitor. It is commonly synthesized from ethionic acid through a series of chemical reactions.

Here's an overview of the production process:

a. **Preparation of Ethionic Acid (2-Hydroxyethanesulfonic Acid):** Ethionic acid is typically prepared by the sulfonation of ethylene oxide. Ethylene oxide reacts with sulfur trioxide (SO_3) to form ethionic acid:
b. $C_2H_4O+SO_3 \rightarrow C_2H_5O_4S$

c. **Neutralization with Sodium Hydroxide:** The ethionic acid produced in the first step is then neutralized with sodium hydroxide (NaOH) to form sodium ethylene sulfonate and water: $C_2H_5O_4S + NaOH \rightarrow C_2H_4O_3SNa + H_2O$

d. **Crystallization and Purification:** The resulting sodium ethylene sulfonate solution can be concentrated and then subjected to crystallization to obtain pure sodium ethylene sulfonate crystals. Various purification techniques may be employed to ensure the final product meets the required specifications.

e. **Drying and Packaging:** The purified sodium ethylene sulfonate crystals are dried to remove any remaining moisture and then packaged for distribution and use in various industrial processes.

f. **Industrial Applications:** Sodium ethylene sulfonate finds widespread use in the following applications:

    a. Polymerization: It serves as a monomer in the production of various polymers, including polyethylene sulfonate, which is used in detergent formulations, water treatment, and as a dispersant in industries such as textiles, ceramics, and paper manufacturing.

    - Surfactants: It acts as a surfactant and wetting agent in formulations for detergents, cleaners, and personal care products.
    - Corrosion Inhibition: SES is utilized as a corrosion inhibitor in various industrial processes, particularly in cooling water systems and metalworking fluids.

APPLICATION OF SULPHUR TRIOXIDE-SHEET 2

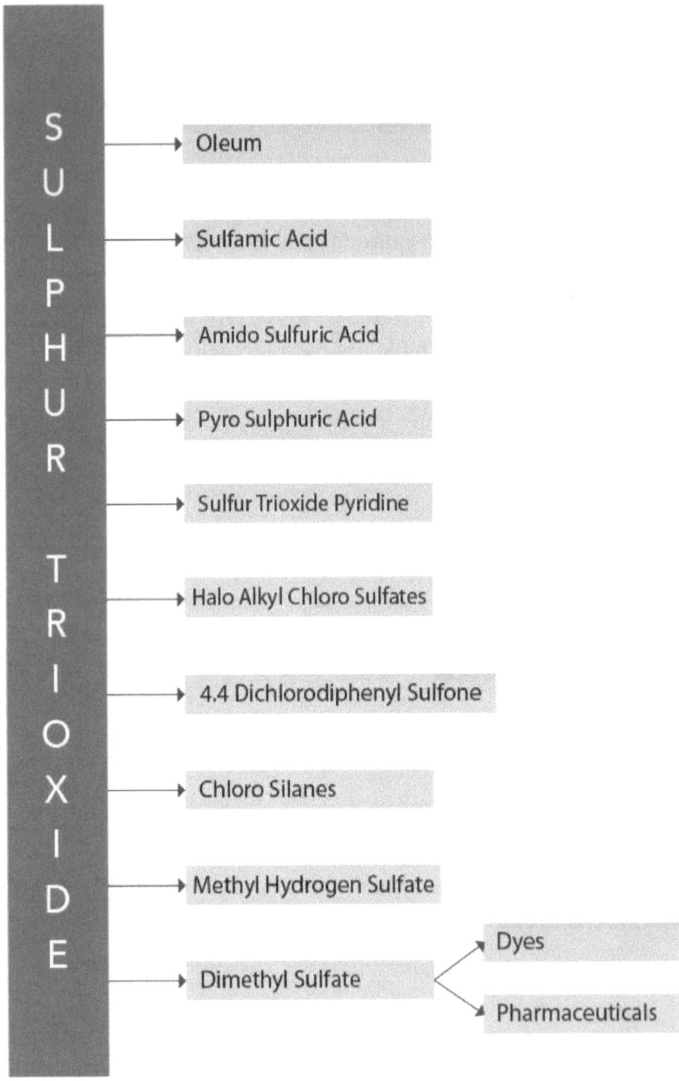

10. Oleum

Oleum, also known as fuming sulphuric acid or sulphuric acid anhydride, is a highly concentrated form of sulphuric acid (H_2SO_4). It is synthesized by dissolving sulphur trioxide (SO_3) in concentrated sulphuric acid.

Here's an overview of the production process and some common applications of oleum:

a. Production Process

- Contact Process: The predominant method involves the contact process. Sulphur dioxide (SO_2) is oxidized to sulphur trioxide (SO_3) using a catalyst like vanadium pentoxide (V_2O_5). The resulting SO3 is then dissolved in concentrated sulphuric acid (H_2SO_4) to yield oleum.
- Double Absorption Process: An alternative method entails absorbing sulphur trioxide gas directly into sulphuric acid. The typical chemical reaction involved is: $SO_3 + H_2SO_4 \rightarrow H_2S_2O_7$ (oleum)

b. Uses of Oleum

- Sulfonating Agent: Oleum serves as a crucial sulfonating agent in the production of detergents, dyes, and various organic compounds. Through sulfonation, it introduces sulfonic acid groups ($-SO_3H$) into hydrocarbons, facilitating the production of alkylbenzenes for detergents and dye intermediates.
- Catalyst: Oleum acts as a catalyst or catalyst precursor in esterification and alkylation processes. For instance, it aids in the production of ethylbenzene, a precursor to styrene used in polystyrene plastics manufacturing.
- Desiccant: With excellent dehydrating properties, oleum removes water from substances, making it valuable in the production of chemicals like nitrocellulose and other cellulose esters. It also assists in gas drying and purification processes.
- Laboratory Reagent: In analytical chemistry, oleum serves as a laboratory reagent, facilitating the preparation of specific acid mixtures or serving as a source of sulphur trioxide in various reactions.
- Sulphuric Acid Production: Oleum acts as a precursor to concentrated sulphuric acid. Diluting oleum with water allows the production of concentrated sulphuric acid, with the sulphuric acid concentration controlled by adjusting the water-to-oleum ratio.

11. Sulfamic Acid

Sulfamic acid (H_3NSO_3) is a versatile industrial chemical widely utilized in various applications, including cleaning agents, descaling agents, pH control, and dye and pigment production. One prevalent method for its production involves the reaction between sulphur trioxide (SO_3) and ammonia (NH_3).

Here's an overview of the production process and the diverse applications of sulfamic acid:

a. **Production of Sulfur Trioxide (SO_3):** Sulfur trioxide is typically manufactured via the contact process, where sulphur dioxide (SO_2) undergoes oxidation with oxygen (O_2) in the presence of a catalyst, often vanadium pentoxide (V_2O_5). The reaction proceeds as follows: $2SO_2 + O_2 \leftrightarrows 2SO_3$
b. **Reaction of Sulphur Trioxide (SO_3) with Ammonia (NH_3):** Sulfamic acid is synthesized by the direct reaction of sulphur trioxide with ammonia gas. This reaction occurs in the gas phase, usually within a dedicated reactor vessel, and is represented by the chemical equation: $SO_3 + NH_3 \rightarrow H_3NSO_3$
c. **Purification and Isolation:** Following the reaction, the resultant mixture comprises sulfamic acid, unreacted ammonia, and various impurities. Purification steps, such as cooling, filtration, and concentration, are typically employed to remove impurities and isolate pure sulfamic acid.
d. **Applications of Sulfamic Acid:** Sulfamic acid serves diverse roles across multiple industries and sectors, including:

- Cleaning Agents: It is a fundamental component in numerous household and industrial cleaning products, including toilet bowl cleaners, metal cleaners, and descaling agents.
- pH Control: Sulfamic acid is utilized as a pH adjuster in various processes, such as water treatment and chemical synthesis.
- Descaling Agents: Its effectiveness in removing scales and deposits from equipment and surfaces makes it invaluable for descaling applications in boilers, heat exchangers, and pipes.
- Dye and Pigment Production: Sulfamic acid is crucial in the synthesis of dyes and pigments, particularly in the production of azo dyes.

Moreover, sulfamic acid serves as a precursor to sweet-tasting compounds. Its reaction with cyclohexylamine, followed by the addition of NaOH, yields sodium cyclamate. Additionally, sulfamates have found applications in designing various therapeutic agents, including antibiotics, anti-cancer drugs, anti-epileptic drugs, and weight loss drugs.

Production Methods and Uses of Sulphamic Acid

Sulphamic Acid, also known as Amidosulfuric acid, is a colorless and water-soluble compound. It melts at 205 °C before decomposing at higher temperatures to water, sulphur trioxide, sulphur dioxide, and nitrogen.

Let's explore alternative production methods and diverse applications of Sulphamic Acid:

a. **Production:** Sulphamic Acid can be synthesized through a two-step reaction process involving sulphur trioxide (SO_3) and ammonia (NH_3):

- Formation of Ammonium Bisulfate (NH_4HSO_4): Sulphur trioxide (SO_3) reacts with ammonia (NH_3) and water (H2O) to produce ammonium bisulfate

  $SO_3 + NH_3 + H_2O \rightarrow NH_4HSO_4$
- Conversion to Sulphamic Acid: Ammonium bisulfate (NH4HSO4) further reacts with ammonia to yield Sulphamic Acid:

  $NH_4HSO_4 + NH_3 \rightarrow NH_2SO_3H + H_2O$

b. **Uses:** Sulphamic Acid exhibits versatile applications across various industries:

- Cleaning Agent: Widely employed as a cleaning agent and descaler, Sulphamic Acid effectively removes hard water scale, rust, and mineral deposits from surfaces and equipment.
- Metal Surface Treatment: Utilized for cleaning and etching metals, especially for preparing surfaces prior to plating, coating, or painting, enhancing adhesion of coatings and ensuring a clean surface.
- Laboratory Applications: Finds utility in laboratory procedures as an acid catalyst in organic synthesis reactions, a strong acid for adjusting pH, and a cleaning agent for laboratory glassware.
- Descaling Agent: Effective for descaling boilers, heat exchangers, condensers, and other industrial equipment prone to mineral deposits and scale accumulation, thus maintaining equipment efficiency.
- Bleaching Agent: Used in certain industries as a bleaching agent for textiles and paper products, efficiently removing color impurities and stains without causing significant damage.
- Dye and Pigment Production: Valued in the production of dyes and pigments, serving as a precursor or intermediate compound in the synthesis of various coloring agents.

12. AmidoSulphuric Acid

The synthesis of amido sulphuric acid entails the reaction between sulphur trioxide (SO_3) and ammonia (NH_3), resulting in the formation of amido sulphuric acid (HSO_3NH_2).Chemical Equation:

$SO_3 + 2NH_3 \rightarrow HSO_3NH_2 + NH_4HSO_4$

a. **Detailed Process:**

- Preparation of Sulphur Trioxide (SO_3): Sulphur trioxide is commonly manufactured by oxidizing sulphur dioxide (SO_2) with a catalyst, such as vanadium pentoxide (V_2O_5). $2SO_2 + O_2 \rightarrow 2SO_3$
- Reaction with Ammonia: Under controlled conditions, sulphur trioxide reacts with ammonia to yield amido sulphuric acid (HSO_3NH_2) alongside ammonium bisulphate (NH_4HSO_4), as a secondary product.
  $SO_3 + 2NH_3 \rightarrow HSO_3NH_2 + NH_4HSO_4$
- Reaction parameters are optimized to enhance the production yield of amido sulphuric acid.
- Isolation and Refinement: Following the reaction, the resultant mixture undergoes processing to separate and refine amido sulphuric acid from the ammonium bisulphate by-product. Techniques such as crystallization or filtration are employed for this purpose.
- Characterization and Quality Assurance: The isolated amido sulphuric acid undergoes meticulous characterization to verify its chemical composition and purity. Stringent quality control measures ensure conformity to predefined standards for diverse applications.

b. **Applications of AmidoSulphuric Acid:**

- Laboratory Reagent: Widely utilized as a reagent in laboratory environments for various chemical reactions and analytical procedures.
- Industrial Utility: Employed across industries including pharmaceuticals, dyes, and textiles as a chemical intermediate or catalyst in organic synthesis.
- Sulfonation Reactions: Functions as a sulfonating agent in organic chemistry reactions, particularly in the synthesis of sulfonic acids.

Environmental Impact Mitigation: Implementation of effluent treatment and waste management protocols is essential to mitigate environmental repercussions and ensure compliance with regulatory standards governing the handling and disposal of chemical by-products.

13. Pyrosulphuric Acid

Pyrosulphuric acid, also known as fuming sulphuric acid or oleum, is a highly corrosive and reactive liquid. It is commonly produced by combining sulphur trioxide (SO_3) with concentrated sulphuric acid (H_2SO_4). The reaction between these two compounds is highly exothermic and should be handled with caution due to the release of heat and the generation of toxic gasses.

a. Production:

- Sulphur Trioxide Generation: Sulphur trioxide (SO_3) can be generated by several methods, including the contact process, the chamber process, or the lead chamber process. The most common method today is the contact process, where sulphur dioxide (SO_2) is oxidized in the presence of a catalyst, such as vanadium pentoxide (V_2O_5), and oxygen (O_2) at high temperatures.
- Combining SO_3 with H_2SO_4: Once sulphur trioxide is produced, it is carefully mixed with concentrated sulphuric acid (H_2SO_4). The reaction between SO_3 and H_2SO_4 is highly exothermic and should be controlled to prevent overheating. The resulting mixture is pyrosulphuric acid or fuming sulphuric acid.

b. Use:

- Sulfonation Reactions: Pyrosulphuric acid is commonly used in the chemical industry for sulfonation reactions, where an organic compound is reacted with pyrosulphuric acid to introducc a sulfonic acid group ($-SO_3H$). Sulfonic acids are versatile compounds used in the production of dyes, detergents, pharmaceuticals, and other organic compounds.
- Desiccant: Pyrosulphuric acid has strong dehydrating properties, and it is used as a desiccant (drying agent) for various applications. It can absorb moisture from gases, liquids, and solids, making it useful in certain chemical processes where water can interfere.
- Catalyst and Reagent: In some chemical reactions, pyrosulphuric acid can act as a catalyst or a reagent. Its strong acidity can facilitate certain reactions or promote the formation of specific products.

14. Sulphur Trioxide Pyridine Complex

Sulphur trioxide pyridine (also known as pyridine complex of sulphur trioxide or pyridine-sulphur trioxide complex) is a compound formed by reacting pyridine (C_5HN) with sulphur trioxide (SO_3). The reaction between pyridine and sulphur trioxide leads to the formation of a stable complex. Here's a brief overview of its production and uses:

a. **Production:** The production of sulphur trioxide pyridine involves the reaction between pyridine and gaseous sulphur trioxide. The reaction typically takes place under carefully controlled conditions, often at low temperatures. The resulting product is a stable adduct of pyridine and sulphur trioxide.

b. **Uses:**

- Reagent in Organic Synthesis: Sulphur trioxide pyridine complex is commonly used as a reagent in organic synthesis. It serves as a convenient source of sulphur trioxide in various reactions. The complex can be easily handled and provides a milder and more controllable sulphur trioxide source compared to using gaseous SO_3.
- Production of Pharmaceuticals and Agrochemicals: The complex can be employed in several reactions to introduce sulfonic acid groups into organic molecules, which are often found in pharmaceutical and agrochemical compounds.
- Manufacturing of Dyes, Pigments, and Colorants: Sulphur trioxide pyridine complex is utilized in the manufacturing of dyes, pigments, and other colorants. It can be involved in reactions to introduce sulfonic acid groups into aromatic compounds, resulting in the formation of water-soluble dyes and pigments.
- Polymerization Reactions: The complex can be employed as a catalyst or co-catalyst in certain polymerization processes, contributing to the synthesis of specific polymers.

It's important to note that the handling and use of sulphur trioxide pyridine complex should be done with caution, as both pyridine and sulphur trioxide can be hazardous substances. Proper safety measures and protocols should be followed when working with these compounds.

15. Haloalkyl Chlorosulphates

Haloalkyl chlorosulphates, also referred to as chlorosulfonic acid esters, constitute a class of organic compounds synthesized from the reaction of sulphur trioxide (SO_3) with haloalkanes. These compounds, denoted by the general structure R-OSO_2Cl, where R represents an alkyl group, are pivotal in diverse chemical processes. Here's an overview of their production and applications:

a. **Production:**

- Reacting SO_3 with Chloroalkane: Sulphur trioxide (SO_3) is reacted with a chloroalkane, an alkane compound containing at least one chlorine atom. This reaction is meticulously controlled, typically involving low temperatures and precise addition rates, to ensure the desired product formation.
- Formation of Haloalkyl Chlorosulphate: The interaction between SO_3 and the chloroalkane leads to the substitution of the chlorine atom with the sulfonic acid group (SO_3H), resulting in the formation of the haloalkyl chlorosulphate compound (R-OSO_2Cl). Specific reaction conditions and reactants may vary based on the targeted Haloalkyl chlorosulphate.

b. **Uses:**

- Sulfation Reactions: Haloalkyl chlorosulphates serve as sulfating agents in chemical reactions, enabling the introduction of the sulfonic acid group (SO_3H) into organic compounds. This sulfation process alters the chemical and physical properties of the target molecules, facilitating the synthesis of desired products.
- Coupling Reactions: These compounds participate in coupling reactions with various nucleophiles, such as amines or alcohols, to forge new carbon-nitrogen or carbon-oxygen bonds. Such reactions are prevalent in the synthesis of pharmaceuticals and other organic compounds.
- Cross-coupling Reactions: Haloalkyl chlorosulphates act as coupling partners in cross-coupling reactions like the Buchwald-Hartwig reaction or the Ullmann reaction. These reactions enable the formation of carbon-carbon bonds, crucial for synthesizing complex organic molecules.
- Functional Group Transformations: Haloalkyl chlorosulphates undergo diverse chemical transformations, including substitution reactions, rearrangements, and conversions to other functional groups. These transformations offer versatility in organic synthesis, facilitating the preparation of diverse chemical structures.

Haloalkyl chlorosulphates play a vital role in organic synthesis, particularly as intermediates for pharmaceuticals, agrochemicals, and specialty chemicals. Their versatility and reactivity make them indispensable in modern chemical processes.

16. Dichlorodiphenyl Sulfone

Dichlorodiphenyl sulfone (DCDPS), though not directly derived from sulphur trioxide (SO_3), is typically synthesized from other precursor compounds. Here's an overview of its production and applications:

a. **Synthesis of Dichlorodiphenyl Sulfone:**

- DCDPS is synthesized through the sulfonation of chlorobenzene with sulphuric acid, often with various additives to enhance the formation of the 4,4′-isomer. Alternatively, it can be produced by chlorination of diphenyl sulfone.
- The most prevalent method involves the reaction of 4,4'-dichlorodiphenyl sulfide with an oxidizing agent like sodium hypochlorite (NaOCl) or hydrogen peroxide (H_2O_2) in the presence of a catalyst.
- The reaction proceeds as follows:
  4,4'-dichlorodiphenyl sulfide + Oxidizing agent → Dichlorodiphenyl sulfone
- This reaction converts the sulfide group (-S-) in the starting material to a sulfone group ($-SO_2-$) in the product, yielding dichlorodiphenyl sulfone.

b. **Uses of Dichlorodiphenyl Sulfone: DCDPS finds diverse applications across industries:**

- Polymers: Widely utilized as a monomer or co-monomer in the production of high-performance polymers like polysulfones and polyether sulfones. These polymers boast excellent thermal stability, chemical resistance, and mechanical properties, making them ideal for aerospace, automotive, electronics, and medical applications.
- Pharmaceuticals: Serves as an intermediate in pharmaceutical synthesis, acting as a building block for drugs and bioactive molecules.
- Organic Synthesis: Used in various organic synthesis reactions as a source of the sulfone functional group. The sulfone group's versatility enables it to participate in nucleophilic substitutions, oxidations, and cross-coupling reactions, expanding its applications in organic chemistry.
- Solvents: Acts as a solvent or solvent additive in specific industrial processes due to its solvency power and high thermal stability.

It's essential to note that the utilization of dichlorodiphenyl sulfone varies based on industry and application requirements.

17. Chloro Silanes

Inorganic chemistry encompasses chlorosilanes, a group of reactive, chlorine-containing compounds related to silane (SiH_4) and pivotal in numerous chemical processes.

a. **Production of Chlorosilanes:** Chlorosilanes are commonly synthesized from other precursor compounds, notably silicon and chlorine sources.
b. **Synthesis Process:** Typically, chlorosilanes are prepared by reacting silicon with chlorine or hydrogen chloride (HCl) gas. This reaction, occurring at high temperatures and often catalyzed, follows a general equation:
    a. $Si + xCl_2$ or $xHCl \rightarrow RnSiCl_4\text{-}n + (4\text{-}x)HCl$
    b. In this equation, "R" denotes an organic group, "n" represents the number of organic groups attached to the silicon atom, and "x" signifies the number of chlorine or hydrogen chloride molecules used. Adjusting these parameters yields different chlorosilanes.
c. **Uses of Chlorosilanes:** Chlorosilanes play vital roles across various industries:

- Silicone Production: Serving as crucial intermediates, chlorosilanes participate in the production of silicones. Through hydrolysis, they react with water to yield silanols, which further condense to form silicone polymers. Silicones have broad applications in automotive, construction, electronics, and medical sectors.
- Chemical Synthesis: Chlorosilanes act as versatile reagents in organic synthesis, undergoing substitution reactions to introduce silicon and organic moieties into molecules. They are particularly valuable in synthesizing organosilicon compounds used in coatings, adhesives, and specialty chemicals.
- Semiconductor Industry: Certain chlorosilanes like trichlorosilane ($SiHCl_3$) are integral in producing ultra-pure silicon for semiconductors. Trichlorosilane serves as a precursor in the chemical vapor deposition (CVD) process, depositing thin silicon films for semiconductor devices.
- Catalysts and Crosslinking Agents: Chlorosilanes function as catalysts or crosslinking agents in various reactions. Dichlorodimethylsilane $(CH_3)_2SiCl_2$, for instance, acts as a catalyst in producing polyethylene foam.

Chlorosilanes' versatility underscores their significance in numerous industrial processes and applications.

18. Methyl Hydrogen Sulphate

Methyl hydrogen sulphate (CH_3HSO_4) is a compound not directly derived from sulphur trioxide (SO_3). Instead, it is typically synthesized through an alternative process.

a. **Production of Methyl Hydrogen Sulphate:** Methyl hydrogen sulphate is commonly generated by the reaction between methanol (CH_3OH) and concentrated sulphuric acid (H_2SO_4).

b. **Synthesis Reaction:** $CH_3OH + H_2SO_4 \rightarrow CH_3HSO_4 + H_2O$
In this chemical equation, methanol reacts with sulphuric acid, yielding methyl hydrogen sulphate and water as products.

c. **Uses of Methyl Hydrogen Sulphate:** Methyl hydrogen sulphate finds versatile applications across various industries:

- Chemical Intermediates: It serves as a vital intermediate in synthesizing diverse chemicals. Acting as a starting material or reactant, it contributes to the production of pharmaceuticals, dyes, agrochemicals, and other specialty chemicals.
- Esterification Reactions: Methyl hydrogen sulphate functions as an esterifying agent in organic synthesis. Through reactions with alcohols, it forms esters. Under appropriate conditions, these esters can be converted back into their respective alcohols, providing flexibility in chemical transformations.
- Catalyst and Reagent: In certain chemical reactions, methyl hydrogen sulphate acts as a catalyst or reagent. It facilitates reactions such as acylation, alkylation, and esterification, enhancing reaction efficiency and product yields.
- Electroplating: Employed as an electrolyte additive in electroplating processes, methyl hydrogen sulphate enhances the efficiency and quality of electroplated metal coatings, contributing to improved surface finishes and corrosion resistance.
- Laboratory Reagent: Methyl hydrogen sulphate serves as a valuable reagent in laboratory settings for conducting various chemical reactions and synthesis procedures, supporting research and experimentation endeavors.

Methyl hydrogen sulphate's diverse applications underscore its significance across multiple industrial and scientific domains.

19. Dimethyl Sulphate

Dimethyl sulphate (DMS) is a crucial methylating agent in organic synthesis, vital for various industrial and laboratory applications.

a. **Production of Dimethyl Sulphate:** Dimethyl sulphate is synthesized through the reaction between sulphur trioxide (SO_3) and methanol (CH_3OH) in the presence of a catalyst, typically sulphuric acid (H_2SO_4).
Synthesis Reaction: $SO_3 + 2(CH_3OH \rightarrow (CH_3)2SO_4 + H_2O$
Here, sulphur trioxide reacts with methanol to yield dimethyl sulphate and water.

b. **Uses of Dimethyl Sulphate:** Dimethyl sulphate finds diverse applications across different industries:

- Chemical Intermediates: Serving as a crucial intermediate, DMS contributes to synthesizing various organic compounds, including pharmaceuticals, dyes, and pesticides. Its ability to introduce methyl groups into molecules during chemical reactions is particularly valuable.
- Methylation Reactions: Widely used as a methylating agent, DMS transfers methyl groups to substrates like alcohols, amines, and phenols. These methylation reactions play a pivotal role in organic synthesis, allowing modification of molecular properties and creating new compounds.
- Industrial Catalysts: Dimethyl sulphate acts as a catalyst in specific industrial processes, such as resin, plastic, and synthetic fiber manufacture. Its catalytic properties promote desired chemical reactions, enhancing production efficiency.
- Laboratory Reagent: Valued as a reagent in laboratory settings, DMS participates in various chemical reactions and synthesis procedures. Its ability to introduce methyl groups to specific molecules makes it a versatile tool in organic chemistry research.

19a. Production and Application of Dimethyl Sulfate in Dye Manufacturing

Dimethyl sulfate (DMS), represented chemically as $(CH_3)\ _2SO_4$, is a versatile compound extensively employed in various industrial processes, notably in dye manufacturing. Below is an overview of its production and utilization in dye production:

a. **Production Process:** Dimethyl sulfate is typically synthesized through the reaction between sulfuric acid (H_2SO_4) and methanol (CH_3OH), often catalyzed by substances like oleum or sulfur trioxide (SO_3).
   $CH_3OH + H_2SO_4 \rightarrow (CH_3)\ 2SO_4 + H_2O$
b. **Key Steps: Esterification of Methanol with Sulfuric Acid:** Methanol undergoes an esterification reaction with sulfuric acid, resulting in the formation of dimethyl sulfate and water.
c. **Application in Dye Manufacturing:** Dimethyl sulfate finds diverse applications in dye production, including:

- Dye Synthesis: Acting as a methylating agent, dimethyl sulfate facilitates the synthesis of specific dye types by introducing methyl groups into target chemical compounds. This modification alters their properties, enhancing their dyeing characteristics.
- Color Enhancement: Dimethyl sulfate is utilized to modify the color and hue of dyes through chemical alterations to their molecular structure. This process enables the creation of a wide array of vibrant and durable dye formulations.
- Introduction of Functional Groups: Dimethyl sulfate enables the integration of functional groups into dye molecules, improving their affinity for various substrates and overall performance in dyeing processes.

Dimethyl sulfate plays a crucial role in dye manufacturing, facilitating the synthesis and enhancement of dyes with desirable properties. By serving as a methylating agent and enabling molecular modifications, dimethyl sulfate contributes to the production of high-quality dyes utilized across diverse industries. However, ensuring its safe handling and compliance with safety regulations are imperative to safeguard the well-being of workers and environmental integrity.

19b. Dimethyl Sulfate Production and Applications in Pharmaceuticals

Dimethyl sulfate (DMS) is synthesized by reacting sulfur trioxide (SO_3) with methanol (CH_3OH).

Chemical Equation: $SO_3 + 2CH_3OH \rightarrow (CH_3O)\ 2SO_2$

a. **Detailed Process:**

- Preparation of Sulfur Trioxide (SO_3): Sulfur trioxide is typically obtained by oxidizing sulfur dioxide (SO_2) with a suitable catalyst, such as vanadium pentoxide (V_2O_5).
  $2SO_2 + O_2 \rightarrow 2SO_3$
- Reaction with Methanol: Sulfur trioxide reacts with methanol under controlled conditions to yield dimethyl sulfate ($(CH_3O)\ 2SO_2$).
  $SO_3 + 2CH_3OH \rightarrow (CH_3O)\ 2SO_2$
  The reaction parameters are adjusted to maximize the dimethyl sulfate yield.
- Purification and Refinement: Dimethyl sulfate undergoes purification via techniques like distillation or extraction to remove impurities.
- Characterization and Quality Control: Purified dimethyl sulfate is subjected to rigorous characterization to ensure compliance with pharmaceutical standards.

b. **Pharmaceutical Applications:**

- Methylation Reagent: Dimethyl sulfate serves as a versatile methylation reagent in organic synthesis, especially in pharmaceutical manufacturing. It introduces methyl groups into organic molecules, altering their properties or biological activity.
- Drug Synthesis: Dimethyl sulfate is used in synthesizing various pharmaceutical compounds, including intermediates, active pharmaceutical ingredients (APIs), and specialty chemicals for drug formulations.
- Research and Development: Pharmaceutical R&D labs employ dimethyl sulfate for modifying nucleic acids and proteins and synthesizing pharmaceutical reference standards.
- Environmental Impact: Effluent treatment and waste management protocols are essential to minimize environmental contamination and ensure compliance with regulatory requirements governing hazardous chemical handling and disposal.

Summary

Sulphur trioxide is a highly reactive compound consisting of three oxygen atoms bonded to a sulphur atom. It holds paramount importance in various industrial applications. Here's a concise overview:

a. **Production:** Sulphur trioxide is primarily produced through the contact process, involving the reaction of sulphur dioxide (SO_2) with oxygen (O_2) in the presence of a catalyst like vanadium pentoxide (V_2O_5). Alternatively, direct oxidation of elemental sulphur can also yield sulphur trioxide.

b. **Chemical Properties:** While colourless under standard conditions, sulphur trioxide readily reacts with atmospheric moisture to form sulphuric acid (H_2SO_4). It exhibits high solubility in water, leading to the formation of fuming sulphuric acid or oleum, known for its extreme corrosiveness.

c. **Uses:**

- Sulphuric Acid Production: A vital precursor, sulphur trioxide is instrumental in sulphuric acid (H_2SO_4) production, extensively used in various industries for manufacturing fertilizers, detergents, and batteries.
- Sulfonation Reactions: Sulphur trioxide plays a pivotal role in sulfonation reactions, crucial for producing detergents, surfactants, and dyes. It introduces sulfonic acid groups into molecules, enhancing their properties.
- Catalysts and Chemical Synthesis: It serves as a catalyst or reactant in diverse chemical processes, including the production of plastics, pharmaceutical intermediates, and specialty chemicals.
- Desiccant and Dehydrating Agent: Owing to its strong water affinity, sulphur trioxide finds application as a desiccant and dehydrating agent, commonly employed in drying gasses and solvents.
- Laboratory Reagent: In laboratory settings, sulphur trioxide derivatives function as reagents for specific chemical reactions or analytical techniques.

REFERENCES:

1. Greenwood, Norman N.; Earnshaw, Alan (1997). Chemistry of the Elements (2nd ed.). Butterworth-Heinemann. pp. 703–704. ISBN 978-0-08-037941-8.
2. Hermann Müller "sulphuric Acid and sulphur Trioxide" in Ullmann's Encyclopedia of Industrial Chemistry, Wiley-VCH, Weinheim. 2000
3. Yoshikubo, K.; Suzuki, M. (2000). "Sulfamic Acid and Sulfamates". Kirk-Othmer Encyclopedia of Chemical Technology. Brauer, George (1963). "Handbook of Preparative Inorganic Chemistry"
4. Speight, James G. (2017-01-01), Speight, James G. (ed.), "Chapter Three - Industrial Inorganic Chemistry", Environmental Inorganic Chemistry for Engineers, Butterworth-Heinemann, pp. 111–169, ISBN 978-0-12-849891-0, retrieved 2021-10-26
5. Considine, Douglas M. (1974). Chemical and Process Technology Encyclopedia. McGraw-Hill. pp. 1070–1.
6. Page, P. C. B.; Wilkes, R. D.; Reynolds, D. (1995). "Alkyl Chalcogenides: sulphur-based Functional Groups". In Ley, Steven V. (ed.). Synthesis: Carbon with One Heteroatom Attached by a Single Bond. Comprehensive Organic Functional Group Transformations. Elsevier.
7. Maynard, G. D. (2001). "sulphuryl Chloride". Encyclopedia of Reagents for Organic Synthesis. John Wiley & Sons
8. Merck Index, 11th Edition, 9625
9. Datenblatt der Oriental Union Chemical Co., Sodium Isethionate – An Eco-friendly Surfactant for Industrial Applications, (PDF; 90 kB), retrieved 17 August 2012.
10. Winum, J. Y.; Scozzafava, A.; Montero, J. L.; Supuran, C. T. (2005). "Sulfamates and their therapeutic potential". Medicinal Research Reviews. 25 (2): 186–228
11. Kovacic, Peter; Brace, Neal O. (1954). "Chlorination of Aromatic Compounds with Metal Chlorides". Journal of the American Chemical Society. 76 (21): 5491–5494

Chapter 4: The Evolution and Properties of Sulphuric Acid

HISTORY

Sulphuric acid's journey spans centuries, from its discovery by the alchemist Jabir ibn Hayyan in the 8th century to its large-scale production in the 18th century by Joseph Priestley. The chemical's significance grew exponentially in the 19th century with the advent of the Solvay process, replacing the Leblanc process and cementing its pivotal role in industrial processes, including the production of fertilizers, dyes, detergents, and petroleum refining.

DESCRIPTION

Sulphuric acid, with its chemical formula H_2SO_4, stands as a highly corrosive and potent mineral acid. Its molecular structure, depicted as HO-S-OH, embodies its formidable nature, being a colorless, dense, and oily liquid, soluble in water, and possessing a distinctive pungent odor. Notably, it serves as a premier dehydrating agent and oxidizing agent, underpinning its multifaceted industrial applications.

PROPERTIES

Sulphuric acid, an inorganic compound comprised of hydrogen, sulfur, and oxygen, epitomizes chemical reactivity and corrosiveness. With a chemical formula H_2SO_4, it exists as a dense, corrosive liquid, characterized by its highly oxidizing nature and tendency to react violently with various substances.

Physical Properties:

- **Appearance**: Sulphuric acid is typically a colorless to slightly yellowish liquid. However, it can appear dark brown due to impurities.
- **State**: It is commonly found in a liquid state at room temperature and pressure.
- **Odor**: Sulphuric acid has no distinctive odor when pure. However, impurities may impart a characteristic smell.
- **Density**: It is a dense liquid with a density of around 1.84 g/cm^3.

- **Viscosity**: Sulphuric acid has a relatively high viscosity, meaning it flows more slowly compared to water.
- **Solubility**: It is highly soluble in water, and its solubility increases with higher temperatures.
- **Boiling Point**: Sulphuric acid has a boiling point of approximately 337°C (638°F) at standard atmospheric pressure.

Chemical Properties:

- **Acidity**: Sulphuric acid is a strong acid and dissociates completely in water to produce hydronium ions (H_3O+) and sulphate ions (SO_4^{2-}).
- **Dehydration**: It is an excellent dehydrating agent, capable of removing water molecules from other substances.
- **Oxidizing Agent**: Sulphuric acid is a powerful oxidizing agent, capable of oxidizing various substances, including non-metals and some metals.
- **Reactivity with Metals**: It reacts with metals to produce hydrogen gas and metal sulphates. The reaction rate varies depending on the reactivity of the metal.
- **Reaction with Bases**: Sulphuric acid reacts vigorously with bases to form salts and water. This reaction is highly exothermic.
- **Electrolytic Properties**: It can conduct electricity when dissociated in water, making it useful in various electrochemical processes.
- **Reaction with Organic Compounds**: Sulphuric acid can react with organic compounds, often leading to dehydration reactions or the addition of sulphonic acid groups.

These physical and chemical properties make sulphuric acid a versatile compound with numerous industrial applications, but its handling requires caution due to its corrosive and hazardous nature.

Sources Of Sulphuric Acid

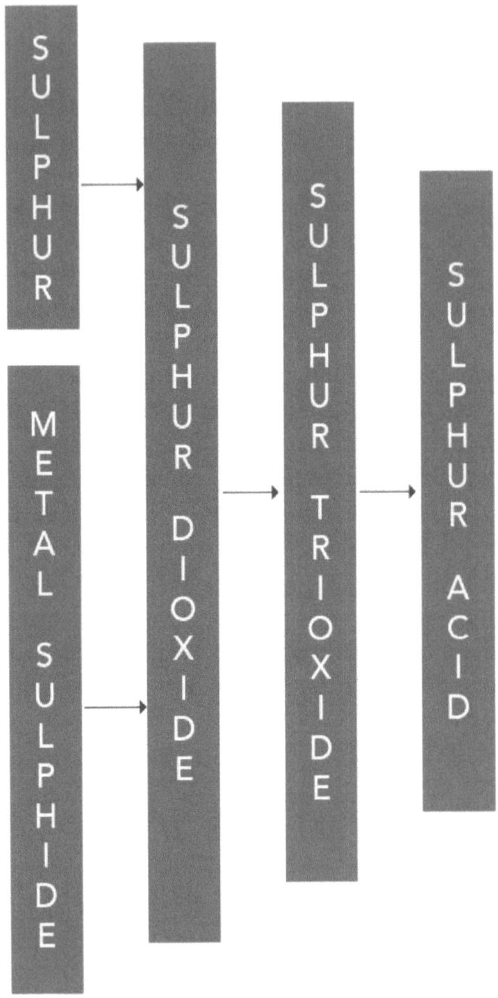

Sulphuric acid, a vital industrial chemical, finds extensive use across various sectors, from fertilizer manufacturing to mineral processing and chemical synthesis. Its production from elemental sulphur involves intricate processes, primarily employing the contact method or the wet sulphuric acid process.

a. Contact Process:

- Sulphur Combustion: Elemental sulphur undergoes combustion to produce sulphur dioxide: $S\ (s) + O_2 \rightarrow SO_2$
- Oxidation to Sulphur Trioxide: Sulphur dioxide reacts with oxygen in the presence of a vanadium(V) oxide catalyst to yield sulphur trioxide: $2SO_2 + O_2 \rightleftarrows 2SO_3$
- Oleum Formation: Sulphur trioxide is absorbed into 97–98% sulphuric acid, forming oleum: $H_2SO_4 + SO_3 \rightarrow H_2S_2O_7$
- Dilution: Oleum is diluted with water to obtain concentrated sulphuric acid: $H_2S_2O_7 + H_2O \rightarrow 2H_2SO_4$

b. Wet Sulphuric Acid Process:

This method involves the direct dissolution of sulphur trioxide in water, albeit infrequently practiced due to the exothermic nature of the reaction, necessitating condensation and separation of the resulting hot aerosol of sulphuric acid.

Key Steps:

- Sulphur Trioxide Production: Sulphur dioxide reacts with oxygen over a vanadium pentoxide catalyst at around 450°C and 0.5–1.0 atm pressure: $SO_2 + O_2 \rightarrow 2SO_3$
- Absorption in Water: Highly reactive sulphur trioxide is absorbed in water to form sulphuric acid, a process managed in absorption towers to dissipate the substantial heat generated: $SO_3 + H_2O \rightarrow H_2SO_4$
- Dilution: The concentrated sulphuric acid is diluted by controlled addition of water, with careful heat management to prevent safety hazards: $H_2SO_4 + H_2O \rightarrow H_2SO_4(aq)$

APPLICATION OF SULPHURIC ACID-SHEET 1

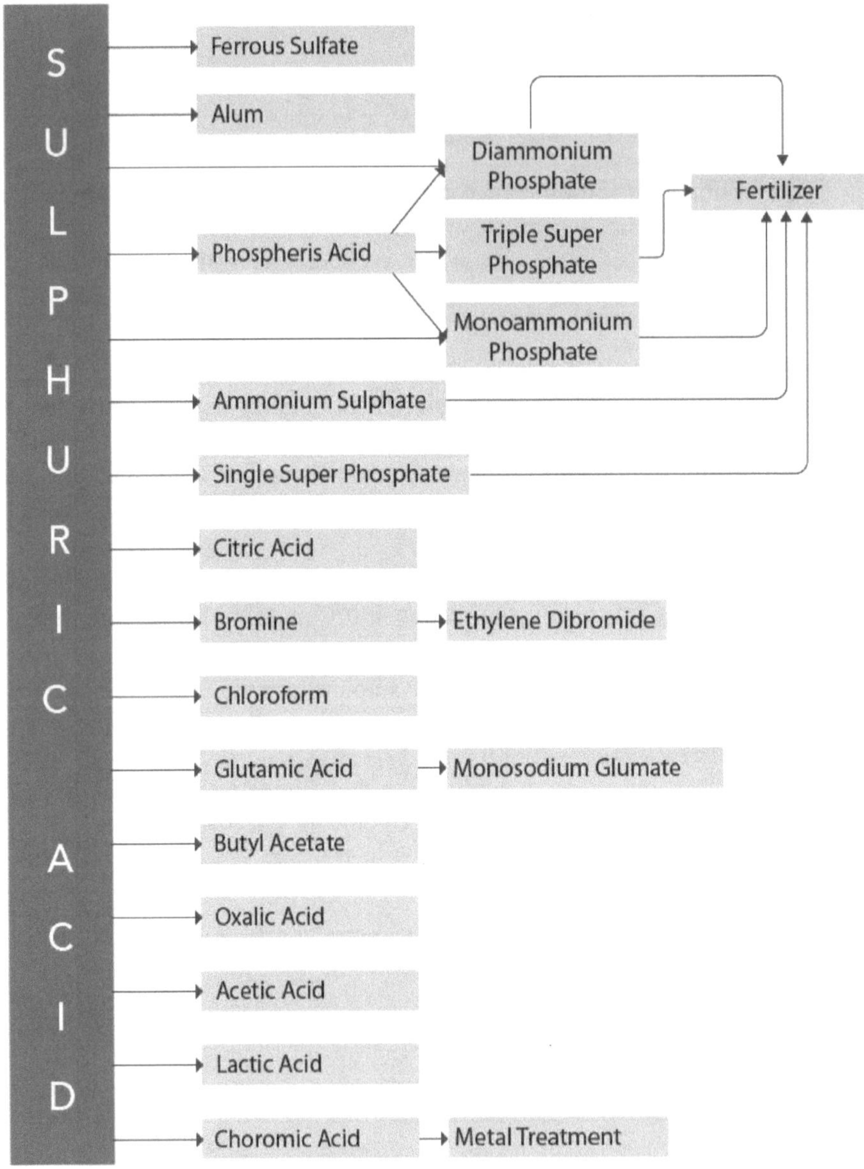

1. Ferrous Sulphate

Ferrous sulphate, also referred to as iron(II) sulphate, is a crucial industrial compound synthesized from sulphuric acid (H_2SO_4) and iron(II) oxide (FeO). The chemical reaction between these components yields ferrous sulphate alongside water: $FeO + H_2SO_4 \rightarrow FeSO_4 + H_2O$

a. The production process involves a series of chemical reactions:

- Iron oxide reacts with sulphuric acid to form ferric sulphate and water: $Fe_2O_3 + H_2SO_4 \rightarrow Fe_2(SO_4)_3 + H_2O$
- Ferric sulphate further reacts with iron to yield ferrous sulphate: $Fe_2(SO_4)_3 + Fe \rightarrow 3FeSO_4$
- Overall Equation: $Fe_2O_3 + 3H_2SO_4 + Fe \rightarrow 3FeSO_4 + 3H_2O$

Typically, iron is introduced to the reaction mixture to facilitate the reduction of ferric sulphate to ferrous sulphate, which is then purified and dried to obtain the final product.

b. Ferrous sulphate boasts a multitude of industrial applications:

- Wastewater Treatment: It serves as a coagulant for water purification processes.
- Chemical Synthesis: It acts as a reducing agent in various chemical reactions.
- Agriculture: Ferrous sulphate is utilized as a fertilizer and soil amendment to supply iron to plants, improving growth and development.
- pH Adjustment: It lowers the pH of alkaline soils, enhancing nutrient availability for plants.
- Leather Tanning: It is employed in the tanning process for leather.
- Textile Industry: Ferrous sulphate acts as a mordant in dyeing textiles.
- Construction: It stains concrete and certain stones, imparting a yellowish rust color
- Animal Feed and Human Health: Ferrous sulphate serves as an iron supplement in animal feed and is used in treating iron-deficiency anemia in humans.

2. Alum

Alum, scientifically known as potassium aluminum sulphate, is a valuable compound synthesized from sulphuric acid (H_2SO_4), potassium sulphate, and either aluminum hydroxide [$Al(OH)_3$] or aluminum oxide (Al_2O_3).

a. **The production process entails a series of reactions:**

- Aluminum hydroxide reacts with sulphuric acid to yield aluminum sulphate and water:
- $2Al(OH)_3 + 3H_2SO_4 \rightarrow Al_2(SO_4)_3 + 6H_2O$
- Potassium sulphate is then added to the solution, resulting in the formation of alum:
- $Al_2(SO_4)_3 + K_2SO_4 + 24H_2O \rightarrow K_2SO_4.Al_2(SO_4)_3.24H_2O$ (alum)
  Overall Reaction:
  $2Al + 3H_2SO_4 + K_2SO_4 + 24H_2O \rightarrow K_2SO_4.Al_2(SO_4)_3.24H_2O + 3H_2$

b. **Alum boasts diverse industrial applications:**

- Water Treatment: It serves as a coagulant to purify water by causing impurities to coagulate and settle.
- Baking Powder: Alum acts as a component in baking powder to provide leavening and regulate acidity.
- Fireproofing: It is utilized in fireproofing materials for wood and textiles due to its flame-retardant properties.
- Textile Industry: Alum functions as a mordant in dyeing textiles, enhancing color fastness and durability.
- Leather Tanning: It is employed in the tanning process to stabilize the structure of leather and improve its quality.
- Ceramics and Paper Production: Alum contributes to the production of ceramics and paper by enhancing strength and durability.
- Food Preservation: Alum is used in food preservation methods such as pickling and canning to inhibit bacterial growth and maintain food quality.

3. Phosphoric Acid

Phosphoric acid, a vital chemical compound, is synthesized through a multi-step process involving the reaction of sulphuric acid with ground phosphate rock. The typical reaction unfolds as follows:

Phosphate rock ($Ca_3(PO_4)_2$) + Sulphuric Acid (H_2SO_4) $\rightarrow$ Phosphoric Acid (H_3PO_4) + Calcium Sulphate ($CaSO_4$)

The phosphoric acid produced undergoes purification and concentration in large-scale processing plants before finding application in various industries.

a. **Agricultural Sector:** Phosphoric acid serves as a crucial fertilizer, supplying plants with essential phosphorus. Phosphorus, integral to plant DNA, facilitates energy transfer and storage in plants, fostering healthy growth and development.

b. **Food Industry:** In the food sector, phosphoric acid functions as an acidulant and flavor enhancer in soft drinks, sports beverages, and processed foods. Additionally, it acts as a sequestrant, preventing the undesirable formation of mineral deposits by inhibiting mineral ion binding.

c. **Chemical Production:** Phosphoric acid finds utility in the synthesis of diverse chemicals, including detergents, cosmetics, and cleaning agents. It contributes as a cleaning agent, pH adjuster, or as a precursor in the manufacture of various chemical compounds.

While phosphoric acid is generally deemed safe for use in food and industrial contexts, improper handling can lead to skin irritation or other health complications. Thus, strict adherence to safety protocols and handling procedures is imperative to mitigate any potential risks associated with its utilization.

4. Diammonium Phosphate (DAP) Derived from Phosphoric Acid

Diammonium phosphate (DAP) is synthesized through a reaction between phosphoric acid (H_3PO_4) and ammonia (NH_3), yielding the compound $(NH_4)_2HPO_4$. This crucial reaction underscores the importance of sulphuric acid in the production chain, as it facilitates the generation of phosphoric acid, a precursor to DAP:

Phosphate rock ($Ca_3(PO_4)_2$) + Sulphuric Acid (H_2SO_4) → Phosphoric Acid (H_3PO_4) + Calcium Sulphate ($CaSO_4$)

Diammonium phosphate (DAP) finds diverse applications across various sectors:

a. **Agricultural Sector:** DAP serves as a pivotal fertilizer, offering plants a vital blend of nitrogen and phosphorus, essential for robust growth and development.

b. **Fire Retardant:** Utilized as a fire retardant in wood products, DAP effectively lowers the combustion temperature, decreases weight loss rates, and enhances residue production, bolstering fire resistance.

c. **Food Industry:** Within the food sector, DAP acts as a yeast nutrient in bread-making and serves as a component in some food additives and animal feed formulations.

d. **Industrial Applications:** DAP plays a vital role in the production of particleboard, medium-density fibreboard, and other wood products, where its flame-retardant properties enhance safety and durability.

While DAP is generally deemed safe for industrial and agricultural usage, caution must be exercised to prevent potential toxicity if ingested in large quantities. Thus, adherence to proper safety protocols during handling and application is essential to mitigate any associated risks.

4a. Triple Super Phosphate (TSP) Fertilizer from Phosphoric Acid

Triple superphosphate (TSP) fertilizer is manufactured through a process involving the treatment of rock phosphate with sulphuric acid, leading to the formation of phosphoric acid. Subsequently, this phosphoric acid undergoes further reaction to yield TSP, a potent source of phosphorus essential for robust plant growth and development in agriculture. The utilization of TSP extends to various agricultural applications due to its efficacy in providing vital nutrients to crops.

Production Process: The synthesis of TSP involves the following reaction:

$Ca_3(PO_4)_2 + 2H_3PO_4 \rightarrow 2Ca(H_2PO_4)_2 + 3CaSO_4$

TSP, also referred to as Concentrated Superphosphate, boasts a significantly higher concentration of effective ingredient P_2O_5 compared to Single Superphosphate (SSP), exceeding 40%.

Applications and Uses:

a. **Agricultural Fertilizer:** TSP serves as a crucial fertilizer in agriculture, supplying crops with essential phosphorus nutrients vital for various physiological processes. Phosphorus plays a pivotal role in root growth, seed development, and flower production, contributing to enhanced overall plant vigor and yield.
b. **Soil Conditioners:** TSP can be incorporated into soil as a soil conditioner to enhance its structure and fertility. By improving soil tilth and nutrient retention capacity, TSP fosters an optimal growth environment conducive to robust plant growth and development.
c. **Plant Growth Regulators:** TSP finds utility as an ingredient in some plant growth regulators, which are chemicals utilized to regulate and optimize the growth and development of plants. Through targeted application, TSP aids in modulating plant growth patterns and optimizing crop yield potential.
d. **Animal Feed:** In certain cases, TSP is incorporated into animal feed formulations to provide essential phosphorus nutrients to livestock, supporting their overall health and growth.

4b. Monoammonium Phosphate (MAP) from Phosphoric Acid

Monoammonium phosphate (MAP) stands as a potent fertilizer synthesized from the chemical reaction between phosphoric acid and ammonia. This compound serves as a highly concentrated reservoir of phosphorus, a vital nutrient indispensable for fostering robust plant growth and development.

Industrially, the preparation of monoammonium phosphate (MAP) entails the exothermic fusion of ammonia and phosphoric acid in precise proportions, as depicted by the equation: $NH_3 + H_3PO_4 \rightarrow NH_4H_2PO_4$. Upon completion of the reaction, crystalline MAP precipitates, ready for application.

MAP finds versatile utility across various domains:

a. **Agriculture:** MAP serves as a fundamental fertilizer, furnishing plants with the requisite phosphorus essential for their optimal growth and development. By ensuring an adequate supply of phosphorus, MAP contributes significantly to enhancing crop yields and improving overall agricultural productivity.
b. **Horticulture:** In horticultural practices, MAP plays a crucial role in fostering healthy root development, stimulating robust flower formation, and promoting the prolific production of fruits in plants. Its application aids in nurturing vibrant and thriving plant ecosystems, conducive to bountiful harvests.
c. **Soil Amendment:** As a potent source of phosphorus, MAP serves as an effective soil amendment agent, bolstering the phosphorus content in soil and enhancing its fertility. By fortifying soil with essential nutrients, MAP facilitates the creation of an optimal growth environment for plants, fostering lush vegetation and robust agricultural yields.

It is imperative to exercise responsible usage practices when employing MAP and adhere to recommended application rates meticulously. By exercising caution and ensuring judicious utilization, potential adverse environmental repercussions such as excessive nutrient runoff into waterways can be mitigated effectively.

5. Ammonium Sulphate (AMS) as Fertilizer

Ammonium sulphate (AMS) stands as a crucial fertilizer derived from the chemical fusion of sulphuric acid and ammonia. Comprising 21% nitrogen and 24% sulphur, it serves as a dual-source of essential nutrients vital for fostering robust plant growth and development. Upon dissolution in water within agricultural settings, AMS undergoes dissociation into ammonium ions (NH_4^+) and sulphate ions (SO_4^{2-}).

The chemical synthesis of AMS entails the following reaction:
$2NH_3 + H_2SO_4 \rightarrow (NH_4)_2SO_4$

AMS finds widespread utility across diverse domains:

a. **Agriculture:** AMS serves as a pivotal fertilizer, furnishing plants with indispensable nitrogen and sulphur nutrients crucial for their growth and development. Additionally, it functions as an agricultural spray adjuvant, facilitating the efficacy of water-soluble insecticides, herbicides, and fungicides. By binding iron and calcium cations present in both water and plant cells, it aids in optimizing nutrient uptake and utilization.
b. **Horticulture:** In horticultural applications, AMS plays a vital role in fostering robust plant growth, stimulating healthy root development, and enhancing flower and fruit production. Its judicious application contributes to nurturing thriving plant ecosystems conducive to flourishing vegetation.
c. **Soil Amendment:** AMS serves as a potent agent for augmenting the nitrogen and sulphur content in soil, thereby enhancing soil fertility. By bolstering soil nutrient levels, it creates an optimal growth environment for plants, fostering robust agricultural yields and sustainable crop production.

It is imperative to exercise responsible usage practices when employing AMS and adhere to recommended application rates meticulously. By exercising caution and ensuring judicious utilization, potential adverse environmental repercussions such as excessive nutrient runoff into waterways can be mitigated effectively.

Furthermore, it is noteworthy that airborne particles originating from evaporated ammonium sulphate contribute to approximately 30% of fine particulate pollution globally. Thus, prudent measures should be adopted to mitigate potential environmental impacts associated with AMS usage.

6. Single super phosphate production

Single superphosphate (SSP), commonly referred to as simple superphosphate, is a crucial fertilizer derived from the reaction between sulphuric acid (H_2SO_4) and rock phosphate. This reaction results in the formation of single superphosphate, a compound essential for agricultural practices and various industrial applications.

The chemical reaction between rock phosphate and sulphuric acid is represented as follows:

$[Ca_3(PO_4)_2]. CaF_2+4H_2SO_4$ (70%) $\rightarrow 3CaH_4(PO_4)_2. 3(CaSO_4)+2HF$

Since phosphate rock naturally contains calcium fluoride (CaF_2), its reaction with sulphuric acid results in the generation of hydrogen fluoride (HF) fumes. These fumes are subsequently eliminated through various chemical processes to ensure safety and environmental compliance.

Single superphosphate finds diverse applications, including:

a. **Agriculture:** SSP serves as a vital source of phosphorus, an indispensable nutrient for plant growth. It is applied to soil to enhance phosphorus levels, thereby promoting robust plant development and increased crop yields. Additionally, it serves as a yeast nutrient in bread-making and is incorporated into certain food additives and animal feed formulations.
b. **Production of Other Phosphatic Fertilizers:** SSP plays a pivotal role as an ingredient in the production of other phosphatic fertilizers, such as triple superphosphate and diammonium phosphate. These fertilizers offer distinct phosphorus concentrations and are tailored to meet specific agricultural requirements.

While single superphosphate is generally regarded as safe for use in both industrial and agricultural settings, caution must be exercised to prevent potential toxicity risks, especially when dealing with large quantities. Adherence to proper safety protocols and handling procedures is paramount to mitigate the risk of injury or environmental harm.

7. Citric Acid

Citric acid, an organic compound renowned for its colorless nature, is naturally abundant in citrus fruits. While it isn't directly synthesized from sulphuric acid (H_2SO_4), the latter plays a pivotal role in the production process of citric acid, particularly through fermentation.

Production Process: Citric acid is typically produced via fermentation, where sugar undergoes transformation with the assistance of a specific mold known as Aspergillus niger. Throughout this process, sulphuric acid serves as a vital component, regulating pH levels and facilitating the purification of the resultant citric acid.

Applications and Uses:

a. **Food and Beverage Industry:** Citric acid finds extensive application as a flavor enhancer and preservative in various food and beverage products. Its characteristic tartness enhances the taste profiles of candies, soft drinks, and processed fruits, while also serving to prolong their shelf life.
b. **Cleaning Agent:** Renowned for its adeptness in removing mineral deposits and stains, citric acid serves as a versatile cleaning agent. It features prominently in household cleaners, dishwashing detergents, and descaling agents, effectively tackling stubborn residues and stains.
c. **Personal Care Products:** Within the realm of personal care, citric acid assumes multiple roles. It functions as an exfoliant in facial scrubs, aiding in

the removal of dead skin cells, and as a pH adjuster in shampoos, conditioners, and soaps, ensuring optimal product efficacy and skin compatibility.

In conclusion, citric acid stands as a versatile compound with widespread applications across various industries. By adhering to proper safety protocols and handling techniques, the multifaceted benefits of citric acid can be harnessed effectively while minimizing the risk of harm to individuals and the environment.

8. Bromine

Bromine (Br_2) isn't directly derived from sulphuric acid (H_2SO_4), but sulphuric acid plays a crucial role in the production of certain bromine compounds.

Production Process: Bromine is typically produced via the electrolysis of seawater or brine, which naturally contains dissolved bromides. During this electrolysis process, sulphuric acid can serve as the electrolyte, facilitating the extraction of bromine from the brine solution.

Applications and Uses:

a. **Flame Retardants:** Bromine serves as a vital component in flame retardants, effectively reducing the flammability of materials and inhibiting the spread of fire. Its incorporation into various products, such as textiles, plastics, and insulation materials, enhances their fire-resistant properties, making them safer for use in numerous applications.

b. **Pharmaceuticals and Agricultural Chemicals:** Bromine finds extensive utilization as a foundational material in the synthesis of several key compounds, including bromoform and ethylene bromide, which are integral components in the production of pharmaceuticals and agricultural chemicals. These compounds play pivotal roles in various processes and formulations within these industries.

c. **Water Treatment:** In water treatment applications, bromine serves as a potent disinfectant, effectively eliminating bacteria, viruses, and other harmful microorganisms present in water sources. Its efficacy extends to controlling algae growth in swimming pools, hot tubs, and spas, ensuring hygienic and safe recreational environments for users.

In summary, bromine's diverse applications underscore its importance across various industries, from enhancing fire safety to safeguarding public health through water disinfection. By prioritizing safety measures and responsible handling practices, the beneficial attributes of bromine can be harnessed effectively while minimizing associated risks.

8a. Ethylene Dibromide: Production from Bromine

Ethylene dibromide (EDB) is an organobromine compound predominantly synthesized rather than occurring naturally, albeit trace amounts are found in oceanic environments, possibly originating from algae and kelp. This dense, colorless liquid emits a subtle, sweet odor detectable at 10 ppm and serves as a widely employed fumigant.

Production Process: EDB is manufactured from bromine (Br_2) through a reaction involving ethylene. This process, facilitated by a catalyst, yields EDB as the end product.

Applications and Uses:

a. **Industrial Intermediates:** EDB serves as a crucial intermediate in the chemical industry, contributing to the production of various other chemicals, including alkylates utilized as additives in gasoline formulations. Its role in this capacity underscores its significance in enabling the synthesis of diverse chemical compounds.
b. **Pesticide and Fumigant:** EDB finds extensive utilization as a pesticide and fumigant across agricultural and industrial domains. Its efficacy in controlling pests, such as insects and mites infesting stored grains, underscores its importance in safeguarding agricultural produce. Additionally, EDB is employed for pest control in soil and on crops, further highlighting its utility in agricultural applications.

In conclusion, while ethylene dibromide plays a vital role in various industrial and agricultural applications, its usage necessitates careful consideration of safety protocols and adherence to regulatory guidelines to mitigate potential risks to human health and the environment. By promoting sustainable practices and exploring alternative solutions, the adverse impacts associated with EDB can be effectively addressed.

9. Chloroform:

Chloroform, also recognized as Trichloromethane ($CHCl_3$), is industrially synthesized through a process involving the heating of a chlorine mixture with either methyl chloride (CH_3Cl) or methane (CH_4). This method operates at temperatures ranging from 400 to 500 °C, facilitating free radical halogenation reactions that gradually chlorinate the precursor compounds.

a. **Initial Reactions:**
   - $CH_4 + Cl_2 \rightarrow CH_3Cl + HCl$
   - $CH_3Cl + Cl_2 \rightarrow CH_2Cl_2 + HCl$
   - $CH_2Cl_2 + Cl_2 \rightarrow CHCl_3 + HCl$

b. Further Chlorination:

$CHCl_3 + Cl_2 \rightarrow CCl_4 + HCl$

The resulting mixture comprises chloromethanes, including chloromethane, methylene chloride (dichloromethane), chloroform, and carbon tetrachloride (tetrachloromethane). These compounds can be separated via distillation. Hydrogen chloride gas, a byproduct of this process, is subsequently isolated.

Applications:

Medical and Pharmaceutical Industry:

- Historically, chloroform was used as an anesthetic in medical procedures. However, due to its potential adverse effects, its medical use has significantly declined.
- It serves as a precursor in the synthesis of various pharmaceutical compounds.

Solvent: Chloroform acts as a solvent in laboratory settings and industrial processes, particularly for dissolving organic compounds. Its ability to dissolve a wide range of substances makes it valuable in chemical synthesis and extraction processes.

Production of Other Chemicals: Chloroform serves as a key intermediate in the production of other chemicals, including fluorocarbons, dyes, and pesticides.

Miscellaneous Uses:

- It finds applications in the manufacture of adhesives, rubber products, and plastics.
- In the past, chloroform was used as a solvent in the production of early refrigerants, although its usage in this regard has declined due to environmental concerns.

In conclusion, while chloroform remains relevant in certain industrial processes and applications, its usage requires stringent adherence to safety protocols and regulatory guidelines to mitigate potential health and environmental risks associated with its handling and disposal. Efforts to explore alternative substances and practices further contribute to the responsible management of chloroform in various industries.

The process of producing chloroformic acid involves a chemical reaction between chloroform and sulphuric acid, resulting in the formation of chloroformic acid. The reaction can be represented as follows:

$$\underset{Chloroform}{\underline{CHCl_3}} + \underset{Sulphuric\ Acid}{\underline{H_2SO_4}} \rightarrow \underset{Chloroformic\ Acid}{\underline{CHCl_3H_2SO_4}}$$

Chloroform, characterized by its colorless, nonflammable nature and sweet, ether-like odor, was historically employed as an anesthetic. However, due to concerns regarding

its potential toxicity and carcinogenic properties, its use in this capacity has significantly declined.
Presently, chloroform finds primary application as a solvent, particularly in the pharmaceutical, pesticide, and chemical manufacturing industries. Its versatility makes it valuable in various processes, including the production of refrigerants and as a cleaning agent in electronics manufacturing.

It is crucial to acknowledge the potential adverse effects of chloroform on both human health and the environment. Regulatory measures are in place in many jurisdictions to control its usage due to these concerns. Therefore, a comprehensive assessment of the safety and environmental implications associated with chloroform utilization is imperative before its incorporation into any process or application.

10. Glutamic Acid

Glutamic acid, an α-amino acid crucial for protein biosynthesis and a primary excitatory neurotransmitter in vertebrates, holds significance across various sectors. While it's not directly synthesized from sulphuric acid, glutamic acid is naturally present in protein-rich foods and can be industrially produced through fermentation processes involving microorganisms.

Reaction: The initial step in the production process involves the following reaction:
Glutamine + $H_2SO_4 \rightarrow$ α-ketoglutaric acid

Glutamic acid serves as a pivotal ingredient in the food industry, notably as a flavor enhancer. It constitutes a vital component of monosodium glutamate (MSG), enhancing the taste profile of diverse food products ranging from soups and sauces to snacks and processed meats.

Moreover, glutamic acid finds utility in various other domains:

a. **Nutritional Supplementation:** Occasionally employed as a nutritional supplement, glutamic acid provides the body with additional amino acids, aiding in diverse physiological processes.
b. **Biodegradable Plastics:** Glutamic acid serves as a foundational element in the manufacture of biodegradable plastics, contributing to sustainable material solutions.
c. **Cosmetics:** In the cosmetic industry, glutamic acid features as an ingredient in skincare and haircare formulations, offering potential benefits for personal care regimes.

10a. Production and Applications of Monosodium Glutamate (MSG)
Monosodium glutamate (MSG), a sodium salt derived from glutamic acid, naturally occurs in certain foods like tomatoes and cheese. Renowned for its ability to enhance savory flavors, MSG is a staple in cooking, replicating the meaty, savory taste found in dishes like stews and soups.

MSG is primarily manufactured from glutamic acid, typically through fermentation processes utilizing microorganisms like bacteria or fungi. This method yields large quantities of glutamic acid, subsequently purified and crystallized to obtain MSG. The predominant production method involves bacterial fermentation akin to vinegar or yogurt production, where Corynebacterium species ferment ammonia and carbohydrates from sources like sugar beets or molasses to excrete amino acids, eventually yielding L-glutamate.

As a flavor enhancer, MSG plays a crucial role in the food industry, enriching the taste profile of savory foods such as soups, sauces, snacks, and processed meats. Beyond flavor enhancement, MSG finds application in various domains:

a. **Seasoning:** Widely embraced in diverse cuisines, MSG serves as a popular seasoning, particularly prominent in Asian culinary traditions.
b. **Food Industry Ingredient:** Incorporated into an extensive array of food products, MSG contributes to enhancing flavors and improving palatability.
c. **Cosmetics:** MSG occasionally features in cosmetic formulations, including skincare and haircare products, potentially offering benefits in personal care regimes.

11. Butyl Acetate

Butyl acetate, a colorless liquid with a fruity aroma, is synthesized by reacting butanol and acetic acid in the presence of sulphuric acid as a catalyst. This reaction, typically conducted at elevated temperatures under controlled conditions, ensures optimal yields of butyl acetate.

Due to its favorable characteristics such as low toxicity, volatility, and reactivity, butyl acetate finds extensive utility across various industries:

a. **Paint and Coatings:** It serves as a solvent in the formulation of paints, lacquers, and coatings, facilitating smooth application and enhancing finish quality.
b. **Perfumes and Fragrances:** Butyl acetate acts as a key ingredient in the production of perfumes and fragrances, contributing to their distinctive scents.

c. **Cellulose Acetate Production:** As a solvent in cellulose acetate production, it enables the creation of a versatile plastic utilized in textiles, films, and photographic materials.
d. **Cleaning Agent:** Its solvent properties make butyl acetate effective for cleaning metals, plastics, and electronic components, removing contaminants efficiently.

Despite its widespread use and versatility, careful consideration of the safety and environmental impact of butyl acetate is essential prior to its application. Certain uses may pose risks to human health and the environment, necessitating thorough evaluation and adherence to safety protocols.

12. Oxalic Acid

Oxalic acid, a colorless crystalline organic acid, is synthesized from the reaction between formic acid and sulphuric acid. This reaction, conducted at elevated temperatures with an excess of sulphuric acid, ensures the complete conversion of formic acid to oxalic acid.

Reaction: The initial step in the process involves the following reaction:
Glycolic acid + $2H_2SO_4 \rightarrow$ Oxalic acid + $2SO_2 + 4H_2O$

Oxalic acid boasts diverse applications across industries owing to its unique properties:

a. **Cleaning and Descaling:** It effectively removes stains and scale from various surfaces, including metal, ceramics, and glass.
b. **Dyeing and Printing:** As a reducing agent, oxalic acid finds use in the dyeing and printing of textiles.
c. **Pharmaceutical Production:** It serves as an intermediate in the synthesis of various pharmaceuticals, such as antacids, laxatives, and antioxidants.
d. **Leather Industry:** Oxalic acid is employed for bleaching leather and eliminating impurities.
e. **Food Industry:** Utilized as a food preservative and flavor enhancer, oxalic acid contributes to food quality and longevity.

13. Acetic Acid in the Presence of Sulphuric Acid

Acetic acid is synthesized in the presence of sulphuric acid through a multi-step process called the carbonylation of methanol. This intricate process involves several reactions, beginning with the conversion of methanol to methyl acetate, followed by hydrolysis to yield acetic acid.

Here's an overview:

a. **Preparation of Methanol:** Methanol (CH_3OH) is typically prepared from synthesis gas, a mixture of carbon monoxide and hydrogen, using a catalyst such as copper-zinc-alumina.

b. **Carbonylation of Methanol:** Methanol undergoes a reaction with carbon monoxide (CO) in the presence of a catalyst, often a metal complex like rhodium or iridium. Sulphuric acid may serve as a catalyst or promoter in this step. The carbonylation reaction yields methyl acetate (CH_3COOCH_3), an ester: $CH_3OH + CO \rightarrow CH_3COOCH_3$

c. **Hydrolysis of Methyl Acetate:** Methyl acetate is subsequently hydrolyzed to form acetic acid (CH_3COOH), with water (H_2O) as a reactant: $CH_3COOCH_3 + H_2O \rightarrow CH_3COOH + CH_3OH$

d. **Separation and Purification:** Acetic acid is separated from the reaction mixture using techniques like distillation or extraction. Further purification steps may be employed to achieve the desired purity level.

In summary, while sulphuric acid may function as a catalyst or promoter in the carbonylation reaction, it does not directly convert into acetic acid. Instead, it facilitates the conversion of methanol into methyl acetate, which is subsequently hydrolyzed to yield acetic acid.

14. Lactic Acid in the Presence of Sulphuric Acid

Lactic acid, a colorless to pale yellow liquid with a sour taste and characteristic odor, can be derived from sugar fermentation or the chemical oxidation of lactose or other sugars, often utilizing sulphuric acid as a catalyst.

Applications of Lactic Acid:

a. **Food and Beverage Industry:** Lactic acid serves as a versatile ingredient in food and beverages, functioning as a souring agent, preservative, and pH adjuster in products like bread, yogurt, cheese, and pickled vegetables.

b. **Personal Care Products:** In skincare, haircare, and deodorants, lactic acid's mild exfoliating and moisturizing properties make it a popular choice. It helps in maintaining skin hydration and texture.

c. **Pharmaceuticals:** Lactic acid finds application in pharmaceuticals, contributing to the production of lactates used in intravenous fluids and lactic acid derivatives employed as active ingredients in various medications. It is also utilized in topical preparations and cosmetics for acidity adjustment and its disinfectant and keratolytic attributes.

d. **Biodegradable Plastics Production:** Lactic acid is a key component in the manufacturing of polylactic acid (PLA), a biodegradable plastic extensively used in packaging, disposable tableware, and medical implants.
e. **Production of Other Chemicals:** As a precursor, lactic acid is involved in the synthesis of chemicals like calcium lactate, utilized as a food additive and firming agent, and lactide, essential in PLA production.

Despite its myriad applications, it's crucial to handle lactic acid with care due to its potential skin and eye irritant properties. Proper safety precautions, including protective clothing, adequate ventilation, and avoidance of skin contact, are imperative when working with this compound.

15. Chromic Acid Derived from Sulphuric Acid

Chromic acid, an inorganic compound composed of chromium, oxygen, and hydrogen, manifests as a dark, purplish-red, odorless, sand-like solid powder. When dissolved in water, it exhibits strong acidic properties. Chromic acid exists in two forms: monochromic acid and dichromic acid.

Chromic Acid Production: Chromic acid (H_2CrO_4), a potent oxidizing agent, is synthesized through the reaction of sulphuric acid and chromium trioxide.

Applications of Chromic Acid:

a. **Electroplating:** Employed as an electrolyte in metal electroplating processes, chromic acid facilitates the deposition of chromium onto metal surfaces, yielding a durable, corrosion-resistant coating.
b. **Metal Cleaning and Pickling:** Chromic acid serves as an effective agent for cleaning metal surfaces and eliminating rust, scale, and impurities, preparing them for subsequent processing or painting.
c. **Glass Etching:** In glass etching applications, chromic acid selectively removes surface layers of glass, enabling the creation of decorative or functional patterns on glass surfaces.
d. **Organic Synthesis:** Chromic acid finds utility as an oxidizing agent in the synthesis of diverse organic compounds, including dyes, pharmaceuticals, and resins, contributing to the production of various essential materials.

15a. Metal Treatment Products Utilizing Chromic Acid

Chromic acid plays a pivotal role in various metal treatment processes, enhancing surface characteristics and performance across diverse applications:

a. **Metal Surface Preparation:** Chromic acid serves as a potent etching in metal surface preparation, effectively eliminating surface contaminants like oils, grease, and oxides. This meticulous cleaning facilitates superior adhesion for subsequent treatments such as plating, painting, or bonding.

b. **Electroplating:** Utilized as an electrolyte in electroplating, chromic acid coats metal components like car parts, jewelry, and household items with a resilient layer of chromium. This chromium coating imparts remarkable hardness and corrosion resistance, thereby augmenting both the durability and aesthetic appeal of the metal.
c. **Passivation:** In stainless steel processing, chromic acid acts as a vital passivating agent, adept at removing surface contaminants and enhancing corrosion resistance. This process fortifies the stainless steel against corrosion, ensuring prolonged structural integrity and longevity.
d. **Anodizing:** Chromic acid finds application in the anodizing of aluminum components, where it facilitates the formation of a protective oxide layer on the surface. This oxide layer enhances the wear resistance and corrosion resistance of aluminum, elevating its durability and performance in various environments.
e. **Cleaning:** For comprehensive metal cleaning tasks, including engine components, gears, and machinery, chromic acid proves invaluable. Its ability to swiftly dissolve grease, oils, and other contaminants ensures optimal performance and efficiency of metal components, contributing to their prolonged functionality.

In each of these applications, chromic acid demonstrates its versatility and efficacy in enhancing the quality and longevity of metal components, thereby underlining its indispensable role in the realm of metal treatment.

APPLICATION OF SULPHURIC ACID-SHEET 2

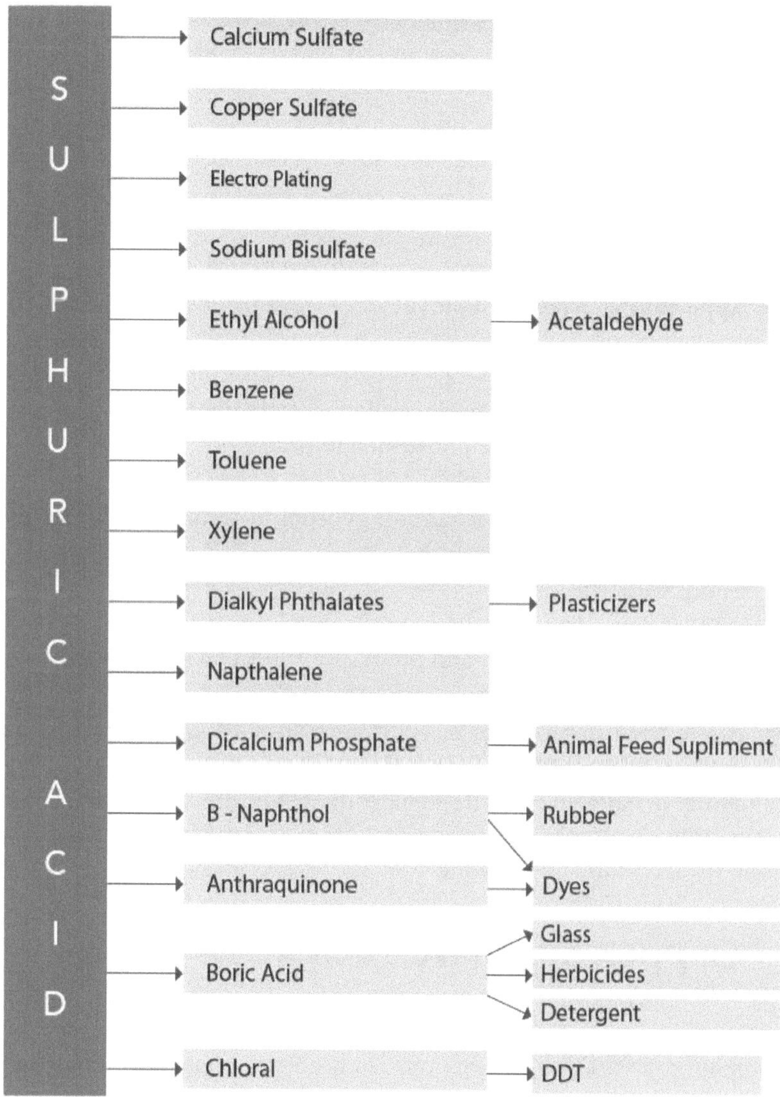

16. Calcium Sulphate

Calcium sulphate, known as plaster of Paris or gypsum in its hydrated form, is a versatile inorganic compound extensively utilized across various industries.

Here's a closer look at its production from sulphuric acid and its manifold uses:
Production Process: Calcium sulphate is synthesized from sulphuric acid through neutralization reactions with calcium-containing compounds, such as calcium hydroxide or calcium carbonate:

(a) $Ca(OH)_2 + H_2SO_4 \rightarrow CaSO_4 + 2H_2O$

(b) $CaCO_3 + H_2SO_4 \rightarrow CaSO_4 + CO_2 + H_2O$

Products and Their Uses:

a. **Gypsum:**
- Application: Widely employed as a fertilizer, soil conditioner, and in the production of plaster, wallboard, and construction materials.

 Anhydrite:
- Application: Occurs naturally as deposits in mineral veins and serves as a filler and extender in various industrial applications.

b. **Plaster of Paris:**
- Application: Derived by heating gypsum, it forms a moldable paste used for casting and modeling purposes.

c. **Food Additive:**
- Application: Acts as a firming agent, emulsifier, and thickener in food products like tofu, beer, and processed fruits and vegetables.

d. **Medical Applications:**
- Application: Utilized as a wound dressing, filling material in dentistry, and injectable bone substitute in orthopedic surgery.

The suitability of calcium sulphate for specific applications is contingent upon its form (e.g., anhydrite, gypsum), purity, and particle size, ensuring optimal performance and functionality across diverse sectors.

17. Copper Sulphate

Copper sulphate, a versatile compound pivotal across various sectors, is synthesized from sulphuric acid through reactions with copper-containing compounds, such as copper oxide or copper carbonate:

Production Process:
(a) $CuO + H_2SO_4 \rightarrow CuSO_4 + 2H_2O$
(b) $CuCO_3 + H_2SO_4 \rightarrow CuSO_4 + CO_2 + H_2O$

Key Products and Their Uses:

a. **Fungicide:**
- Application: Widely utilized in agriculture as a fungicide to combat various plant diseases effectively.

b. **Algaecide:**
- Application: Employed to control algae proliferation in water bodies like lakes and reservoirs.

c. **Herbicide:**
- Application: Serves as a non-selective herbicide for weed control in non-crop areas.

d. **Wood Preservation:**
- Application: Utilized as a preservative to safeguard wood against rot and decay, enhancing its longevity.

e. **Electroplating:**
- Application: Acts as a source of copper ions in electroplating processes, facilitating the deposition of a thin copper layer on substrates.

f. **Blue Vitriol:**
- Application: Known colloquially as "blue vitriol" due to its distinctive blue crystalline appearance. Historically used as a dye, pigment, and in copper electroplating.

18. Electroplating

Sulphuric acid stands as a common electrolyte in electroplating processes, facilitating the deposition of metal ions onto substrates, thereby forming a thin, even layer of metal.

The products and applications of electroplated materials span various sectors, including:

a. **Decorative Coatings:**
- Application: Electroplated materials are utilized decoratively, exemplified by the electroplating of gold or silver onto jewelry, tableware, and assorted objects, elevating their aesthetic appeal.

b. **Protective Coatings:**

- Application: Serving as a shield against corrosion, wear, and environmental influences, electroplated materials like nickel or chromium safeguard metal items from rust and degradation, prolonging their lifespan.

c. **Electronic Components:**

- Application: In the realm of electronics, electroplated materials find utility in crafting components such as printed circuit boards, where meticulously deposited metal layers establish crucial electrical connections.

d. **Automotive Components:**

- Application: Electroplated materials contribute to automotive manufacturing, notably in the fabrication of catalytic converters. Here, a thin layer of precious metal is applied to catalyze chemical reactions, enhancing performance and durability.

e. **Industrial Applications:**

- Application: Industries rely on electroplated materials for diverse needs, ranging from electrical connectors to battery components. The uniform deposition of metal layers ensures optimal functionality in various industrial settings.

These applications underscore the versatility and indispensability of electroplating processes utilizing sulphuric acid. By bestowing enhanced functionality, durability, and aesthetic appeal, electroplated materials serve as integral components across a spectrum of industries, embodying the seamless fusion of innovation and practicality.

19. Sodium Bisulphate

Sodium bisulphate, an acid salt derived from the partial neutralization of sulphuric acid by sodium base equivalents, manifests as a dry granular product, exhibiting hygroscopic tendencies in its anhydrous state.

This compound is crafted from sulphuric acid through neutralization reactions with sodium hydroxide:

Reaction: $NaOH + H_2SO_4 \rightarrow NaHSO_4 + H_2O$

Diverse industries benefit from sodium bisulphate, finding utility in an array of applications:

a. **Cleaning Agent:**

- Usage: In the food, beverage, and pharmaceutical sectors, sodium bisulphate serves as a potent cleaning agent, effectively eliminating mineral deposits and stains from equipment, ensuring hygiene and safety standards.

b. **Pool pH Decreaser:**

- Usage: Employed in swimming pools, sodium bisulphate assumes the role of a pH decreaseor, lowering both pH and total alkalinity in the water. This action mitigates scaling and corrosion risks, preserving the integrity of pool surfaces and equipment.

c. **Food Additive:**

- Usage: Sodium bisulphate functions as a pivotal food additive, primarily as a pH adjuster. By modulating pH levels, it safeguards the color and flavor of food products, enhancing their shelf life and sensory attributes.

d. **Textile Industry:**

- Usage: Within the textile realm, sodium bisulphate contributes to pH adjustments in dye baths, ensuring optimal dyeing conditions. Moreover, it aids in equipment maintenance by effectively removing mineral deposits, bolstering operational efficiency.

e. **Industrial Applications:**

- Usage: Embraced across diverse industrial domains, sodium bisulphate finds multifaceted applications as a cleaning agent, pH adjuster, and sulphate ion source, catering to varied operational needs.

20. Ethyl Alcohol

Ethyl alcohol, or ethanol, embodies a volatile, flammable liquid, characterized by its colorless hue, wine-like aroma, and distinct taste. Renowned as both a psychoactive recreational drug and the quintessential component of alcoholic beverages, ethanol serves a myriad of purposes.

Naturally occurring through the fermentation of sugars by yeasts or via petrochemical processes like ethylene hydration, ethanol is a cornerstone of various industries. In petrochemical pathways, ethylene undergoes indirect hydration by reacting with concentrated sulphuric acid to yield ethyl sulphate, subsequently hydrolyzed to regenerate sulphuric acid and produce ethanol.

Ethyl Alcohol Products and Their Uses:

a. **Beverage Alcohol:**

- Utility: Ethyl alcohol forms the crux of alcoholic beverages, including beer, wine, and spirits, contributing to their intoxicating allure and sensory profiles.

b. **Fuel:**

- Utility: Employed as a fuel, either in its pure form or blended with gasoline, ethanol powers internal combustion engines in automobiles, marine vessels, and power generators, offering an eco-friendly alternative.

c. **Solvent:**

- Utility: Ethyl alcohol serves as a versatile solvent in diverse industrial and chemical applications. It finds application in the production of personal care items, cosmetics, pharmaceuticals, and cleaning agents, facilitating dissolution and formulation processes.

d. **Antiseptic:**

- Utility: With its potent antibacterial and antiviral properties, ethyl alcohol emerges as a pivotal antiseptic agent, employed to sterilize skin and surfaces, safeguarding against microbial contaminants.

e. **Chemical Intermediate:**

- Utility: Ethyl alcohol assumes the role of a chemical intermediate in the synthesis of various compounds. Notably, it serves as a precursor to ethyl acetate, a key constituent in adhesives and coatings production, contributing to diverse industrial applications.

In essence, ethyl alcohol epitomizes versatility, permeating various sectors with its multifaceted utility. From the intoxicating elixirs of the beverage industry to the eco-conscious fuels propelling modern transportation, ethanol stands as a testament to innovation and adaptability, enriching lives and industries alike.

20a. Acetaldehyde

Acetaldehyde, a pivotal chemical compound, stems from the oxidation of ethyl alcohol. Its versatile nature renders it indispensable across numerous industries.

Acetaldehyde Products and Their Uses:

a. **Flavor & Fragrance:**

- Utility: Acetaldehyde finds application as a flavor and fragrance component, enriching food and cosmetic formulations with its distinctive aroma and taste.

b. **Intermediate Chemical:**

- Utility: Serving as a cornerstone in chemical synthesis, acetaldehyde acts as an intermediate for various compounds. Notably, it facilitates the production of acetic acid, a vital ingredient in vinegar manufacturing, and ethyl acetate, essential for adhesives and coatings.

c. **Fuel:**

- Utility: Embracing a role in energy production, acetaldehyde serves as a fuel source for generating heat and electricity, contributing to diverse industrial and domestic applications.

d. **Preservative:**

- Utility: Leveraging its preservative properties, acetaldehyde safeguards food products from spoilage while preserving their flavor and aroma, enhancing their shelf life and quality.

e. **Industrial Applications:**

- Utility: Acetaldehyde serves as a linchpin in various industrial processes. It acts as a precursor for the synthesis of resins, plastics, and synthetic fibres, underpinning the production of a myriad of essential materials.

In essence, acetaldehyde emerges as a multifaceted compound, permeating diverse sectors with its array of applications. From enhancing the sensory experience in food and cosmetics to fueling energy production and driving industrial processes, acetaldehyde stands as a testament to its versatility and indispensability in modern society.

21. Benzene

Benzene, a fundamental component of petroleum, stands as a cornerstone in the realm of petrochemicals. Recognized for its aromaticity due to cyclic pi bonds, benzene emerges as a colorless, highly flammable liquid boasting a distinct sweet aroma, notably contributing to the fragrance of gasoline.

While theoretically, benzene can be derived from sulphuric acid via the cumene process, involving the catalyzed reaction of phenol with propene, most commercially, it is obtained through crude petroleum oil refinement.

Benzene Products and Their Diverse Applications:

a. **Plastics and Rubber Manufacturing:**

- Utility: Benzene serves as a vital precursor in the production of plastics and rubber, playing a pivotal role in the synthesis of various polymers essential for manufacturing a wide array of products.

b. **Dyes Production:**

- Utility: Leveraging benzene's chemical properties, it acts as a key ingredient in the synthesis of dyes, facilitating the creation of vibrant pigments utilized across numerous industries

c. **Detergent Formulation:**

- Utility: Benzene contributes to the production of detergents, enabling the formulation of effective cleaning agents essential for household and industrial applications.

d. Pharmaceutical Manufacturing:

- Utility: In pharmaceuticals, benzene serves as a foundational building block, fostering the creation of diverse medicinal compounds crucial for healthcare and therapeutic treatments.

e. Solvent in Chemical Processes:

- Utility: Benzene's solvent properties find extensive use in chemical manufacturing processes, facilitating the dissolution and extraction of various substances, including fuels like gasoline.

Expanding the horizon of benzene's utility unveils a plethora of valuable products derived through strategic chemical transformations:

a. Benzenesulfonic Acid:

- Synthesis Reaction: $C_6H_6 + H_2SO_4 \rightarrow C_6H_5SO_3H + H_2O$
- Benzene undergoes a transformation to yield benzenesulfonic acid through sulfonation, a pivotal reaction in organic chemistry.
- Utility: Benzenesulfonic acid finds application in various industrial processes, including as a catalyst and intermediate in the synthesis of detergents, dyes, and pharmaceuticals.

b. Nitrobenzene:

- Synthesis Reaction: $C_6H_6 + HNO_3/H_2SO_4 \rightarrow C_6H_5NO_2 + H_2O/H_2SO_4$
- Through nitration, benzene transforms into nitrobenzene, a chemical compound with diverse applications.
- Utility: Nitrobenzene serves as a key intermediate in the production of various chemicals, including aniline, which in turn is utilized in the synthesis of dyes, pharmaceuticals, and rubber chemicals.

22. Toluene

Toluene, a pivotal aromatic hydrocarbon, can be synthesized through various routes, including methods involving sulfuric acid. One such pathway involves the treatment of methylbenzene with sulfuric acid to achieve toluene. While this process has been historically employed, it's essential to note that contemporary methods may prioritize alternative synthetic routes for toluene production.

Chemical Formula and Physical Characteristics: Toluene, represented by the chemical formula C_7H_8 or $C_6H_5CH_3$, manifests as a colorless liquid with distinctive characteristics reminiscent of paint thinners. Its solubility in water is limited, contributing to its classification as an organic solvent.

Applications of Toluene: The versatility of toluene extends across various industrial domains, underpinning its significance in numerous applications:

a. **Solvent in Paints and Coatings:** Toluene serves as a solvent in the formulation of paints, coatings, and varnishes, contributing to their consistency and facilitating application.
b. **Chemical Feedstock:** As a precursor in organic synthesis, toluene acts as a vital feedstock for the production of numerous chemicals, including benzene, phenol, and trinitrotoluene (TNT).
c. **Fuel Additive:** Toluene finds utility as a blending component in gasoline, enhancing its octane rating and combustion properties.
d. **Industrial Cleaning Agent:** Due to its strong solvent properties, toluene is utilized in industrial cleaning applications, aiding in the removal of grease, oil, and adhesive residues.
e. **Polymer Production:** Toluene serves as a key component in the manufacture of various polymers and plastics, contributing to the synthesis of materials like polyurethane and polystyrene.
f. **Pharmaceuticals:** In the pharmaceutical industry, toluene finds application in the synthesis of drugs and pharmaceutical intermediates, facilitating the production of medications for diverse therapeutic purposes.

Toluene Derivatives

Sulphuric acid facilitates the conversion of toluene into various valuable derivatives, expanding its utility in diverse applications:

a. **Formation of p-Toluenesulfonic Acid:** Through the reaction with sulphuric acid, toluene undergoes transformation into p-toluenesulfonic acid, denoted by $C_7H_7SO_3H$, presenting a versatile precursor in chemical synthesis.
b. **Generation of Nitrotoluene:** Interaction with nitric acid or a mixture of nitric acid and sulphuric acid yields nitrotoluene, marked by the formula $C_7H_7NO_2$, showcasing its potential as an intermediate in organic synthesis.

Versatility and Applications: Toluene, as a foundational chemical, serves as the building block for numerous industrially significant products:

a. **Feedstock for Chemical Synthesis:** Toluene serves as a pivotal starting material for the synthesis of diverse chemicals, including benzene, xylene, and various other organic compounds, contributing to the production of plastics, pharmaceuticals, and industrial chemicals.

b. **Solvent in Product Formulation:** With its effective solvency properties, toluene finds extensive use as a solvent in the formulation of paints, inks, adhesives, and other industrial products, facilitating their production and application.
c. **Fuel Component:** Toluene serves a dual role in the fuel industry, both as an integral component of gasoline, enhancing its octane rating, and as a standalone fuel with notable combustion properties.
d. **Industrial Processes and Materials Production:** Toluene plays a crucial role in the manufacturing of explosives, foam plastics, and various industrial materials, underscoring its significance in process industries.

Toluene is naturally present in crude petroleum oil and is also a by-product of coal coke production. Purification techniques such as distillation and solvent extraction refine toluene for industrial use.

23. Xylene

Xylene, an essential organic compound, arises from the substitution of two hydrogen atoms with methyl groups in a benzene ring. This colorless, flammable liquid, with its slightly greasy texture, holds significant industrial significance, characterized by the chemical formula $C_6H_4(CH_3)_2$ or $^-CH_3{}^-CH_3$ or C_8H_{10}.

a. Production via Sulphuric Acid: Xylene is synthesized efficiently from sulphuric acid through the xylene isomerization process. This technique involves utilizing mixed xylenes, a blend of xylene isomers, as reactants alongside sulphuric acid catalysts. The resultant reaction yields a mixture of ortho-, meta-, and para-xylene, constituting a pivotal step in xylene production.
b. Industrial Scale and Significance: With several million tons manufactured annually, xylene production forms an integral component of the BTX aromatics (benzene, toluene, and xylenes) extracted from catalytic reforming products, notably reformate. Notably, some of the world's largest xylene plants operate in Singapore, underscoring its global importance in industrial processes.
c. Versatile Applications: Xylene serves as a foundational chemical in diverse industrial applications, facilitating the production of a myriad of products:
   - Raw Material for Product Synthesis: Xylene functions as a fundamental starting material for the synthesis of various products, including plastics, resins, synthetic fibers, rubber, dyes, and detergents, contributing to the manufacturing of everyday essentials across industries.

- Solvent in Product Formulation: Its exceptional solvent properties render xylene invaluable in formulating paints, inks, adhesives, and other industrial products, enhancing their application and performance characteristics.
- Petrochemical Feedstock: Within the petrochemical sector, xylene serves as a key feedstock for the production of additional chemicals, facilitating the synthesis of a wide array of compounds critical for industrial processes.
- Fuel Component: Furthermore, xylene finds application as a component in fuel formulations, contributing to its utilization as a fuel additive within specific contexts.

Xylene, synthesized efficiently through sulphuric acid-mediated processes, stands as a cornerstone in industrial chemistry, exemplifying its versatility and indispensability across numerous sectors. From its role as a raw material in product synthesis to its utility as a solvent and fuel component, xylene continues to play a pivotal role in advancing industrial processes and product development worldwide.

24. Dialkyl Phthalates

Dialkyl phthalates, renowned for their plasticizing properties, serve as vital additives in plastics, augmenting their flexibility, transparency, and overall durability. Among these, Diethyl Phthalate (DEP) exemplifies a prominent member, with its chemical formula epitomizing its molecular structure.

a. **Production via Sulphuric Acid:** The synthesis of dialkyl phthalates commences with the utilization of sulphuric acid in conjunction with phthalic anhydride and alcohols. The ensuing reactions orchestrate the transformation of precursor compounds into the desired dialkyl phthalates:
    - Formation of Phthalic Anhydride: Ortho-xylene undergoes oxidation to yield phthalic anhydride in a meticulously controlled reaction, setting the stage for subsequent transformations.
    Reaction:-
    $$\frac{C_6H_4(CH_3)_2}{Xylene} + 3O_2 \rightarrow \frac{C_6H_4(CO)_2O}{Phthalic\ anhydride} + 2H_2O$$
    - Conversion to Dialkyl Phthalates: Phthalic anhydride engages with methanol in a transformative encounter, generating dimethyl phthalate (DMP), a quintessential member of the dialkyl phthalate family.

b. **Versatile Applications:** Dialkyl phthalates play multifaceted roles across various industries, contributing significantly to product formulations and manufacturing processes:

- Plasticizer in Polymer Production: Renowned for their plasticizing prowess, dialkyl phthalates serve as indispensable additives in the fabrication of flexible and semi-rigid plastics, prominently PVC and other vinyl polymers. By enhancing flexibility and resilience, they mitigate the risk of cracking and breakage, elevating the performance and longevity of plastic products.
- Solvent in Formulation Processes: Leveraging their solvency properties, dialkyl phthalates find application as solvents in the synthesis of inks, dyes, and personal care products. Their compatibility with various formulations facilitates the production of vibrant pigments and stable emulsions essential for diverse applications.
- Intermediates in Chemical Synthesis: Serving as pivotal intermediates, dialkyl phthalates contribute to the synthesis of fragrances and fragranced products, amplifying olfactory experiences across personal care and household realms.

9a. Plasticizers: Dialkyl Phthalates

Dialkyl phthalates represent a significant category of plasticizers, inherently designed to enhance the pliability and versatility of various materials. As quintessential plasticizers themselves, they play a pivotal role in augmenting the flexibility and processability of polymers and plastics.

a. **Production and Mechanism:** Derived from meticulous synthesis processes, dialkyl phthalates exhibit inherent plasticizing properties that stem from their molecular structure. By reducing the glass transition temperature and enhancing flexibility, they imbue materials with desirable traits essential for diverse applications.
b. Dialkyl phthalates find extensive application across a myriad of industries, owing to their exceptional plasticizing capabilities. Some key applications include:
    - PVC Products: Integral to the fabrication of PVC products, such as pipes, flooring materials, inflatable products, and wire and cable coatings, dialkyl phthalates optimize the mechanical properties and processability of these materials, ensuring durability and resilience.
    - Rubber Products: In the rubber industry, dialkyl phthalates serve as indispensable additives in the production of gloves, seals, hoses, and other rubber-based products, facilitating enhanced flexibility and performance.
    - Adhesives: Leveraging their compatibility and plasticizing prowess, dialkyl phthalates find utility in adhesives, particularly pressure-

sensitive variants, where they contribute to adhesive strength and flexibility.

- Paints and Coatings: Within the realm of paints and coatings, dialkyl phthalates play a dual role, serving as solvents and additives to enhance workability and flexibility. Their incorporation ensures optimal coating properties and improved durability of the finished product.
- Personal Care Products: Recognized for their biocompatibility and versatility, dialkyl phthalates feature prominently in personal care formulations, including lotions, perfumes, and hair care products, where they contribute to texture, fragrance dispersion, and overall product performance.

25. Naphthalene

Naphthalene stands as a pivotal organic compound and polycyclic aromatic hydrocarbon, manifesting as a white crystalline solid with a distinct odor detectable even at minute concentrations. Its molecular structure comprises a fused pair of benzene rings, epitomizing its status as an aromatic hydrocarbon.

Production Method:The synthesis of naphthalene from sulphuric acid is facilitated through the sulphuric acid digestion process. During this chemical digestion, coal tar or petroleum undergoes a transformative reaction with sulphuric acid, yielding a composite mixture of chemical components, among which naphthalene prominently features.

Naphthalene embodies versatility, finding application across diverse industrial sectors and consumer domains. Some notable applications include:

a. **Precursor for Phthalic Anhydride:** Serving as a fundamental precursor, naphthalene catalyzes the synthesis of phthalic anhydride. This compound, in turn, is pivotal in the production of plasticizers, dyes, and pigments, underpinning various manufacturing processes.
b. **Moth Repellent:** Renowned for its efficacy as a moth repellent, naphthalene assumes a crucial role in safeguarding textiles and garments from moth infestation. Its volatile nature and characteristic odor deter moth larvae, making it a staple in mothballs and other repellent formulations.
c. **Raw Material for Naphthalene Sulfonic Acid:** Naphthalene serves as a primary raw material in the synthesis of naphthalene sulfonic acid. This derivative finds utility as a detergent builder and as a precursor in the production of an array of chemical compounds, further underscoring naphthalene's intrinsic value.

d. **Fuel Source:** Traditionally employed as a fuel source, particularly in the form of naphthalene balls, naphthalene finds application in heating homes and buildings, predominantly in certain geographical regions where this practice prevails.

26. Dicalcium Phosphate

Dicalcium phosphate, also known as dibasic calcium phosphate, emerges as a crucial compound utilized across various industries for its calcium and phosphorus content. Its production involves the neutralization of calcium hydroxide with phosphoric acid, resulting in the precipitation of the dihydrate as a solid. Alternatively, the anhydrous form can be obtained through precipitation at 60°C:

The synthesis of dicalcium phosphate entails the following chemical reaction:
$H_3PO_4 + Ca(OH)_2 \rightarrow CaHPO_4 + 2H_2O$

Phosphoric acid, a key precursor, is derived from the reaction between sulphuric acid (H_2SO_4) and naturally occurring phosphate rock. Notably, this reaction yields calcium sulphate ($CaSO_4$), commonly known as gypsum.

Dicalcium phosphate finds diverse applications across several industries, owing to its versatile properties:

a. **Livestock Nutrition:** Dicalcium phosphate serves as a vital food additive in the livestock industry, providing a rich source of calcium and phosphorus for farm animals. It plays a crucial role in promoting bone health, aiding in the growth and development of livestock.

b. **Food Industry:** Within the food industry, dicalcium phosphate functions as a calcium supplement, enhancing the nutritional value of various food products. Moreover, it serves as an anticaking agent in powdered food formulations, ensuring product consistency and quality.

c. **Medical Sector:** In the medical field, dicalcium phosphate features prominently in the production of dental and orthopedic implants. It is utilized as a component of bone cement, facilitating the fixation of implants and promoting bone regeneration in surgical procedures.

d. **Fertilizer and Flame Retardants:** Dicalcium phosphate finds application in the production of fertilizers, where it contributes to soil enrichment and plant growth. Additionally, it serves as a component in flame retardant formulations, enhancing fire resistance in various materials.

25a. Dicalcium Phosphate in Animal Feed

Dicalcium phosphate serves as a prevalent animal feed supplement, enriched with vital nutrients like calcium and phosphorus crucial for the well-being of livestock and poultry. These nutrients play pivotal roles in fostering robust bone structure, dental health, and overall physiological functions, encompassing muscle contraction, nerve transmission, and enzymatic activity.

Production and Utilization: In the realm of animal husbandry, dicalcium phosphate finds extensive application as a dietary supplement for various farm animals, including cattle, swine, poultry, and fish. Its incorporation into feed formulations is geared towards meeting the specific nutritional requirements of these animals. By serving as a readily available and cost-effective source of calcium and phosphorus, dicalcium phosphate aids in maintaining the optimal calcium-to-phosphorus ratio within feed rations.

The utilization of dicalcium phosphate in animal feed confers several noteworthy advantages:

a. **Enhanced Nutritional Balance:** Dicalcium phosphate facilitates the attainment of a balanced nutrient profile in animal diets, ensuring that livestock and poultry receive adequate levels of calcium and phosphorus essential for their growth and vitality.
b. **Improved Feed Efficiency:** Incorporating dicalcium phosphate into feed formulations can enhance feed efficiency, promoting optimal nutrient utilization and metabolic processes within the animals' bodies. This, in turn, contributes to improved growth rates and overall productivity.
c. **Support for Health and Productivity:** The provision of essential nutrients through dicalcium phosphate supplementation supports the overall health and well-being of livestock and poultry. Stronger bones, healthier teeth, and enhanced physiological functions translate to improved animal welfare and productivity on the farm.

27. β-Naphthol

β-Naphthol, a fluorescent colorless crystalline compound, boasts solubility in various solvents like simple alcohols, ethers, and chloroform. Serving as a crucial intermediate, it finds extensive application in the production of dyes and assorted compounds.

Production Process: The synthesis of β-Naphthol from sulphuric acid entails a multi-step procedure, primarily utilized in azo dye manufacturing.

This method involves sulfonation of naphthalene followed by subsequent hydrolysis:

a. **Sulfonation of Naphthalene:** Naphthalene undergoes sulfonation with concentrated sulphuric acid to yield naphthalene sulfonic acid under reflux conditions: $C_{10}H_8 + H_2SO_4 \rightarrow C_{10}H_7SO_3H + H_2O$
b. **Formation of β-Naphthol:** The naphthalene sulfonic acid is subjected to hydrolysis in the presence of a strong base like sodium hydroxide, resulting in the liberation of β-Naphthol: $C_{10}H_7SO_3H + NaOH \rightarrow C_{10}H_7OH + NaHSO_4$
c. **Isolation and Purification:** Following the reaction, β-Naphthol is separated from the reaction mixture via filtration or extraction. Subsequent purification steps, such as recrystallization, ensure the attainment of high-purity β-Naphthol.

β-Naphthol serves as a pivotal building block in the synthesis of azo dyes, offering versatile coloration options across various industries:

a. **Textile Dyeing:** Azo dyes derived from β-Naphthol are extensively employed in coloring textiles ranging from cotton and wool to synthetic fibres.
b. **Paper Dyeing:** These dyes find application in the paper industry, imparting vibrant hues to diverse paper products including newsprint, tissue, and paperboard.
c. **Leather Dyeing:** β-Naphthol-based azo dyes are utilized in the leather industry to color leather goods like shoes, bags, and upholstery.
d. **Other Applications:** The versatility of azo dyes extends to various other sectors, including the production of food and cosmetic colorants, as well as certain medical applications.

27a. Rubber Additives Utilizing β-Naphthol

Natural rubber, typically sourced from the sap of rubber trees (Hevea brasiliensis), undergoes processing to yield the rubber material used in various applications.

While β-Naphthol isn't directly involved in rubber production, it serves as a crucial chemical intermediate in the manufacturing of additives that enhance rubber properties.

a. β-Naphthol is utilized as a key component in the synthesis of additives aimed at improving rubber characteristics:
    - Accelerators: β-Naphthol acts as a reactant in the production of accelerators, which expedite the curing process of rubber during manufacturing. These additives enhance the speed and efficiency of rubber vulcanization.
    - Antioxidants: β-Naphthol serves as a precursor in the synthesis of antioxidants, which shield rubber from degradation caused by factors like heat, light, and environmental exposure. These additives prolong the lifespan and durability of rubber products.

b. Applications of Rubber: Rubber finds extensive usage across diverse sectors, owing to its versatile properties:
    - Tires: Rubber serves as the primary material in the production of tires for vehicles, bicycles, and aircraft, providing traction and durability on various surfaces.
    - Footwear: Rubber is employed in footwear manufacturing, particularly for crafting durable and slip-resistant soles for shoes and other footwear types.
    - Conveyor Belts: Rubber is integral to the fabrication of conveyor belts utilized in industries like mining and agriculture, facilitating the efficient transportation of materials.
    - Vibration Isolation: Rubber is utilized to isolate vibrating machinery from surrounding structures, minimizing noise and structural damage.
    - Seals and Gaskets: Rubber seals and gaskets, fabricated using β-Naphthol-derived additives, find application in automotive, aerospace, and electronics industries for ensuring leak-proof connections.
    - Other Applications: Rubber is further utilized in the production of hoses, gloves, toys, and various other consumer and industrial products, benefiting from its flexibility, resilience, and chemical resistance.

27b. Dyes Synthesis Leveraging β-Naphthol

β-Naphthol serves as a fundamental component in the synthesis of azo dyes, a prominent class of synthetic colorants widely employed across various industries.

a. **Utilization in Dyes Synthesis:** β-Naphthol acts as a pivotal precursor in the manufacturing process of azo dyes, facilitating the creation of an extensive array of color variations through reactions with diazonium salts.
b. **Applications of Azo Dyes:** The versatile nature of azo dyes, synthesized with β-Naphthol, finds application in diverse sectors:
    - Textile Dyeing: Azo dyes impart vibrant hues to a broad spectrum of textiles, encompassing cotton, wool, silk, and synthetic fibres, enhancing the aesthetic appeal of garments and fabrics.
    - Paper Coloring: Azo dyes are employed in the coloring of various paper products, ranging from newsprint to tissue paper and paperboard, contributing to the visual appeal and distinctiveness of printed materials.
    - Leather Dyeing: Azo dyes are utilized in the dyeing process of leather goods, including shoes, bags, and upholstery, offering a diverse palette of colors to meet aesthetic preferences.

- Other Applications: Azo dyes extend their utility beyond conventional sectors, finding usage in the production of food and cosmetic colorants, where they enhance the visual appeal of consumables. Additionally, they play a role in specific medical applications, contributing to diagnostic and therapeutic formulations.

28. Anthraquinone

Anthraquinone stands as a vital aromatic organic compound, serving as a foundational component in the creation of numerous dyes and pigments. With its distinct properties, it finds utility in various industrial sectors.

a. **Production of Anthraquinone:** Anthraquinone is synthesized through the oxidation of anthracene, a process catalyzed by sulfuric acid. Alternatively, commercial production involves the condensation of benzene and phthalic anhydride. Anthraquinone's molecular structure, represented by the chemical formula $C_{14}H_8O_2$, exhibits a characteristic arrangement, forming a yellow, highly crystalline solid.

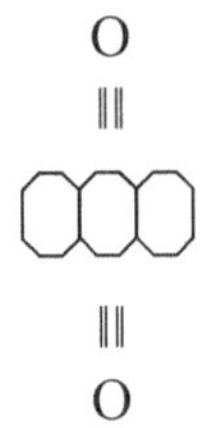

b. **Applications of Anthraquinone:**
   - Dyes and Pigments: Anthraquinone serves as a fundamental building block in the synthesis of dyes and pigments. It plays a pivotal role in the creation of azo dyes, which impart vibrant colors to textiles, paper, and various materials. Additionally, it contributes to the production of pigments utilized in inks, paints, and plastics.
   - Pharmaceuticals: Certain derivatives of anthraquinone feature prominently in pharmaceutical formulations, particularly in laxatives. Their efficacy in promoting bowel movement makes them valuable active ingredients in medicinal products aimed at addressing gastrointestinal issues.
   - Intermediates: Anthraquinone serves as a versatile intermediate in chemical synthesis, facilitating the production of a diverse range of compounds. It acts as a precursor in the creation of pigments, dyes, and pharmaceuticals, contributing to the synthesis of various functional chemicals.

- Energy Production: Anthraquinone finds application in the field of renewable energy, particularly in solar cells. Its role as a sensitizer aids in the conversion of sunlight into electrical energy, enhancing the efficiency of solar cell technologies.

28a. Anthraquinone-Based Dyes

Anthraquinone serves as a crucial precursor in the production of diverse dyes, offering a spectrum of vibrant colors for various applications across industries.

Production of Anthraquinone Dyes: Anthraquinone acts as a foundational building block in the synthesis of different types of dyes, including vat dyes, direct dyes, and pigments. These dyes are synthesized through tailored chemical processes that leverage anthraquinone's unique properties.

Types of Anthraquinone Dyes:

a. **Vat Dyes:** Anthraquinone serves as a primary ingredient in the creation of vat dyes, which find extensive use in coloring textiles, paper, and other materials. Vat dyes, known for their water-insolubility, undergo reduction during application and subsequently oxidize to impart insoluble hues to the substrate.
b. **Direct Dyes:** Anthraquinone contributes to the formulation of direct dyes, essential for coloring textiles, paper, and other materials. Direct dyes, characterized by their water-solubility, can be directly applied to the substrate, providing a versatile coloring solution.
c. **Pigments:** Utilizing anthraquinone as a precursor, pigments are synthesized for applications in inks, paints, and plastics. These pigments offer an extensive color palette and exhibit excellent stability, enhancing the visual appeal of various products.

Applications of Anthraquinone-Based Dyes:

a. **Textile Dyeing:** Anthraquinone-based dyes are integral to the textile industry, offering a broad spectrum of colors for cotton, wool, silk, and synthetic fibres. These dyes impart vibrant and long-lasting hues to fabrics, enhancing their aesthetic appeal.
b. **Paper Dyeing:** Anthraquinone-based dyes play a vital role in paper dyeing applications, contributing to the production of newsprint, tissue paper, and paperboard with vivid and durable colors.
c. **Leather Dyeing:** Anthraquinone-based dyes find application in leather dyeing processes, enabling the coloring of leather products such as shoes, bags, and upholstery with rich and enduring shades.
d. **Other Applications:** Beyond textiles, paper, and leather, anthraquinone-based dyes are utilized in diverse sectors. They are employed in the production of

food and cosmetic colorants, contributing to the visual appeal of various consumables. Additionally, these dyes find application in specific medical contexts, serving diagnostic and therapeutic purposes.

29. Boric Acid

Boric acid, known for its colorless crystalline form, is a versatile compound with various applications across industries. Its soluble nature in water and weak acidic properties make it a valuable component in numerous products.

Production of Boric Acid: Boric acid can be synthesized through a straightforward process involving the reaction of borax with sulfuric acid. This chemical reaction yields boric acid along with sodium sulfate and water as byproducts.
Chemical Equation:

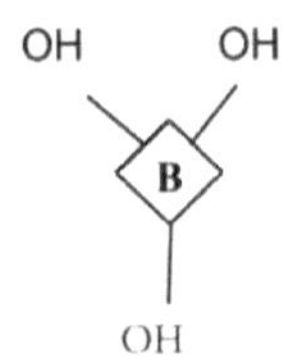

Chemical Equation:

$$\frac{Na_2B_4O_7.10(H_2O)}{Borax} + \frac{H_2SO_4}{Sulphuric\ Acid} \rightarrow \frac{4(H_3BO_3)}{Boric\ Acid} + Na_2SO_4 + 5H_2O$$

Key Applications of Boric Acid:

a. **Preservative:** Boric acid serves as an effective preservative, safeguarding cereals such as rice, wheat, and millets against harmful fungi and insects. Its inclusion in these products extends their shelf life by inhibiting microbial growth.
b. **Pesticide:** Widely utilized as an insecticide and roach killer, boric acid offers potent pest control capabilities. Its formulation targets insects while posing minimal risk to humans and pets, making it a preferred choice for pest management.
c. **Medical Use:** Boric acid exhibits antiseptic and mild anti-inflammatory properties, rendering it suitable for medical and veterinary applications. It finds use in eyewashes and topical creams, where its gentle yet effective nature aids in treating various ailments.

d. **Textile Industry:** In the textile sector, boric acid serves as a flux in soldering applications and as a fireproofing agent. Its ability to enhance solder flow and impart flame-retardant properties to textiles makes it indispensable in these processes.
e. **Glass Manufacturing:** Boric acid acts as a flux in the production of specialized glass types like borosilicate glass, renowned for its exceptional thermal resistance. Its inclusion in glass formulations facilitates melting and enhances the properties of the final product.
f. **Nuclear Power Plants:** Boric acid finds utility in nuclear reactors as a control rod absorber due to its neutron-absorbing capabilities. This application ensures the safe and efficient operation of nuclear power plants by regulating reactor performance.
g. **Household Cleaning:** Due to its mild antiseptic properties, boric acid is incorporated into household cleaning products like laundry boosters. Its inclusion aids in disinfection and odor elimination, contributing to a hygienic living environment.

29a. Boric Acid in Glassmaking

Boric acid serves as a crucial component in the manufacturing process of specific glass types, notably borosilicate glass. As a flux, it plays a pivotal role in reducing the melting temperature of glass batches, facilitating easier processing and resulting in superior quality glass products.

Production of Borosilicate Glass with Boric Acid: The integration of boric acid into glass production involves meticulous blending with other raw materials to form a glass batch. By carefully controlling the composition and proportion of ingredients, manufacturers ensure the desired properties of borosilicate glass, such as thermal resistance and durability.

Key Uses of Borosilicate Glass:

a. **Laboratory Glassware:** Borosilicate glass finds extensive application in laboratory settings, where its exceptional thermal and chemical resistance make it ideal for manufacturing test tubes, flasks, beakers, and other scientific apparatus.
b. **Cookware:** Borosilicate glass is favored in cookware manufacturing, particularly for oven-safe bakeware, owing to its remarkable resistance to thermal shock. This property ensures safe and reliable performance, even under extreme temperature variations.
c. **Lighting Fixtures:** The superior thermal stress resistance of borosilicate glass renders it suitable for use in various lighting applications, including street lights and other fixtures. Its efficient light transmission properties contribute to enhanced illumination efficiency.

d. **Art and Craft:** Artists and craftsmen utilize borosilicate glass for creating intricate beads and other decorative items due to its ability to withstand thermal and chemical stresses, ensuring durability and longevity of the finished products.

In addition to its role as a flux in borosilicate glass production, boric acid finds utility in enhancing the strength and durability of other glass varieties, such as tempered glass and fiberglass, further expanding its application scope in the glass industry.

29b. Herbicides with Boric Acid

While boric acid is not commonly utilized as the primary active ingredient in herbicides, it serves as a supplementary component in certain formulations. Its role often involves stabilizing pH levels or modifying viscosity to optimize the performance of herbicidal products.

Production of Herbicides with Boric Acid: Herbicides containing boric acid are formulated through precise blending of active ingredients, adjuvants, and carriers. Boric acid is integrated into herbicide formulations to enhance their efficacy and ensure optimal application results.

Key Uses of Boric Acid in Herbicides:

a. **Weed Control in Agriculture:** Boric acid may be incorporated into herbicide formulations designed for weed control in agricultural settings. By adjusting the pH and viscosity of the herbicidal solution, boric acid contributes to the effective dispersion and adherence of the active ingredients onto targeted weed species.

b. **Horticultural and Landscaping Applications:** In horticulture and landscaping, herbicides containing boric acid may be utilized to manage unwanted vegetation in ornamental gardens, turf, and non-crop areas. These formulations help control weed growth while maintaining the aesthetic appeal of cultivated spaces.

29c. Detergents with Boric Acid

Boric acid serves as a versatile ingredient in the formulation of detergents and cleaning products, contributing to their efficacy in removing dirt, stains, and residues. Its multifunctional properties make it a valuable component in various cleaning formulations.

Production of Detergents with Boric Acid: Detergents containing boric acid are formulated through meticulous blending of active ingredients, surfactants, and additives. Boric acid is incorporated into detergent formulations to enhance their cleaning capabilities and to provide additional benefits.

Key Uses of Boric Acid in Detergents:

a. **Laundry Detergents:** Boric acid acts as a builder in laundry detergents, aiding in the removal of dirt, stains, and grime from fabrics. Its ability to soften water helps prevent mineral buildup and improves detergent performance, resulting in cleaner and brightener laundry.
b. **Household Cleaning Products:** Boric acid finds application in various household cleaning products, such as all-purpose cleaners, where it assists in loosening and lifting dirt and stains from different surfaces. Its mild abrasive properties make it effective for scrubbing surfaces without causing damage.
c. **Dishwashing Detergents:** In dishwashing detergents, boric acid plays a crucial role in eliminating food residues, grease, and stubborn stains from dishes and cookware. Its alkalinity helps break down organic matter, facilitating thorough cleaning and leaving dishes sparkling clean.

30. Chloral

Chloral, with the chemical formula $Cl_3C\text{-}CHO$ or C_2HCl_3O, is synthesized through the chloral hydrate synthesis reaction, involving the chlorination of acetaldehyde. This process, catalyzed by concentrated sulfuric acid, yields chloral hydrate, which is subsequently dehydrated to produce chloral. Commercially, chloral serves various purposes owing to its chemical properties and versatility.

Production Process of Chloral:

1. Chloral Hydrate Synthesis: Acetaldehyde undergoes chlorination with three equivalents of hydrogen chloride in the presence of concentrated sulfuric acid, resulting in the formation of chloral hydrate.

Reaction: $\frac{H_3C-CHO}{Acetaldehyde} + \frac{3Cl}{Chlorine} + \frac{H_2O}{Water} \rightarrow \frac{Cl_3C-CH(OH)_2}{Chloral\ Hydrate} + \frac{3HCl}{Hydrochloric\ Acid}$

2. Dehydration to Chloral: Chloral hydrate is dehydrated using concentrated sulfuric acid, leading to the production of chloral.

Reaction: $\frac{Cl_3C-CH(OH)_2}{Chloral\ Hydrate} + \frac{H_2SO_4}{Cone\ Sulphuric\ Acid} \rightarrow \frac{Cl_3C-CHO}{Chloral} + \frac{(H_2SO_4, H_2O)}{Dilute\ Sulphuric\ Acid}$

Applications of Chloral:

a. **Sedative and Hypnotic Drug:** Chloral hydrate, derived from chloral, has historically been utilized as a sedative and hypnotic drug. However, its use has diminished over time due to safety concerns and the availability of safer alternatives.

b. **Laboratory Reagent:** In laboratory settings, chloral hydrate finds utility as a dehydrating agent for organic compounds. Its ability to remove water molecules makes it valuable in various chemical processes.

c. **Solvent for Organic Compounds:** Chloral hydrate serves as a solvent for a wide range of organic compounds, including fats, waxes, and resins. Its solvent properties make it useful in research, experimentation, and industrial applications.

30a. DDT

DDT, or dichlorodiphenyltrichloroethane, emerged in the 1940s as a synthetic insecticide and found widespread use in agricultural and household pest control. However, growing concerns regarding its environmental persistence and adverse effects on non-target organisms led to regulatory restrictions on its application in the 1970s.

a. DDT Production Process:
   - Formation of Chloral Hydrate: Chloral, derived from trichloroacetaldehyde, reacts with sodium hydroxide to produce chloral hydrate.
   - Reaction with Chlorobenzene: Chloral hydrate undergoes a reaction with chlorobenzene in the presence of sulfuric acid to yield DDT.

b. Applications of DDT:
   - Insecticide: DDT gained prominence as a potent insecticide, effectively combating agricultural pests and household insects. Its chemical properties enabled long-lasting pest control measures.
   - Veterinary Use: DDT has been employed as a veterinary drug for the management of lice and ticks in livestock. Its insecticidal properties extended to animal health applications, contributing to parasite control efforts.
   - Evolution of DDT Usage: Despite its effectiveness in pest management, the widespread adoption of DDT faced scrutiny due to environmental concerns. Restrictions on its agricultural use ensued as alternative pest control strategies emerged. The adverse ecological impacts associated with DDT prompted shifts towards more sustainable and environmentally friendly pest management practices.

APPLICATION OF SULPHURIC ACID-SHEET 3

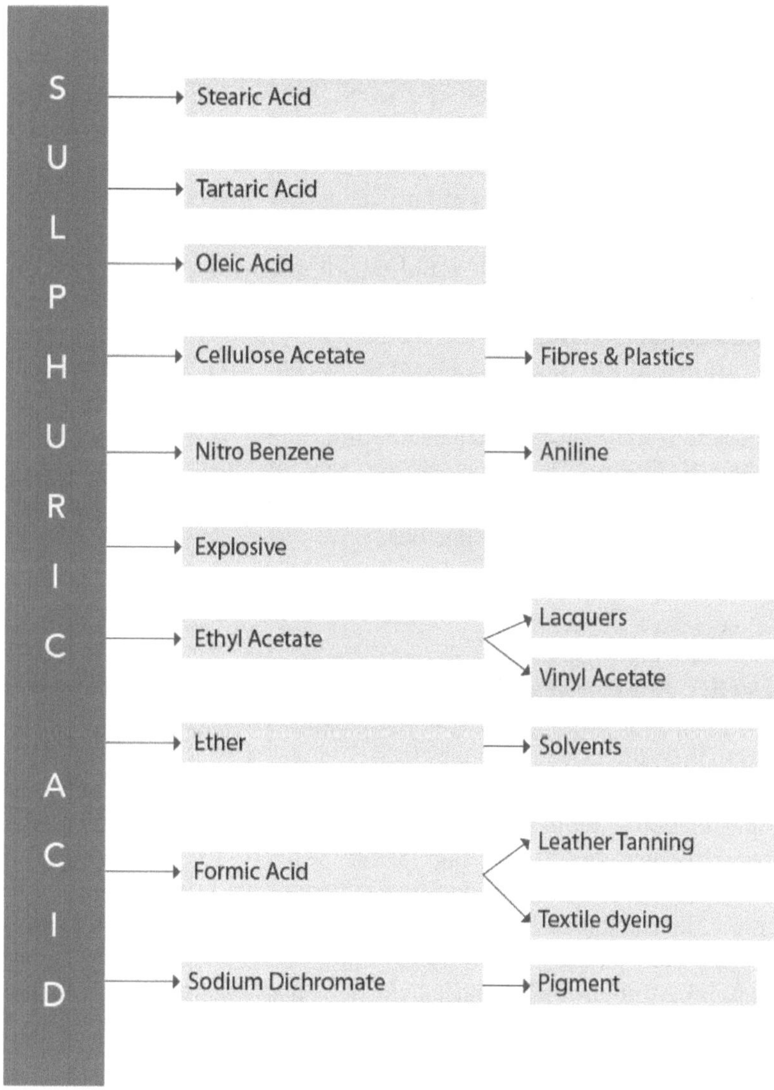

31. Stearic Acid

Stearic acid is a saturated fatty acid commonly found in animal and vegetable fats. It is used in a variety of industries, including cosmetics, pharmaceuticals, food, and personal care products. While stearic acid is not directly produced from sulfuric acid, sulfuric acid can be used in processes related to its production. Here's an overview:

Production of Stearic Acid:

a. **Saponification or Hydrolysis of Fat:** Stearic acid is often obtained by the hydrolysis or saponification of natural fats or oils. In this process, fats or oils are reacted with a strong base (such as sodium hydroxide or potassium hydroxide) to form soap (salts of fatty acids) and glycerol.
b. **Acidification to Obtain Fatty Acids:** The soap produced from the saponification process is then acidified with a strong mineral acid, typically sulfuric acid, to break down the salts of fatty acids back into their respective free fatty acids.
c. **Separation and Purification:** The mixture obtained from acidification is then subjected to separation techniques such as distillation or precipitation to isolate stearic acid from other fatty acids and impurities.
d. **Further Refining:** The crude stearic acid obtained may undergo further refining processes, such as filtration or crystallization, to enhance its purity and remove any remaining impurities.

Use of Stearic Acid:

a. **Cosmetics and Personal Care Products:** Stearic acid is commonly used in the production of cosmetics, including creams, lotions, and soaps, as it acts as an emulsifier, thickener, and stabilizer.
b. **Pharmaceuticals:** In pharmaceuticals, stearic acid is used as an excipient in the formulation of tablets and capsules, providing lubrication and aiding in the uniform distribution of active ingredients.
c. **Food Industry:** Stearic acid and its derivatives are used in the food industry as emulsifiers, stabilizers, and texture modifiers in products such as margarine, chocolate, and confectionery.
d. **Industrial Applications:** Stearic acid finds applications in various industrial processes, including the production of candles, plastics, rubber, and metal processing.
e. While sulfuric acid may not directly participate in the production of stearic acid, it is often used in the chemical industry for various purposes, including

as a catalyst, dehydrating agent, or in neutralization reactions, which may indirectly support processes involved in stearic acid production.

32. Tartaric Acid

Tartaric acid, a crystalline organic acid, is naturally present in various fruits, notably grapes, as well as bananas, tamarinds, and citrus fruits. Its salt, potassium bitartrate or cream of tartar, forms during fermentation and serves culinary and industrial purposes. This acid plays a vital role in food preparation, pharmaceuticals, cosmetics, and various industrial processes.

Tartaric Acid Production:

a. **Extraction from Fruit Juices:** Tartaric acid is extracted from grape juice concentrate through a reaction with sulfuric acid, yielding tartaric acid, sulfur dioxide, and water.

Applications of Tartaric Acid:

a. **Food and Beverages:**

- Flavoring Agent: Tartaric acid imparts a tangy flavor and acts as an acidulant in food and beverages. Its significance in winemaking lies in adjusting acidity levels.
- Leavening Agent: In baking, tartaric acid, often combined with sodium bicarbonate, forms baking powder, contributing to leavening.

b. **Pharmaceuticals:**

- Effervescent Tablets: Tartaric acid finds utility in effervescent tablets, enhancing their disintegration and dissolution rates.
- Chiral Building Block: It serves as a chiral building block in synthesizing various pharmaceutical compounds.

c. **Cosmetics:**

- pH Balancing: Tartaric acid is incorporated into personal care products like skin and hair care items for its pH-balancing properties.
- Antioxidant: Its antioxidant attributes contribute to skincare formulations.

d. **Photography:**

- Stabilizing Agent: In photographic film production, tartaric acid acts as a stabilizer for silver halides, aiding in image formation.

e. **Metal Finishing:**

- Chelating Agent: Tartaric acid serves as a chelating agent in metal finishing processes, facilitating impurity removal from metal surfaces.

33. Oleic Acid

Oleic acid is a monounsaturated omega-9 fatty acid commonly found in various vegetable oils, such as olive oil, canola oil, and sunflower oil. It is widely used in the food, pharmaceutical, and cosmetic industries. While sulfuric acid isn't typically directly involved in the production of oleic acid, it may be used in certain processes related to its extraction or synthesis. Here's an overview:

Production of Oleic Acid:

a. **Extraction from Vegetable Oils:** The primary source of oleic acid is vegetable oils, particularly those rich in monounsaturated fats. Extraction methods, such as pressing or solvent extraction, are used to isolate oleic acid along with other fatty acids from the oil.
b. **Hydrolysis:** In some cases, oleic acid can be obtained through the hydrolysis of triglycerides present in vegetable oils. In this process, the oil is treated with water and a catalyst (such as sodium hydroxide) to break down the triglycerides into glycerol and free fatty acids, including oleic acid.
c. **Acidification or Neutralization:** Sulfuric acid may be used in the acidification or neutralization step to convert the soaps formed during the hydrolysis process back into free fatty acids. This step helps to separate and purify the oleic acid from other components.
d. **Separation and Purification:** The mixture obtained from acidification or neutralization is then subjected to separation techniques, such as distillation or solvent extraction, to isolate oleic acid from other fatty acids and impurities.
e. **Refining:** The crude oleic acid obtained may undergo further refining processes, such as filtration or crystallization, to enhance its purity and remove any remaining impurities.

Uses of Oleic Acid:

a. **Food Industry:** Oleic acid is commonly used in the food industry as a cooking oil, salad dressing, and food additive. It serves as an emulsifier, flavor enhancer, and stabilizer in various food products.
b. **Cosmetics and Personal Care Products:** In cosmetics and personal care products, oleic acid is used in the formulation of creams, lotions, and soaps due to its emollient properties. It helps to soften and moisturize the skin and hair.
c. **Pharmaceuticals:** Oleic acid is used as an excipient in pharmaceutical formulations, particularly in topical creams and ointments, as it enhances the absorption of active ingredients through the skin.

d. **Industrial Applications:** Oleic acid finds applications in various industrial processes, including the production of lubricants, surfactants, and biodiesel.

While sulfuric acid may not directly participate in the production of oleic acid, it is commonly used in the chemical industry for various purposes, such as in neutralization reactions or as a catalyst, which may indirectly support processes involved in oleic acid production.

34. Cellulose Acetate

Cellulose acetate, a versatile material, finds application in various industries, serving as a film base in photography, a component in coatings, and a material for eyeglass frames. Additionally, it is utilized as a synthetic fiber in manufacturing cigarette filters and playing cards.

Production Process:

a. **Chemical Reaction:**
   Acetic Anhydride + Sulphuric Acid → Cellulose Acetate + Acetic Acid
   Steps:
b. **Dissolution of Cellulose:** Cellulose is dissolved in a mixture of sulphuric acid and acetic anhydride, yielding cellulose acetate and acetic acid.
c. **Purification:** The cellulose acetate solution undergoes purification to remove residual sulphuric acid and acetic anhydride. Neutralization with a basic substance, like sodium hydroxide, ensures the solution's stability.
d. **Formation of Product:** The purified solution is extruded through a spinneret to create fibers or cast into sheets and other shapes for plastic products.
e. **Drying Process:** The fibers or molded products are dried to eliminate any remaining solvent, resulting in a robust and durable end product.

Applications of Cellulose Acetate:

a. **Photography:**
- Film Base: Cellulose acetate serves as a reliable film base in photography, providing a stable support for images.

b. **Coatings:**
- Component in Coatings: Its inclusion in coatings enhances their properties, such as adhesion and durability.

c. **Eyewear:**
- Frame Material: In the eyewear industry, cellulose acetate is prized for its lightweight yet sturdy nature, making it an ideal choice for eyeglass frames.

d. **Synthetic fiber:**

- Cigarette Filters: As a synthetic fiber, cellulose acetate is commonly used in cigarette filters to trap particulates and reduce harmful substances in smoke.
- Playing Cards: Its use in playing cards ensures durability and longevity, providing a smooth and flexible material for cardstock.

4a. Production and Applications of Fibre & Plastics from Cellulose Acetate

Cellulose acetate, a synthetic polymer derived from cellulose, offers versatility in fiber and plastic production. Originating from plant cell walls, cellulose serves as the foundation for this adaptable material.

a. **Fiber Production:**
    - Cellulose acetate is dissolved in a solvent like acetone before being extruded through a spinneret, resulting in fiber formation.
    - Upon cooling and solidification, the fibers exhibit strength, flexibility, and lightness, making them ideal for various applications including clothing, upholstery, and cigarette filters.

b. **Plastic Production:**
    Molding and Shaping:
    - In plastic production, cellulose acetate undergoes molding into diverse shapes such as films, sheets, and molded objects.
    - Its transparency, durability, and moisture resistance render it suitable for packaging, optical applications, and photographic film.

c. **Unique Properties and Applications:**
    - Versatility in Use: Cellulose acetate's unique properties make it indispensable in fibre and plastic manufacturing, offering flexibility and adaptability across industries.
    - Fashion and Apparel:
        - Similar to rayon, cellulose acetate finds application in bridal attire and wedding gowns, owing to its breathable nature and lustrous sheen.
        - Its smooth, satiny texture serves as an excellent alternative to silk, adding elegance to various clothing items.

35. Nitrobenzene

Nitrobenzene, with the chemical formula $C_6H_5NO_2$, is a pale yellow oil with an almond-like odor. It plays a crucial role as a precursor to aniline and finds utility in various industrial applications.

Production Process:

a. Nitration of Benzene:
   - Nitrobenzene is synthesized through the nitration of benzene, employing sulfuric acid as a dehydrating agent.
   - Benzene reacts with nitric acid in the presence of sulfuric acid to yield nitrobenzene and water.
   - Chemical Reaction:
     Benzene + HNO_3 + H_2SO_4 → Nitrobenzene + H_2O + H_2SO_4

$$\frac{C_6H_6}{Benzene} + \frac{HNO_3}{Nitric\ Acid}\ \frac{C^6H^5NO^2}{Nitrobenzene} + \frac{H_2O}{Water}$$

Applications:

a. **Intermediate in Aniline Production:** Nitrobenzene serves as a vital intermediate in the production of aniline, which finds application in dye manufacturing, rubber production, and pharmaceuticals.
b. **Solvent:** Due to its non-polar nature, nitrobenzene functions as a solvent for dissolving organic compounds like resins, fats, and waxes.
c. **Pesticide Manufacturing:** It acts as a raw material in the synthesis of insecticides and fungicides, contributing to pest control in agricultural practices.
d. **Dye Production:** Nitrobenzene is employed as a starting material for the synthesis of various dyes, including aniline dyes, utilized in textile and fabric coloring.

35a. Production and Applications of Aniline

Aniline is a fundamental organic compound and the simplest aromatic amine. It holds substantial industrial significance and serves as a pivotal starting material for various chemical syntheses.

a. **Production Process:** Reduction of Nitrobenzene: Aniline is synthesized through the reduction of nitrobenzene using a suitable reducing agent, such as iron with sulfuric acid or hydrogen gas in the presence of a catalyst.
   Chemical Reaction: Nitrobenzene + $6H_2$ → Aniline + $3H_2O$

Applications:

a. **Dye Production:** Aniline serves as a crucial precursor in the production of diverse dyes, notably aniline dyes, pivotal in textile and leather industries. It plays a significant role in the synthesis of indigo, a primary component in denim dyeing for blue jeans.

b. **Rubber Manufacturing:** Aniline acts as a key starting material in the production of various rubber types, including nitrile rubber and neoprene, contributing to the fabrication of durable and resilient rubber products.
c. **Pharmaceutical Industry:** Aniline serves as a vital starting material in the synthesis of numerous pharmaceuticals, facilitating the production of aniline-based pain relievers and anti-inflammatory medications, among others.
d. **Fuel Additive:** Aniline finds application as an additive in gasoline to enhance its octane rating and mitigate engine knocking, thereby improving engine performance and efficiency.

36. Explosive

Sulphuric acid plays a pivotal role as both a reactant and a catalyst in the production of various explosives, contributing to their synthesis and enhancing their efficacy.

Nitrocellulose

a. **Synthesis:** Nitrocellulose, a combustible and explosive substance, is synthesized through the nitration of cellulose with sulphuric acid acting as a catalyst.
b. **Applications:** Rocket Propellants: Nitrocellulose finds extensive use as a propellant fuel in rocketry, harnessing its explosive properties to propel rockets into space.
    - Smokeless Gunpowder: It serves as a binding agent in the formulation of smokeless gunpowder, providing stability and consistency in firearms.

Nitroglycerin

a. **Synthesis:** Nitroglycerin, renowned for its high explosiveness, is produced via the nitration of glycerol with sulphuric acid as a catalyst.
b. **Applications:**

- Mining Industry: Nitroglycerin is a staple blasting agent in the mining sector, facilitating excavation and extraction operations.
- Dynamite Manufacturing: It serves as a crucial ingredient in the production of dynamite, enhancing its detonation capability for various industrial and construction applications.

37. Ethyl Acetate

Ethyl acetate, an organic compound renowned for its colorless appearance and sweet aroma, finds versatile applications across various industries, owing to its properties as a solvent and flavoring agent.

Synthesis of Ethyl Acetate: Ethyl acetate, with the chemical formula $C_4H_8O_2$, is synthesized through the esterification reaction between ethanol and acetic acid, typically catalyzed by sulphuric acid. This reaction yields ethyl acetate along with water.

The esterification reaction can be represented as follows:

Ethanol + Acetic Acid → Ethyl Acetate + Water

Uses of Ethyl Acetate:

a. **Solvent:** Ethyl acetate serves as a prevalent solvent in various industrial processes, including the manufacturing of inks, adhesives, and paints. Additionally, it facilitates the extraction of essential oils from botanical sources.
b. **Food Flavoring:** Its fruity aroma makes ethyl acetate a desirable flavoring agent in certain food and beverage products, contributing to their sensory appeal.
c. **Pharmaceuticals:** Ethyl acetate's solvent properties render it valuable in pharmaceutical manufacturing, particularly in the production of topical medications and injectables.
d. **Cleaning Agents:** It is utilized as a solvent in nail polish removers and some cleaning agents, owing to its ability to dissolve various organic compounds effectively.

37a. Production and Uses of Lacquers with Ethyl Acetate

Lacquers, revered for their protective and decorative qualities, serve as varnish-like coatings applied to diverse surfaces like wood, metal, and plastic. Ethyl acetate, a solvent known for its efficacy, finds widespread usage in the formulation of certain lacquers.

Synthesis and Composition: Lacquers containing ethyl acetate as a solvent are crafted through precise formulations that blend various ingredients to achieve desired properties. Ethyl acetate contributes to the solubility and application characteristics of the lacquer, facilitating its even spread and rapid drying.

Uses of Lacquers:

a. **Wood Finishing:** Lacquers offer an ideal solution for safeguarding and enhancing the aesthetic appeal of wooden surfaces, ranging from furniture to flooring, providing durability and a glossy finish.
b. **Automotive Industry:** In the automotive sector, lacquers serve as a protective shield, preserving the integrity and appearance of vehicles, including their exteriors and individual components.
c. **Plastic Coating:** Lacquers find application in the protection and embellishment of plastic articles, such as toys, household items, and electronic gadgets, enhancing their durability and visual appeal.
d. **Paper and Printing:** Lacquers are utilized to coat printed materials like books, magazines, and packaging, safeguarding them from wear and tear while imparting a glossy finish.

37b. Vinyl Acetate: Production and Applications

Vinyl acetate ($C_4H_6O_2$) stands as a pivotal organic compound synthesized through the polymerization of ethyl acetate. This reaction, catalyzed by either alkaline or acidic agents, initiates the formation of vinyl acetate polymers, unlocking its versatile applications.

Synthesis and Composition: The synthesis of vinyl acetate involves the polymerization of ethyl acetate, facilitated by catalytic agents that kickstart the polymerization process. This reaction leads to the formation of vinyl acetate, a compound fundamental to various industrial applications.

Applications of Vinyl Acetate:

a. **Adhesives:** Vinyl acetate serves as a fundamental constituent in the production of diverse adhesives, encompassing hot melt adhesives and pressure-sensitive variants, owing to its adhesive properties and compatibility with different substrates.
b. **Coatings:** As a binder, vinyl acetate finds widespread use in coatings production, including paints, inks, and surface coatings for paper and textiles, enhancing adhesion and imparting durability to coated surfaces.
c. **Emulsion Polymers:** Vinyl acetate acts as a monomer in the synthesis of emulsion polymers, notably vinyl acetate-ethylene (VAE) copolymers, prized for their adhesive, coating, and binding properties across various industries.
d. **Photographic Film:** Vinyl acetate features as a component in the formulation of select photographic films, contributing to their structural integrity and photographic properties.

38. Ethers

Sulfuric acid plays a pivotal role as a reactant in the synthesis of various ether compounds, encompassing ethyl ether, diethyl ether, and ethylene glycol. These ethers find diverse applications across industries, owing to their unique properties and versatile nature.

a. **Ethyl Ether:** Ethyl ether emerges as a colorless, highly flammable solvent utilized as a primary material in organic compound synthesis. Additionally, it serves as a fuel for lighters and finds application as a general anesthetic, highlighting its significance in both industrial and medical domains.
b. **Diethyl Ether:** Diethyl ether, sharing similar characteristics with ethyl ether, is another colorless, highly flammable solvent. Employed as a starting material in organic compound synthesis and as a fuel for alcohol burners, its versatility underscores its utility across various applications.\
c. **Ethylene Glycol:** Ethylene glycol, a colorless and virtually odorless liquid, serves as an industrial solvent and coolant. Beyond its cooling properties, it finds utility in the production of polyester fibers and serves as a de-icing agent for airplanes, showcasing its multifaceted applications in diverse industries.

Beyond the aforementioned ethers, sulfuric acid facilitates the synthesis of other ether compounds, such as monoethyl ether, dibutyl ether, and tert-butyl methyl ether. These compounds boast a wide array of applications across the chemical, pharmaceutical, and petrochemical sectors, further underscoring sulfuric acid's indispensable role in ether production.

38a. Solvents Derived from Ethers

Ethers serve as versatile solvents across a spectrum of industrial and commercial domains, offering a range of applications. Here's an insight into some common solvents derived from ethers and their respective uses:

a. **Ethyl Ether:** Primarily utilized as a solvent in organic compound synthesis, ethyl ether also finds application in fuel production, highlighting its role in both chemical processes and energy applications.
b. **Diethyl Ether:** A staple solvent in the chemical and pharmaceutical sectors, diethyl ether is favored for its efficacy in dissolving coatings, resins, and various organic compounds, contributing significantly to industrial processes.
c. **Ethylene Glycol Ethers:** Derived from ethylene glycol, this group of ethers serves as indispensable solvents in the paint and coatings industry. They are also instrumental in synthesizing a diverse range of organic compounds, showcasing their versatility.

d. **Tert-Butyl Methyl Ether (TBME):** TBME emerges as a solvent of choice in the chemical industry, particularly valued for its role in polymer synthesis and the production of various organic compounds, underlining its importance in chemical manufacturing.
e. **Diisopropyl Ether (DIPE):** Widely employed in the paint and coating sector, DIPE functions as a reliable solvent, aiding in the dissolution of coatings. Additionally, it finds utility in oil and fat extraction processes, contributing to diverse industrial applications.
f. **Dibutyl Ether (DBE):** Recognized for its role in polyurethane foam production, DBE serves as a solvent in polymer synthesis and resin manufacture. Its versatility extends to various industrial processes, where it facilitates the production of polymers and related materials.

In summary, the diverse range of ethers as solvents underscores their significance in numerous industrial processes, from chemical synthesis to coatings and polymer production. However, adherence to safety protocols is paramount to ensure safe handling and application in various industrial settings.

39. Formic Acid

Formic acid, with the chemical formula HCOOH or CH_2O_2, stands as the simplest carboxylic acid, pivotal in chemical synthesis and occurring naturally, notably in certain ants. Its esters, salts, and derived anions are termed formats.

Production Process:
Formic acid synthesis from sulfuric acid involves the Ostwald process, comprising the following steps:

a. **Formation of sodium formate:** Carbon monoxide reacts with sodium hydroxide to yield sodium formate:
   $CO + NaOH + H_2O \rightarrow HCOONa$
b. **Conversion to formic acid:** Sodium formate reacts with sulfuric acid to produce formic acid: $2(HCOONa) + H_2SO_4 \rightarrow 2(HCOOH) + Na_2SO_4$

Products and Uses:

a. **Preservative:** Formic acid finds application as a preservative in animal feed, serving to prevent spoilage and curb bacterial proliferation.
b. **Leather Tanning:** In the tanning of leather, formic acid acts as a crucial agent, neutralizing alkaline solutions employed during the process.
c. **Disinfectant:** Due to its antimicrobial properties, formic acid serves as a disinfectant in various industries, particularly in food processing.

d. **pH Adjustment:** Industrial processes, such as textile and dye production, utilize formic acid for pH adjustment in solutions.
e. **Herbicide:** Employed as a herbicide in agriculture, formic acid demonstrates toxicity towards plants, aiding in weed control.
f. **Rubber Production:** Formic acid functions as a coagulating agent in rubber production, facilitating the solidification of rubber latex.
g. **Fuel:** In certain industrial processes, formic acid serves as a fuel, contributing to electricity production. It finds application in fuel cells, both directly in formic acid fuel cells and indirectly in hydrogen fuel cells.
h. In essence, formic acid stands as a multifaceted compound, finding utility across diverse industries, albeit requiring careful management to ensure safe handling and application.

39a. Application of Formic Acid in Leather Tanning

Formic acid plays a vital role in the leather tanning industry, serving as both a neutralizing agent and a key source of hydrogen ions.

Its specific uses encompass several critical aspects of the tanning process:

a. **Neutralization:** Formic acid serves as a crucial neutralizing agent, tasked with balancing the alkalinity inherent in tanning solutions. By adjusting the pH of tanning baths to optimal levels, it facilitates the progression of the tanning process.
b. **Hydrogen Ion Source:** As a source of hydrogen ions, formic acid assumes a pivotal role in leather tanning. These ions contribute significantly to cross-linking tanning agents with collagen fibres in the leather, culminating in the formation of a resilient and stable tanned leather material.
c. **Deliming:** Formic acid functions as a proficient deliming agent during the preparatory stages of hides and skins before tanning. By effectively removing impurities and excess lime from the hide's surface, it primes the material for subsequent tanning processes.
d. **Disinfectant:** Leveraging its antimicrobial properties, formic acid serves as a potent disinfectant in select leather tanning operations. Its application helps thwart the proliferation of bacteria and other microorganisms, ensuring the hygienic integrity of the tanning process.

While formic acid proves invaluable in leather tanning, adherence to stringent safety protocols is imperative. Proper handling and storage procedures are essential to mitigate exposure to the chemical's hazardous properties, safeguarding both personnel and the environment.

39b. Utilizing Formic Acid in Textile Dyeing

In the textile dyeing industry, formic acid plays a pivotal role as both a pH adjuster and a reducing agent.

Its applications are diverse and crucial for ensuring the efficacy and quality of the dyeing process.

a. **pH Adjustment:** Formic acid serves as a fundamental component for pH adjustment in dye baths during textile dyeing. This function is essential for optimizing the absorption of dyes by fibers, thereby facilitating the attainment of desired colors.
b. **Fixation of Dyes:** As a reducing agent, formic acid participates in the fixation of certain direct dyes. By aiding in the bonding of dye molecules to fibres through reduction dyeing, it ensures the durable and vibrant adhesion of colors to textiles.
c. **Scouring:** Formic acid acts as an effective scouring agent, eliminating impurities from fibers before dyeing commences. This scouring process fosters a clean fiber surface, enhancing dye absorption and retention for superior color saturation.
d. **Desizing:** In the desizing phase, formic acid plays a vital role in breaking down sizing agents, such as starch, applied to fabrics. This action facilitates fabric preparation by rendering it more manageable for subsequent dyeing processes.

Formic acid stands as an efficient and indispensable component in textile dyeing operations. However, it is imperative to handle and store it with utmost care due to its hazardous properties. Employing proper safety protocols ensures the safe and controlled utilization of this chemical, safeguarding both personnel and the environment.

40. Sodium Dichromate

Sodium dichromate, an inorganic compound pivotal in the synthesis of various chromium-based materials, finds widespread applications across diverse industries.

Production from Sulphuric Acid:

Sodium dichromate is synthesized from sulphuric acid through a chemical reaction, as depicted below: $Na_2Cr_2O_7 + 2H_2SO_4 \rightarrow Cr_2(SO_4)_3 + Na_2SO_4 + H_2O$

Applications:

a. **Chromium Plating:** Utilized as an electrolyte in chromium plating processes, sodium dichromate facilitates the deposition of chromium layers onto metal surfaces, enhancing their aesthetic appeal and resistance to corrosion.

b. **Dyeing:** Serving as an oxidizing agent, sodium dichromate is instrumental in fabric dyeing, catalyzing oxidation reactions essential for the development of desired dyes on fabrics and other materials.
c. **Leather Tanning:** In the leather tanning process, sodium dichromate functions as a neutralizing agent and a source of chromium ions, facilitating the tanning of hides and skins while imparting durability and resilience to the leather.
d. **Wood Preservation:** Due to its ability to thwart decay and insect damage, sodium dichromate is employed as a wood preservative, safeguarding wooden structures and products against environmental degradation.
e. **Oil Drilling:** Playing a pivotal role as an initiator, sodium dichromate aids in the production of cross-linked polymers crucial in the oil drilling sector. These polymers enhance the viscosity of drilling fluids, optimizing drilling operations.
f. **Water Treatment:** In water treatment applications, sodium dichromate is utilized to eliminate heavy metals and other impurities, ensuring the purity and safety of water resources.

While sodium dichromate boasts numerous industrial applications, it is imperative to acknowledge its highly toxic and carcinogenic nature. Consequently, stringent safety protocols must be observed during its handling and disposal to mitigate environmental and health hazards.

40a. Utilizing Sodium Dichromate as a Pigment

Sodium dichromate, renowned for its vivid yellow hue, serves as a versatile pigment in various industrial applications, imparting vibrancy and durability to a spectrum of products.

a. **Ceramics:** In the ceramics industry, sodium dichromate is employed as a yellow pigment, infusing ceramic creations with a striking and enduring coloration, enhancing their aesthetic appeal.
b. **Paints and Coatings:** Within the realm of paints and coatings, sodium dichromate functions as a prized yellow pigment. Its incorporation adds a lively burst of color to coatings and paints while ensuring resistance to fading, thereby extending the longevity of painted surfaces.
c. **Inks:** Sodium dichromate finds its place in the formulation of inks, where its yellow pigment enhances the visual impact of printed materials. The vibrant color it imparts adds depth and character to printed designs.
d. **Plastics:** As a yellow pigment in plastics manufacturing, sodium dichromate contributes to the production of plastic products with enduring coloration. Its

vibrant hue not only elevates the visual appeal of plastics but also ensures resistance to fading over time, maintaining their aesthetic allure.

By leveraging sodium dichromate as a pigment, industries can infuse their products with a captivating yellow hue that not only enhances visual appeal but also withstands the test of time, ensuring lasting vibrancy in various applications.

APPLICATION OF SULPHURIC ACID-SHEET 4

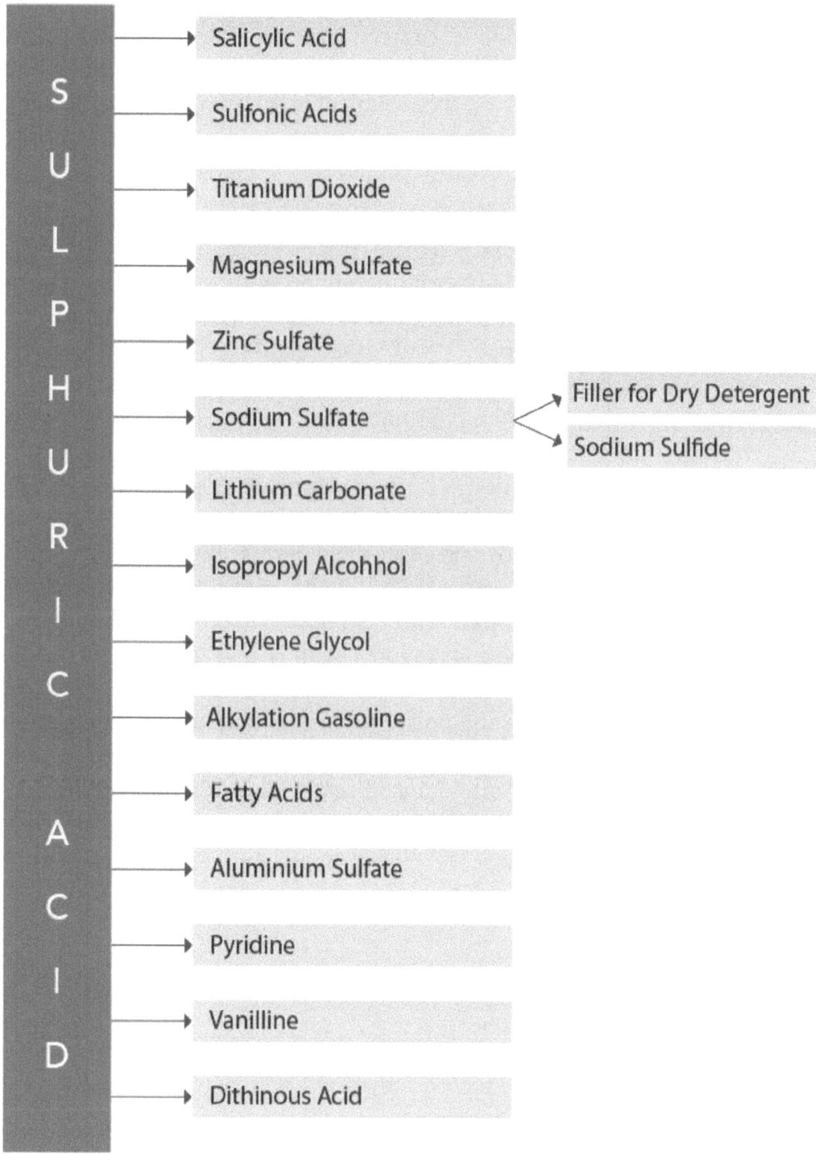

41. Salicylic Acid

Salicylic acid, an organic compound, serves as a pivotal precursor to aspirin and holds multifaceted roles in various domains, from skincare to plant protection.

Production Process: Commercially, salicylic acid is synthesized through a meticulous process. Initially, sodium phenolate, derived from phenol, undergoes a transformation with carbon dioxide under high pressure and temperature conditions (100 atm and 115°C). The ensuing product is then subjected to treatment with sulphuric acid, yielding salicylic acid.

Applications:
Salicylic acid finds extensive use across diverse sectors:

a. **Skincare:** Renowned for its exfoliating properties and pore-cleansing abilities, salicylic acid is a staple ingredient in numerous skincare formulations, including cleansers, toners, and spot treatments. It effectively tackles acne, psoriasis, and warts.
b. **Pharmaceuticals:** In the pharmaceutical realm, salicylic acid plays a crucial role in the production of aspirin, a widely-used analgesic and anti-inflammatory medication. It also features in various other pharmaceutical products such as creams, gels, and ointments tailored for treating skin conditions like dandruff and seborrheic dermatitis.
c. **Hair Care:** Salicylic acid's follicle-unclogging properties make it a favored component in select hair care products, including shampoos and scalp treatments. Its application promotes healthy hair growth and scalp health.
d. **Plant Health:** Functioning as a plant growth regulator, salicylic acid stimulates root growth and bolsters overall plant health. Additionally, it serves as a pesticide, aiding in the control of fungal growth and other plant pests.
e. **Food Preservation:** Salicylic acid lends its preservative prowess to the food industry, inhibiting bacterial and microbial growth in food products.

42. Sulfonic Acid

Sulfonic acids, crucial components in various industries, are synthesized from sulfuric acid through a chemical reaction.

Production Process:
The synthesis of sulfonic acids involves a reaction with sulfuric acid, depicted by the general equation: $R\text{-}H + H_2SO_4 \rightarrow R\text{-}SO_3H + H_2O$

Here, (R) signifies the organic compound. For instance, benzene (C_6H_6) undergoes sulfonation with concentrated sulfuric acid to yield benzenesulfonic acid ($C_6H_5SO_3H$):
$C_6H_6 + H_2SO_4 \rightarrow C_6HSO_3H + H_2O$

Applications: Sulfonic acids find widespread utilization across various sectors:

a. **Detergents:** Sulfonic acids serve as pivotal surfactants in detergent manufacturing, bestowing cleansing and emulsifying properties to products like laundry detergents, dishwashing liquids, and shampoos.
b. **Textile Processing:** In the textile industry, sulfonic acids function as softeners and finishing agents, enhancing fabric texture and bolstering stain resistance.
c. **Oil Refining:** Sulfonic acids act as catalysts in petroleum and oil refining processes, augmenting gasoline and other derivative yields.
d. **Dyes and Pigments:** These acids serve as intermediates in dye and pigment manufacturing, serving as foundational materials for the production of vibrant colorants.
e. **Pharmaceuticals:** In pharmaceutical production, sulfonic acids act as intermediates, facilitating the synthesis of diverse drugs and medicinal products

43. Titanium Dioxide

Titanium dioxide (TiO_2) is a pivotal compound utilized extensively across various industries. While not directly derived from sulphuric acid, it plays a vital role in the production process through the sulphate method, employing ilmenite concentrate (60% TiO_2) as a primary feedstock, treated with Sulphuric Acid to extract and refine Titanium dioxide.

Production Process Overview:
The essential reaction is succinctly illustrated as follows:
$TiO_2 + H_2SO_4 \rightarrow Ti(SO_4)_2 + H_2O$

Key Applications:

a. Pigment: Titanium dioxide serves as a ubiquitous white pigment, enriching various products such as viscose rayon, paints, coatings, plastics, papers, and cosmetics. Its exceptional brightness, whiteness, and superior covering power enhance the aesthetic appeal and functionality of these materials.
b. Sunscreen: Widely employed as a physical sunscreen, titanium dioxide offers effective protection against harmful UV rays in personal care products, safeguarding the skin from sun-induced damage.

c. Food Additive: In the food industry, titanium dioxide acts as a food additive, imparting brightness and whiteness to a diverse array of products, including chewing gums, candies, and powdered sugars.
d. Photocatalysis: Renowned for its photocatalytic properties, titanium dioxide serves as a catalyst for degrading pollutants in air and water when exposed to light, contributing to environmental remediation efforts.
e. Ceramics: In the realm of ceramics, titanium dioxide emerges as a pivotal material, endowing products with exceptional wear resistance and thermal stability.

44. Magnesium Sulphate

Magnesium sulphate, commonly known as Epsom salt, is a compound comprising magnesium, sulphur, and oxygen. It is typically produced from sulphuric acid and magnesium oxide or magnesium hydroxide through chemical reactions, as illustrated below:

$$MgO + H_2SO_4 \rightarrow MgSO_4 + H_2O$$
$$MgCO_3 + H_2SO_4 \rightarrow MgSO_4 + CO_2 + H_2O$$

Key Applications:

a. **Agriculture:** Magnesium sulphate serves as a crucial fertilizer, enriching soil with magnesium—a vital nutrient essential for chlorophyll synthesis and photosynthesis. It aids in correcting magnesium-deficient soils and adjusting pH levels to optimize crop growth and yield.
b. **Medical:** In the medical field, magnesium sulphate finds widespread use. It acts as a laxative to relieve constipation, and is employed in treating heart disease, hypertension, and managing symptoms of pre-eclampsia in pregnant women.
c. **Bathing:** Adding magnesium sulphate to bathwater is a popular practice for alleviating muscle pain and soothing skin irritation, offering a therapeutic and relaxing bathing experience.
d. **Food and Beverage:** As a food additive, magnesium sulphate enhances food stability and texture in processed foods, ensuring quality and shelf life.
e. **Industrial:** In industrial settings, magnesium sulphate contributes to various processes. It aids in paper production, textile manufacturing, and the development of fireproofing materials. Additionally, it facilitates brackish water purification.

45. Zinc Sulphate

Zinc sulphate, a compound comprising zinc, sulphur, and oxygen, is a vital resource with multifaceted applications. It is commonly synthesized through the reaction of sulphuric acid with zinc oxide or zinc carbonate, as demonstrated below:

a. $$\frac{\text{ZnO}}{\text{Zinc Ash or Zinc Dross}} + \frac{H_2SO_4}{Sulphuric\ Acid} \rightarrow \frac{Zn(SO_4)}{Zinc\ Sulphate} + H_2O$$

b. $$\frac{ZnCO_3}{Zinc\ Carbonate} + \frac{H_2SO_4}{Sulphuric\ Acid} \rightarrow \frac{Zn(SO_4)}{Zinc\ Sulphate} + CO_2 + H_2O$$

Zinc sulphate is commercially available in various hydrate forms, including:

1. Monohydrate ($ZnSO_4 \cdot H_2O$)
2. Hexahydrate ($ZnSO_4 \cdot 6H_2O$)
3. Heptahydrate ($ZnSO_4 \cdot 7H_2O$)

Key Applications:

a. **Agriculture:** Zinc sulphate serves as a crucial nutrient in agriculture, enriching soil with zinc to address zinc deficiencies in crops. It promotes plant growth, development, and overall crop yield.

b. **Medical:** In the medical field, zinc sulphate acts as a dietary supplement, aiding in the treatment of zinc deficiencies and supporting immune function. It plays a vital role in maintaining overall health and well-being.

c. **Industrial:** Zinc sulphate finds application in various industrial processes. It is utilized in the production of zinc salts like zinc chloride and zinc acetate, integral components in the manufacturing of textiles, rubber products, and ceramics.

d. **Water Treatment:** Zinc sulphate plays a crucial role in water treatment, particularly in the purification of drinking water. It effectively removes impurities such as iron, manganese, and heavy metals, ensuring water safety and quality.

e. **Mining:** In the mining industry, zinc sulphate is utilized in mineral flotation processes. It enhances the separation of valuable metal minerals, such as lead and copper, from waste minerals, contributing to efficient mineral extraction.

46. Sodium Sulphate

Sodium sulphate, also referred to as sulphate of soda or Glauber's salt, constitutes a compound composed of sodium, sulphur, and oxygen. This white, odorless powder is commonly synthesized through the reaction of sulphuric acid with sodium carbonate, colloquially known as soda ash.

Production and Variants: Sodium sulphate can also be obtained as a by-product during the manufacture of Viscose Rayon. In this process, Viscose is reacted with hot and diluted sulphuric acid, yielding Rayon yarn as the primary product and sodium sulphate as the by-product. Commercially, sodium sulphate salt is termed "Glauber Salt," existing in a crystalline form with water molecules as water of crystallization. Its chemical formula, $[Na_2SO_4(_{10}H_2O)]$, distinguishes it from the anhydrous form of Glauber Salt, $NaSO_4$.

Manufacturing Sodium Hydrogen Sulphate: When sodium sulphate reacts with sulphuric acid, a double displacement reaction occurs, yielding sodium hydrogen sulphate and water: $Na_2SO_4 + H_2SO_4 \rightarrow 2NaHSO_4 + H_2O$
This exothermic reaction can also proceed reversibly, converting sodium hydrogen sulphate back into sodium sulphate and sulphuric acid upon the addition of water.

Diverse Utilizations:

a. **Industrial Applications:** Sodium sulphate serves as a filler in detergent, glass, and textile manufacturing. Additionally, it acts as a desiccant in chemical and pharmaceutical production processes.
b. **Food and Beverage Industry:** Employed as a food additive, sodium sulphate enhances texture and functions as a firming agent for fruits and vegetables.
c. **Agriculture:** As a soil amendment, sodium sulphate improves soil structure and aids in pH adjustment.
d. **Mining:** Sodium sulphate finds utility in mineral flotation processes, facilitating the separation of valuable metal minerals from waste minerals, particularly in lead and copper mining.
e. **Medical Uses:** Sodium sulphate serves as a laxative to alleviate constipation when administered under professional guidance.

46a. Filler in Dry Detergent Formulations from Sodium Sulfate

a. **Inert Nature:** Sodium sulfate boasts a chemically inert nature, ensuring it doesn't react with other detergent components. This characteristic guarantees safety in usage and avoids interference with the cleaning process.

b. **Cost-Effectiveness:** Offering relative affordability compared to many detergent ingredients, sodium sulfate emerges as a cost-effective option for manufacturers, allowing them to maintain competitiveness without compromising quality.
c. **Bulkiness:** Characterized by its crystalline structure, sodium sulfate adds substantial bulk to detergent formulations without inflating production costs. This feature enables manufacturers to scale up production volumes while keeping prices competitive.
d. **Absorbency:** Sodium sulfate's capacity to absorb small amounts of water plays a crucial role in preventing clumping, ensuring the detergent maintains a free-flowing consistency, and facilitating easy dispensing.
e. **pH Buffering:** Acting as a stabilizer, sodium sulfate aids in maintaining the pH balance of detergent solutions over time, ensuring consistent and effective cleaning performance.
f. **Environmental Considerations:** Recognized for its environmentally friendly and biodegradable properties, sodium sulfate aligns with the growing demand for sustainable detergent formulations, making it a preferred choice for environmentally conscious consumers and manufacturers alike.

When integrating sodium sulfate into detergent formulations, manufacturers meticulously consider factors such as particle size, purity, and compatibility with other ingredients to optimize both performance and appearance. Moreover, the quantity of sodium sulfate employed may vary depending on the specific formulation requirements and desired detergent properties.

46b. Sodium Sulphide from Sodium Sulfate

Sodium sulfide, a chemical compound comprising sodium and sulfur, is pivotal in numerous industrial processes. It is synthesized through several methods, including the reaction of sodium hydroxide with excess sulfur, as illustrated below:

$2(NaOH) + 4S \rightarrow 2(Na_2s) + Na_2S_2O_3 + 3(H_2O)$

Alternatively, sodium sulfide can be produced by reducing sodium sulfate with a suitable agent like carbon or hydrogen sulfide gas, resulting in its application across diverse sectors. These reactions typically necessitate elevated temperatures and controlled conditions, commonly executed in industrial furnaces or reactors.

Applications of Sodium Sulfide:

a. **Pulp and Paper Industry:** Sodium sulfide is instrumental in pulp and paper production, where it facilitates lignin dissolution and minimizes paper discoloration, ensuring high-quality paper output.
b. **Textile Manufacturing:** In textile production, sodium sulfide acts as a potent reducing and bleaching agent, aiding in fabric treatment to achieve desired coloration and texture.

c. **Leather Tanning:** Sodium sulfide finds utility in leather tanning processes, effectively removing hair and softening the skin, thus enhancing leather quality.
d. **Dyes and Pigments:** Within the dye and pigment industry, sodium sulfide serves as a crucial reducing and precipitating agent, facilitating the production of vibrant and stable colorants.
e. **Mining Operations:** In mining, sodium sulfide plays a pivotal role in mineral flotation processes, particularly in the extraction of copper, lead, and zinc. It aids in the separation of valuable metal minerals from waste materials, thereby optimizing mineral recovery.

47. Lithium Carbonate

Lithium carbonate stands as a pivotal compound in both pharmaceutical and chemical domains, renowned for its versatility and efficacy. This white, crystalline powder, bearing the chemical formula Li_2CO_3, finds widespread utility across various industries, with its paramount role as a mood stabilizer for individuals grappling with bipolar disorder. Let's delve into its production process from sulphuric acid and its multifaceted applications:

Production Process Overview:

a. **Preparation of Lithium Sulphate:** Lithium-containing ores or minerals undergo treatment with sulphuric acid, yielding lithium sulphate (Li_2SO_4) along with other byproducts.
   Example Reaction: $LiAlSi_2O_6 + H_2SO_4 \rightarrow Li_2SO_4 + Al_2(SO_4)_3 + SiO_2$
b. **Formation of Lithium Hydroxide:** The obtained lithium sulphate undergoes a hydrothermal process, resulting in the production of lithium hydroxide (LiOH) alongside sulphuric acid.
   Example Reaction: $Li_2SO_4 + 2H_2O \rightarrow 2LiOH + H_2SO_4$
c. **Reaction with Carbon Dioxide:** Lithium hydroxide interacts with carbon dioxide (CO_2), leading to the formation of lithium carbonate (Li_2CO_3) and water (H_2O).
   Example Reaction: $2LiOH + CO_2 \rightarrow Li_2CO_3 + H_2O$
d. **Isolation and Purification:** The lithium carbonate product undergoes rigorous isolation and purification procedures, encompassing filtration, precipitation, and drying steps.

Applications of Lithium Carbonate:

a. **Pharmaceuticals:** Primarily renowned as a mood stabilizer for bipolar disorder, lithium carbonate holds a prominent place in psychiatric treatments.

b. **Ceramics and Glasses:** Lithium carbonate acts as a crucial 'flux' agent in ceramic and glass production, facilitating desired melting and fusion properties.
c. **Desiccant:** With its remarkable ability to absorb moisture, lithium carbonate serves as an effective desiccant, combating humidity in diverse environments.
d. **Battery Technology:** A key player in the realm of lithium-ion batteries, lithium carbonate serves as a vital source of lithium ions. These ions enable energy storage and release within battery cathodes, powering numerous electronic devices and vehicles.
e. **Chemical Industry:** Lithium carbonate finds application in producing various lithium salts like lithium bromide and lithium chloride, essential components in heat transfer fluids for air conditioning and refrigeration systems.

In essence, the synthesis and utilization of lithium carbonate epitomizes the amalgamation of scientific innovation and industrial exigencies. From stabilizing moods to revolutionizing battery technology, its impact spans across critical sectors, underscoring its indispensable role in modern society's fabric.

48. Isopropyl Alcohol

Isopropyl alcohol, also recognized as isopropanol, emerges as a clear, flammable liquid with a potent, distinct aroma. Renowned for its multifaceted utility as a solvent, cleanser, and disinfectant, it finds widespread application across diverse industries.

Production Process Overview:

a. **Hydration of Propane:** Isopropyl alcohol is synthesized through the hydration of propane (C_3H_8) in the presence of sulphuric acid (H_2SO_4).

Example Reaction: $C_3H_8 + H_2O$ (in the presence of H_2SO_4) $\rightarrow C_3H_8O$

b. **Chemical Formula:** The chemical equation can also be represented as follows: Example Reaction:

$CH_3CH{=}CH_2 + H_2O$ (in the presence of H_2SO_4) $\rightarrow (CH_3)_2CHOH$

Applications of Isopropyl Alcohol:

a. **Healthcare:** Isopropyl alcohol serves as a crucial disinfectant for sanitizing surfaces and medical equipment, alongside functioning as a skin antiseptic.
b. **Household Cleaning:** Widely utilized for cleaning electronic devices, glass surfaces, and stubborn stains, isopropyl alcohol proves effective in household cleaning chores.
c. **Cosmetics and Personal Care:** It features as an essential ingredient in hand sanitizers and rubbing alcohol, ensuring hygiene and sanitation in personal care products.

d. **Automotive Sector:** Isopropyl alcohol serves as a potent cleaning agent for automotive components and acts as a solvent for various fluids essential for vehicle maintenance.
e. **Printing Industry:** Employed as a solvent for inks and resins, isopropyl alcohol facilitates the printing process by ensuring optimal fluidity and viscosity.
f. **Chemical Synthesis:** Isopropyl alcohol serves as a pivotal starting material for the synthesis of other chemicals, including isopropyl acetate, a solvent utilized in the paint and coatings industry, and 2-propanol, employed both as a solvent and as a raw material for diverse chemical productions.

In essence, isopropyl alcohol emerges as a cornerstone in numerous industrial processes, offering a potent blend of cleansing, disinfecting, and solvent properties. From healthcare to household chores and industrial applications, its versatility underscores its indispensable role in modern-day operations.

49. Ethylene Glycol

Ethylene glycol stands as a clear, colorless, syrupy liquid with a multitude of applications in the chemical industry. Renowned for its role as a raw material in polyester fiber production and as a key component in antifreeze formulations, its chemical composition, and diverse utility underscore its significance in industrial processes.

Production Process Overview:

a. Ethylene Oxide Intermediate: Ethylene glycol is synthesized from ethylene (ethene), employing ethylene oxide (C_2H_4O) as an intermediate.

Example Reaction:

$C_2H_4O + H_2O$ (in the presence of H_2SO_4 catalyst) → $HO\text{-}CH_2CH_2\text{-}OH$

Applications of Ethylene Glycol:

a. **Antifreeze Formulation:** Ethylene glycol serves as the primary constituent in automotive antifreeze and coolant formulations, crucial for preventing engine overheating and freezing in extreme weather conditions.
b. **Polyester fibres and Resins:** As a vital raw material, ethylene glycol contributes significantly to the production of polyester fibres and resins, pivotal in textile manufacturing, packaging materials, and various consumer goods.
c. **Unsaturated Polyester Resins:** These resins find application in fabricating a diverse array of products, including bathtub and shower enclosures, boats, and storage tanks.

d. **Solvent:** Ethylene glycol's solvent properties render it indispensable in industries such as paint and coatings, pharmaceuticals, and personal care, where it serves as a versatile solubilizing agent.
e. **Dehumidifiers:** In select dehumidifiers, ethylene glycol is employed to effectively absorb moisture from the air, facilitating humidity control in enclosed environments.
f. **Personal Care Products:** Ethylene glycol functions as a humectant in certain personal care items like lotions and creams, aiding in skin hydration and moisture retention.

50. Alkylation Gasoline

Alkylation stands as a vital refining process employing sulfuric acid as a catalyst to transform low-value straight-chain hydrocarbons into high-octane, branched-chain molecules. The resultant product, alkylate, serves as a prized gasoline blendstock renowned for its superior octane rating, reduced volatility, and minimal toxicity.

Alkylation Reaction Overview:

a. **Reaction Components:** The alkylation process involves the reaction of an olefin, such as isobutene, with an isoparaffin, such as isohexane, in the presence of sulfuric acid. Example Reaction:
C_4H_8 + C_6H_{14} (in the presence of H_2SO_4 catalyst) → C_7H_{16} + C_3H_6

Uses of Alkylate:

a. **Gasoline Enhancement:** Alkylate significantly elevates the octane rating of gasoline, thereby enhancing its combustion efficiency and reducing engine knocking. Moreover, it aids in minimizing emissions of harmful pollutants like benzene, contributing to cleaner air quality.
b. **Blendstock:** Alkylate serves as a fundamental component in formulating high-quality gasoline products. When blended with other gasoline constituents like reformate and straight-run gasoline, it helps achieve optimal performance and compliance with stringent fuel quality standards.
c. **Feedstock for Chemical Production:** Beyond gasoline production, alkylate finds utility as a feedstock for manufacturing high-grade chemicals, such as alpha-olefins. These compounds serve as crucial building blocks for a diverse range of consumer and industrial products, further accentuating the versatility and value of alkylate.

The incorporation of alkylate in gasoline refining processes plays a pivotal role in elevating the overall quality and performance of gasoline products. By improving combustion efficiency and reducing emissions of harmful pollutants, alkylate contributes to cleaner air and a healthier environment. Its indispensability in the refining industry underscores its high demand among gasoline refiners and marketers, highlighting its significance in advancing sustainable fuel solutions.

In essence, alkylate emerges as a cornerstone in gasoline production, offering a multifaceted approach to enhancing fuel quality, promoting environmental stewardship, and meeting the evolving demands of the modern automotive industry.

51. Fatty Acid

Fatty acids, pivotal constituents of fats and oils, serve as fundamental building blocks with diverse industrial and consumer applications. Their synthesis involves the esterification of carboxylic acids and alcohols facilitated by sulfuric acid as a catalyst:
Esterification Reaction Overview:
RCOOH + R'OH (catalyzed by H_2SO_4) → RCOOR' + H_2O
Here, R and R' denote the alkyl groups of the carboxylic acid and alcohol, respectively.

Applications of Fatty Acids:

a. **Soap and Detergents:** Fatty acids undergo saponification with strong bases like sodium hydroxide, yielding soap. This soap serves as a key component in various cleaning agents, including laundry detergents, dishwashing liquids, and personal care products.
b. **Biodiesel Production:** Fatty acids are pivotal in the production of biodiesel, a renewable fuel offering an eco-friendly alternative to conventional diesel derived from petroleum sources.
c. **Personal Care Products:** Fatty acids feature prominently in personal care formulations, such as lotions, creams, and shampoos, where they contribute to moisturization and nourishment of skin and hair.
d. **Food Industry:** Fatty acids find extensive usage in food products like margarine, shortening, and cooking oils, enhancing flavor profiles and imparting desirable textures.
e. **Animal Feed Supplements:** As constituents of animal feed, fatty acids serve as crucial sources of energy and essential nutrients, fostering the health and growth of livestock and poultry.

Fatty acids play an indispensable role across various industries, underpinning the formulation of a wide array of products essential to everyday life. Their multifaceted applications underscore their significance in supporting human and animal well-being, as well as driving advancements in sustainable fuel solutions and personal care formulations. As key components of diverse formulations, fatty acids continue to contribute to the functionality, efficacy, and overall quality of numerous consumer and industrial products.

52. Aluminum Sulphate

Aluminum sulphate, a soluble compound, serves as a vital coagulating agent, particularly in water purification, wastewater treatment, and paper manufacturing processes. Its anhydrous form, occurring naturally as millosevichite, is relatively rare and found in volcanic environments and coal-mining waste dumps. In various industries, it is often referred to as alum.

This chemical compound, denoted by the formula $Al_2(SO_4)_3$, is synthesized through the reaction between aluminum hydroxide and sulphuric acid:

$$2[Al(OH)_3] + 3(H_2SO_4) \rightarrow Al_2(SO_4)_3 + 6H_2O$$

Applications of Aluminum Sulphate:

Water Treatment: Widely employed as a coagulant in water treatment facilities, aluminum sulphate facilitates the aggregation of suspended particles, enabling their easier removal via filtration. This process enhances water clarity and purity by eliminating impurities.

a. **Pulp and Paper Production:** Aluminum sulphate functions as a flocculant in pulp and paper manufacturing, aiding in the clarification of fibre suspensions and impurities. This improves the efficiency of removing unwanted materials from the production stream.

b. **Textile Printing:** Utilized as a mordant in textile printing processes, aluminum sulphate plays a crucial role in fixing dyes onto fabric, ensuring color fastness and vibrant designs.

c. **Firefighting:** In firefighting applications, aluminum sulphate is utilized as a fire-retardant when applied in an aqueous solution. It forms a protective barrier that suppresses flames and mitigates fire hazards.

d. **Agricultural Use:** Aluminum sulphate serves as a soil amendment in agriculture, aiding in the adjustment of soil pH levels to create a more acidic environment. This is beneficial for crops such as blueberries and azaleas that thrive in acidic soils.

53. Pyridine

Pyridine, a basic heterocyclic organic compound, possesses distinctive properties, including high flammability, weak alkalinity, and a fish-like odor. Widely utilized in agrochemicals, pharmaceuticals, and vitamins, pyridine plays a crucial role in numerous important compounds.

With a chemical formula of C_5H_5N, pyridine is synthesized through a multi-step process involving nitrobenzene and sulphuric acid:

a. **Reduction of Nitrobenzene to Aniline:** Nitrobenzene ($C_6H_5NO_2$) undergoes reduction to form aniline ($C_6H_5NH_2$) using a reducing agent like iron or tin and hydrochloric acid (HCl):
   $C_6H_5NO_2 + 6HCl + 6Fe \rightarrow C_6H_5NH_2 + 3H_2O + 6FeCl_2$
b. **Conversion of Aniline to Phenylhydrazine:** Aniline reacts with excess nitrous acid (HNO_2) to yield phenylhydrazine ($C_6H_5NHNH_2$):
   $C_6H_5NH_2 + HNO_2 \rightarrow C_6H_5NHNH_2 + H_2O$
c. **Conversion of Phenylhydrazine to Pyridine:** Phenylhydrazine reacts with formaldehyde (HCHO) and sulphuric acid (H_2SO_4) to produce pyridine:
   $C_6H_5NHNH_2 + HCHO + H_2SO_4 \rightarrow C_5H_5N + 2H_2O + H_2SO_4 + H_2CO$

Applications of Pyridine:

b. **Pharmaceuticals:** Pyridine serves as an intermediate in synthesizing various pharmaceuticals, including antihistamines, analgesics, and anticonvulsants, contributing significantly to the healthcare industry.
c. **Agrochemicals:** As an intermediate, pyridine is integral in synthesizing agrochemicals such as herbicides and insecticides, supporting agricultural practices worldwide.
d. **Flavor and Fragrance:** The strong, distinctive odor of pyridine finds application in the flavor and fragrance industries, enhancing the sensory appeal of numerous products.
e. **Pesticides:** Pyridine derivatives act as active ingredients in pesticides, playing a vital role in fungicides and insecticides used to safeguard crops and environments.
f. **Solvents:** Known for its polar nature, pyridine serves as a solvent in various chemical reactions and processes, facilitating the dissolution of resins and extraction of oils and fats.

54. Vanilline

Vanillin, a phenolic aldehyde, is renowned as the primary flavor component extracted from vanilla beans. While synthetic vanillin has supplanted natural extracts in many applications, it continues to be a prominent flavoring agent in foods, beverages, and pharmaceuticals.
With a chemical formula of $C_8H_8O_3$, vanillin can be synthesized from various starting materials, including guaiacol and eugenol, with sulphuric acid catalyzing the process:

Vanillin is synthesized from isoeugenol through oxidation, catalyzed by sulphuric acid:
$2C_{10}H_{12}O + 3O_2 \rightarrow 2C_8H_8O_3 + 2H_2O$ (catalyzed by H_2SO_4)

Applications of Vanillin:

a. **Flavor and Fragrance:** Vanillin is extensively used as a flavoring agent in foods and beverages, contributing its characteristic sweet, creamy taste. Additionally, it serves as a key ingredient in perfumes and fragrances, imparting a pleasant aroma.
b. **Pharmaceuticals:** In the pharmaceutical industry, vanillin acts as a flavor enhancer in various formulations such as cough syrups and lozenges, improving palatability and masking undesirable tastes.
c. **Cosmetics:** Vanillin finds application as a fragrance component in cosmetic products like soaps, lotions, and perfumes, enhancing their sensory appeal and providing a soothing aroma.
d. **Aromatherapy:** Due to its calming and relaxing aroma, vanillin is utilized in aromatherapy practices, where it is believed to promote a sense of tranquility and well-being.

55. Dithionous Acid

Dithionous acid, an unstable sulfur oxoacid, finds stability in its salts, known as dithionite. Despite its inherent instability, dithionous acid serves as an intermediate in various chemical processes and holds potential applications in specialized areas.
With the chemical formula $H_2S_2O_4$, dithionous acid is not directly obtained from sulphuric acid. However, sulphuric acid plays a crucial role in its preparation through a multi-step process.

Synthesis Overview:
Dithionous acid is synthesized via a multi-step process, beginning with the reaction of sodium sulfite with sulphuric acid to yield sulphurous acid:
$Na_2SO_3 + H_2SO_4 \rightarrow H_2SO_3 + Na_2SO_4$

Subsequently, sulphurous acid is further reduced to dithionous acid by heating in the presence of a reducing agent, such as iron:
$H_2SO_3 + 2H^+ + 2e^- \rightarrow H_2S_2O_4 + H_2O$

Applications of Dithionous Acid:

a. **Wastewater Treatment:** Dithionic acid holds potential in the treatment of wastewater, where it can effectively remove heavy metal ions, aiding in environmental remediation efforts.
b. **Chemical Synthesis:** Dithionic acid serves as a precursor in the synthesis of various sulfur-containing compounds, including dithionite. These compounds

find applications in bleaching and papermaking processes, contributing to the production of high-quality paper products.

While dithionous acid may not be extensively utilized on its own, its role as an intermediate compound is indispensable in the production of sulfur-containing chemicals. Its applications in specialized areas underscore its importance in various industrial processes, contributing to environmental sustainability and product quality enhancement.

APPLICATION OF SULPHURIC ACID-SHEET 5

56. Iron & Steel Pickling

Pickling, a surface treatment method, serves to eliminate impurities like stains, inorganic contaminants, rust, or scale from various metals, including iron/ferrous metals, copper, precious metals, and aluminum alloys. Typically, a solution known as pickle liquor, containing acid, is employed to achieve this surface purification. The pickling process is extensively utilized in steelmaking to descale or clean steel surfaces.

Iron and steel pickling involves the removal of impurities, particularly rust, from the surface of iron and steel products. Sulphuric acid, renowned for its potent acidic properties, is commonly utilized in this process to facilitate the removal of impurities through chemical reactions. The overall reaction involved in rust removal (pickling) is as follows: Chemical Reaction: $Fe + H_2SO_4 \rightarrow FeSO_4 + H_2$

This reaction yields ferrous sulphate and hydrogen gas, with the latter being safely vented into the atmosphere while the former is either disposed of or recycled for other processes.

Pickling Procedure:

a. **Surface Preparation:** The iron or steel surface undergoes thorough cleaning to eliminate loose debris and contaminants.
b. **Immersion in Acid Bath:** Subsequently, the product is immersed in a bath containing sulphuric acid, typically maintained at temperatures of around 50-60°C. The acid interacts with surface impurities, effectively dissolving them and leaving behind a clean surface.
c. **Rinsing:** After the desired duration of immersion, the product is withdrawn from the acid bath and rinsed meticulously with water to eliminate any residual acid and impurities.
d. **Passivation:** Optionally, a passivation step may be conducted to provide additional protection to the surface and prevent future corrosion. This step typically involves immersing the product in a solution of nitric acid or a similar chemical.

The resultant clean and impurity-free surface achieved through iron and steel pickling is crucial for enhanced corrosion resistance, improved paint adhesion, and better weldability.

57. Cyclic Sulphates

Cyclic sulphates, also referred to as ring-opened sulphates or epoxy sulphates, constitute a class of chemical compounds characterized by a cyclic structure containing a sulphate group ($-SO_4{}^{\wedge}{}_2-$). Typically, they are synthesized from sulfamic acid or sulfamates, with sulphuric acid serving as a catalyst in the process.

Applications:

a. **Plasticizers:** Certain cyclic sulphates function as plasticizers in the production of plastics, rubber, and other polymeric materials. They enhance flexibility and refine the processing characteristics of the final products.
b. **Intermediates:** Cyclic sulphates play a pivotal role as intermediates in the synthesis of various chemicals, including pharmaceuticals and fragrances, contributing to the production of diverse compounds.
c. **Epoxy Resins:** Some cyclic sulphates serve as essential raw materials in the manufacturing of epoxy resins, widely utilized in coatings, adhesives, and composites, owing to their robust properties.
d. **Detergents:** Cyclic sulphates find application as surfactants in detergents, aiding in the emulsification of oils and grease, thereby enhancing the cleaning efficacy of the detergent formulations.

58. Chlorine

Chlorine, a vital chemical, is commonly manufactured from sulphuric acid in a process known as the chlor-alkali process. Below outlines the process and highlights the versatile applications of chlorine:

Chlorine Production Process:

a. **Brine Formation:** Sodium chloride (NaCl) is dissolved in water to create a brine solution.
b. **Electrolysis:** A small amount of sulphuric acid is added to the brine solution to facilitate electrolysis. The solution undergoes electrolysis, wherein an electric current is passed through it, causing the salt (NaCl) to dissociate into its constituent ions: sodium (Na+) and chloride (Cl-).
c. **Chlorine Generation:** At the anode, chloride ions (Cl-) are oxidized to form chlorine gas (Cl_2).

Applications of Chlorine Products:

a. **Water Treatment:** Chlorine serves as a potent disinfectant, extensively utilized for purifying drinking water and treating wastewater, safeguarding public health.
b. **Bleaching:** In industries such as paper and textiles, chlorine is employed for bleaching processes, ensuring the brightness and purity of the final products.
c. **Disinfectants and Sanitizers:** Chlorine-based compounds feature prominently in household and industrial cleaning products, effectively eliminating germs and microbes, thus maintaining hygiene.
d. **Solvents:** Chlorine finds application as a solvent in the production of various chemicals and plastics, facilitating crucial manufacturing processes.
e. **Chemical Production:** Chlorine acts as a crucial reactant in the synthesis of numerous chemicals, including polyvinyl chloride (PVC), a ubiquitous plastic employed across various industries.
f. **Agricultural Chemicals:** In agriculture, chlorine is utilized in the formulation of pesticides and herbicides, aiding in crop protection and ensuring agricultural productivity.

These examples merely scratch the surface of the extensive applications of chlorine products derived from sulphuric acid. From safeguarding public health to enhancing industrial processes, chlorine plays a pivotal role across diverse sectors, underscoring its indispensable nature in modern society.

58a. Chlorine Applications in Pulp and Paper Manufacturing

Chlorine plays a crucial role in various facets of the pulp and paper production process, particularly in bleaching and water treatment. Below are specific applications of chlorine in the pulp and paper industry:

a. **Bleaching:** Chlorine is integral to the bleaching process, effectively eliminating coloration from the pulp and enhancing the brightness of the final paper product. Typically, chlorine gas or chlorine-based bleaches like hydrogen peroxide or sodium hypochlorite are employed for this purpose.
b. **Water Treatment:** Chlorine finds extensive use in treating both process water and effluent water in pulp and paper mills. Its disinfectant properties aid in reducing levels of bacteria, fungi, and other contaminants, thereby ensuring the quality of the end product.
c. **Effluent Treatment: In** the treatment of pulp and paper mill effluent, chlorine is utilized to neutralize harmful pollutants, contributing to the mitigation of the mill's environmental footprint.
d. **Recycled Paper Treatment:** Chlorine is applied in the treatment of recycled paper to facilitate ink and contaminant removal, while also enhancing the brightness and overall quality of the recycled fibres.

These applications underscore the vital role of chlorine in the pulp and paper industry, where it contributes to the production of high-quality paper products characterized by brightness and cleanliness. By leveraging chlorine's efficacy in bleaching and water treatment processes, pulp and paper manufacturers can uphold stringent quality standards while minimizing environmental impact.

38b. Organic Chlorides: Applications and Uses

Organic chlorides, compounds featuring a chloride group (-Cl) bound to a carbon atom, serve diverse functions across industries and commerce. They serve as crucial starting materials or intermediates in the production of various chemicals and materials.

Below are examples of organic chlorides and their applications:

a. **Vinyl Chloride:** Primarily utilized as a precursor for polyvinyl chloride (PVC), vinyl chloride forms the backbone of this extensively employed plastic.
b. **Ethyl Chloride:** Serving as a versatile compound, ethyl chloride finds application as a refrigerant, propellant in aerosol sprays, and solvent in numerous chemical reactions.
c. **Chloroform:** Recognized for its utility as a solvent, chloroform also serves as a foundational material in the synthesis of refrigerants, pharmaceuticals, and dyes.
d. **Methylene Chloride:** Employed as a solvent, cleaning agent, and blowing agent in foam insulation production, methylene chloride boasts diverse applications across industries.
e. **Trichloroethylene:** Renowned for its solvent and cleaning properties, trichloroethylene is particularly favored in the electronics industry for its efficacy.

These examples illustrate the multifaceted roles of organic chlorides, underscoring their significance across industries and applications. Given the variability in molecular structure and properties, organic chlorides cater to a broad spectrum of needs, contributing to diverse industrial processes and product formulations.

59. Cresols

Cresols, also referred to as hydroxytoluene, represent a group of aromatic organic compounds. They occur widely, either naturally or through manufacturing processes. Often existing in solid or liquid states due to their melting points closely aligning with room temperature, cresols, akin to other phenols, undergo slow oxidation upon exposure to air, emitting a distinctive odor reminiscent of "coal tar."

Production Process:

Cresols, with the chemical formula C_7H_8O or alternately $C_6H_4(OH)\text{-}CH_3$, are commonly synthesized through a sulfonation process. This method involves benzene's reaction with sulphuric acid to yield benzenesulfonic acid, subsequently processed to obtain cresols. The reaction unfolds in stages:

a. **Benzene Sulfonation:** Sulphuric acid reacts with benzene to produce benzenesulfonic acid and water.
b. **Benzenesulfonic Acid Hydrolysis:** Hydrolyzing benzenesulfonic acid with water yields cresol and sulphuric acid.
c. **Cresol Distillation and Purification:** Cresol is distilled from the reaction mixture and purified through techniques like crystallization or chromatography.

The yield and purity of cresols hinge on factors like sulphuric acid concentration, reaction duration, and purification methodologies employed.

Cresols find application across various sectors, including:

a. **Disinfectants:** Widely used as disinfectants and antiseptics in healthcare settings, cresols exhibit efficacy against bacteria, fungi, and viruses.
b. **Solvents:** Employed in industrial and chemical processes, cresols serve as solvents in resin, adhesive, and pharmaceutical production.
c. **Pesticides:** Crescent compounds function as active ingredients in certain pesticides and herbicides, effectively controlling pest populations.
d. **Fragrances:** Adding aromatic notes to consumer products like perfumes, soaps, and personal care items, cresols contribute to fragrance formulations.
e. **Industrial Cleaners:** Cresols act as cleaning agents in industrial settings, facilitating equipment and machinery maintenance.

These applications highlight the versatility of cresols, which cater to diverse needs across industries. Their specific properties and uses may vary based on structural nuances and composition.

59a. Phenolic Resins from Cresols

Phenolic resins represent a category of synthetic polymers synthesized from phenol and an aldehyde, commonly formaldehyde. Renowned for their remarkable heat resistance, chemical durability, and electrical insulation capabilities, these resins find extensive application across diverse industries.

Here are several examples showcasing their utility:

a. **Wood Adhesives:** Utilized as adhesives within the wood industry, phenolic resins play a pivotal role in manufacturing plywood, particleboard, and various composite wood products. Their strong bonding properties ensure structural integrity and longevity in wood-based constructions.

b. **Insulation:** In the electrical and electronics sectors, phenolic resins serve as invaluable insulating materials. Their exceptional electrical and heat resistance properties make them indispensable for safeguarding components and equipment against electrical hazards and thermal stress.
c. **Laminates:** Phenolic resins contribute to the production of laminates, including decorative variants used for countertops, tabletops, and wall panels. These laminates offer durability, aesthetic appeal, and resistance to moisture, making them suitable for a wide range of architectural and interior design applications.
d. **Molding Compounds:** As molding compounds, phenolic resins enable the fabrication of diverse products such as gears, bearings, and electrical components. Their ability to maintain dimensional stability, withstand high temperatures, and resist corrosion ensures the durability and reliability of molded parts.
e. **Friction Materials:** In the automotive and industrial sectors, phenolic resins find use as binders in the production of friction materials, including brake linings and clutch plates. Their robust bonding strength and thermal stability contribute to the efficient performance and longevity of friction components.

These examples underscore the versatility and indispensability of phenolic resins across multiple industries, including construction, automotive, electrical and electronics, and consumer goods. Their diverse properties and wide-ranging applications make them a cornerstone of modern manufacturing and engineering endeavors.

60. Phenol

Phenol, also known as carbolic acid, stands as a vital organic compound renowned for its diverse applications across numerous industries. Its synthesis typically involves the reaction of benzene with propylene in the presence of sulfuric acid, yielding a compound vital in various sectors.Let's delve into its production and myriad uses:

Production Process:

Phenol is predominantly synthesized through the cumene process. Benzene and propylene react in the presence of a catalyst, usually phosphoric acid or aluminum chloride, to form cumene. The cumene is then oxidized to cumene hydroperoxide, which upon treatment with sulfuric acid, yields phenol and acetone.

Applications:

a. **Disinfectants:** In healthcare, phenol serves as a potent disinfectant and antiseptic, effectively combating bacteria, fungi, and viruses.
b. **Pharmaceuticals:** It acts as a crucial starting material for various pharmaceuticals, including pain relievers, antiseptics, and local anesthetics.

c. **Adhesives:** Phenol finds application as a component in adhesives, notably in wood industries for manufacturing plywood, particleboard, and composite wood products.
d. **Dyes:** It plays a pivotal role in dye production, serving as a precursor to aniline dyes extensively utilized in textiles, leather, and printing.
e. **Resins:** Phenol serves as a fundamental component in phenolic resin production. These resins, with their exceptional properties, are integral in construction, automotive, electrical and electronics, and consumer goods industries.

These applications merely scratch the surface of phenol's versatility. Its widespread use spans across pharmaceuticals, agriculture, and consumer goods, making it an indispensable compound in modern industrial processes.

61. Hydrochloric Acid

Hydrochloric acid (HCl), a potent inorganic acid, finds its origins in the sulfuric acid-based salt-acid process, also known as the Mannheim process. Here's a comprehensive look at its production and versatile applications:

In the salt-acid process, sulfuric acid reacts with sodium chloride, or common salt, catalyzed by materials like ceramic balls or quartz sand, yielding hydrochloric acid and sodium sulfate. The reversible reaction is favored at higher temperatures. Further, sodium bisulfate is regenerated into sulfuric acid and sodium sulfate.

The synthesis of hydrochloric acid from sulfuric acid and sodium chloride can be summarized as follows:

a. $NaCl+H_2SO_4 \rightarrow HCl+NaHSO_4$
b. $2NaHSO_4 + H_2SO_4 \rightarrow Na_2SO_4 + 2H_2O$

Following the reaction, hydrochloric acid is purified and separated using techniques such as distillation or extraction. The resultant purified hydrochloric acid can be concentrated to the desired strength for industrial applications.

Applications:

a. **Steel Pickling:** In the steel industry, hydrochloric acid plays a pivotal role in removing impurities like rust and scale from steel surfaces.
b. **Food Production:** Hydrochloric acid serves as a crucial food additive, particularly in the production of preservatives, flavorings, and various food products.
c. **Cleaning Agents:** It's widely used in concrete and masonry surface cleaning owing to its potent cleaning properties.

d. **Oil and Gas Production:** Hydrochloric acid finds application in oil and gas production, especially in well stimulation processes.
e. **Water Treatment:** It's instrumental in water treatment processes, aiding in pH level adjustment in drinking water.
f. These applications merely scratch the surface of hydrochloric acid's utility. From metal fabrication to water treatment, its corrosive and potent nature makes it indispensable across various industries. Nonetheless, its handling demands caution due to its highly toxic properties, emphasizing the importance of proper safety protocols.

61a. Hydrochloric Acid in Metal Industry

Hydrochloric acid (HCl) plays a pivotal role in the metal industry, offering diverse applications across various processes.

Here's an insight into its manifold uses:

a. **Steel Pickling:** In the steel sector, hydrochloric acid is instrumental in a process called steel pickling. It's utilized to eradicate impurities like rust and scale from steel surfaces, thereby enhancing the quality and properties of steel products.
b. **Electroplating:** Hydrochloric acid serves as an electrolyte in electroplating procedures, including nickel and chromium plating. This facilitates the deposition of metal ions onto surfaces, enabling the enhancement of durability and aesthetics.
c. **Refining:** For refining metals like copper and zinc, hydrochloric acid is indispensable. It aids in the elimination of impurities, thereby purifying the metal and enhancing its quality for various industrial applications.
d. **Ore Processing:** Hydrochloric acid finds utility in ore processing, particularly in the extraction of precious metals such as gold and silver from ores. Its acidic properties facilitate the dissolution of metal ions from ores, enabling efficient extraction processes.
e. **Cleaning:** In metal surface cleaning, hydrochloric acid is a go-to cleaning agent. Whether it's aluminum or stainless steel surfaces, hydrochloric acid effectively removes contaminants, ensuring the cleanliness and integrity of metal components.

These applications merely scratch the surface of hydrochloric acid's significance in the metal industry. Its potent and corrosive properties make it indispensable across metal processing and refining processes, contributing to the improvement of metal product quality and properties. Nonetheless, it's imperative to exercise caution in handling hydrochloric acid, given its highly toxic nature, to prevent potential harm or accidents.

62. Nitroglycerin

Nitroglycerin, with the chemical formula $C_3H_5N_3O_9$, is a potent explosive liquid crucial in diverse industries like mining, construction, and medicine. Its production involves a catalytic nitration of glycerol with a mixture of concentrated sulfuric acid and nitric acid. Here's an overview of its applications:

In the construction, mining, and demolition sectors, nitroglycerin is a cornerstone explosive material. Its high explosive power makes it ideal for breaking rocks, excavating tunnels, and demolishing structures. Moreover, it serves as a key ingredient in dynamite and other high-performance explosives.

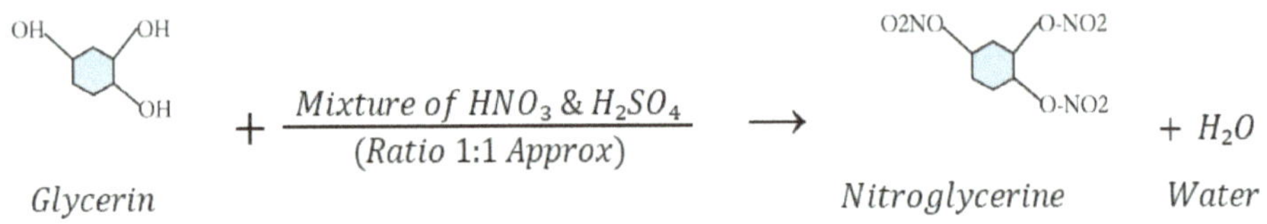

a. **Medical Use:** Beyond its role in explosives, nitroglycerin finds application in medicine as a vasodilator. By widening blood vessels, it improves blood flow and is particularly valuable in treating angina pectoris and various heart conditions.

b. **Propellants:** Nitroglycerin's energetic properties extend to propellants, where it contributes to rocket fuel and other high-energy propellant formulations. Its explosive power makes it invaluable in propelling vehicles and projectiles with precision and efficiency.

c. **Oil and Gas Production:** In the oil and gas industry, nitroglycerin plays a crucial role in stimulating oil and gas wells. By detonating nitroglycerin charges downhole, it creates fractures in reservoir rocks, enhancing fluid flow and increasing production rates.

These applications underscore nitroglycerin's significance across diverse sectors. However, its explosive nature demands careful handling and stringent safety protocols to mitigate risks of accidents and ensure worker safety.

62a. Dynamite from Nitroglycerin

Dynamite stands as a cornerstone in various industries, offering controlled explosive power for essential tasks. Formulated by absorbing nitroglycerin into a porous absorbent material like diatomaceous earth, dynamite undergoes stabilization to enhance safety during handling.

Applications:

a. **Construction and Demolition:** In the construction and demolition sectors, dynamite reigns supreme for heavy-duty tasks like rock breaking and excavation. Its controlled explosive power ensures precision and efficiency, vital for large-scale construction projects and demolitions.
b. **Mining:** Dynamite plays a pivotal role in mining operations, particularly in hard-rock mining for mineral extraction. By effectively breaking through tough geological formations, dynamite enables efficient resource extraction and enhances operational productivity.
c. **Military:** Within military applications, dynamite serves various purposes, including structural demolition of buildings, bridges, and other installations. Its versatility and reliability make it an indispensable tool for military engineering operations.
d. **Civil Engineering:** In civil engineering endeavors, dynamite contributes to critical infrastructure projects like road construction and tunnel excavation. Its controlled explosive power facilitates rapid progress in challenging terrains, ensuring efficient project completion.

63. Tetryl

Tetryl, an explosive compound crucial in various applications, emerges from the reaction between nitric acid and 2,4,6-trinitrotoluene (TNT). Despite its production not directly involving sulfuric acid, the manufacturing process of TNT, a key component of tetryl, commonly utilizes sulfuric acid.

Applications:

Boosting Explosive Power: Primarily serving as a booster or initiator in explosives like dynamite and TNT, tetryl enhances the explosive force, ensuring efficient detonation in various applications.

Military and Industrial Use: Tetryl finds extensive application in military and industrial settings, featuring prominently in explosive devices such as mines, artillery shells, and hand grenades. Its reliable performance and potent explosive capabilities make it indispensable for strategic and industrial purposes.

While tetryl serves as a valuable tool in various industries, its potential hazards underscore the critical importance of adhering to stringent safety measures. Through careful production and responsible handling, tetryl's explosive capabilities can be harnessed effectively while minimizing risks to personnel and the environment

64. Hydrogen Peroxide

Hydrogen peroxide, with the chemical formula H_2O_2, stands as a versatile compound, prized for its oxidizing, bleaching, and antiseptic properties. Its production often involves the utilization of sulfuric acid, although alternative methods exist.

Sulfuric acid serves as a catalyst in the industrial synthesis of hydrogen peroxide. The process typically involves anthraquinone autoxidation, where anthraquinone reacts with hydrogen peroxide in the presence of sulfuric acid, yielding anthraquinone dihydroperoxide. This compound, upon decomposition, releases hydrogen peroxide, anthraquinone, and oxygen.

Applications:

a. **Oxidizer and Bleaching Agent:** Hydrogen peroxide finds extensive use as an oxidizer in various industrial processes, including the bleaching of textiles, paper, and pulp. Its ability to break down chromophores without generating harmful by-products makes it a preferred choice for achieving bright, clean surfaces.

b. **Antiseptic and Disinfectant:** In healthcare settings, hydrogen peroxide serves as a potent antiseptic and disinfectant, effectively eliminating pathogens and promoting wound healing. Its bubbling action upon contact with organic material helps to mechanically dislodge debris and foreign particles.

c. **Additional Applications:** Beyond its primary roles, hydrogen peroxide finds application in diverse fields. It serves as a component in hair dyes, contact lens solutions, and teeth whitening products, owing to its gentle yet effective bleaching properties. Moreover, hydrogen peroxide is utilized in the electronics industry for etching printed circuit boards and as a propellant in rocketry.

d. **Anthraquinone Utilization:** Anthraquinone, a key component in the production process, also finds application in various industries. It serves as a digester additive in paper pulp production and as an electrolyte in flow batteries, facilitating long-term electrical storage.

Hydrogen peroxide, aided by sulfuric acid in its production, stands as a multipurpose compound with widespread applications across industries. Through careful synthesis and diligent utilization, it contributes to diverse processes, ranging from industrial bleaching to healthcare disinfection, underscoring its significance in modern chemistry and technology.

APPLICATION OF SULPHURIC ACID-SHEET 6

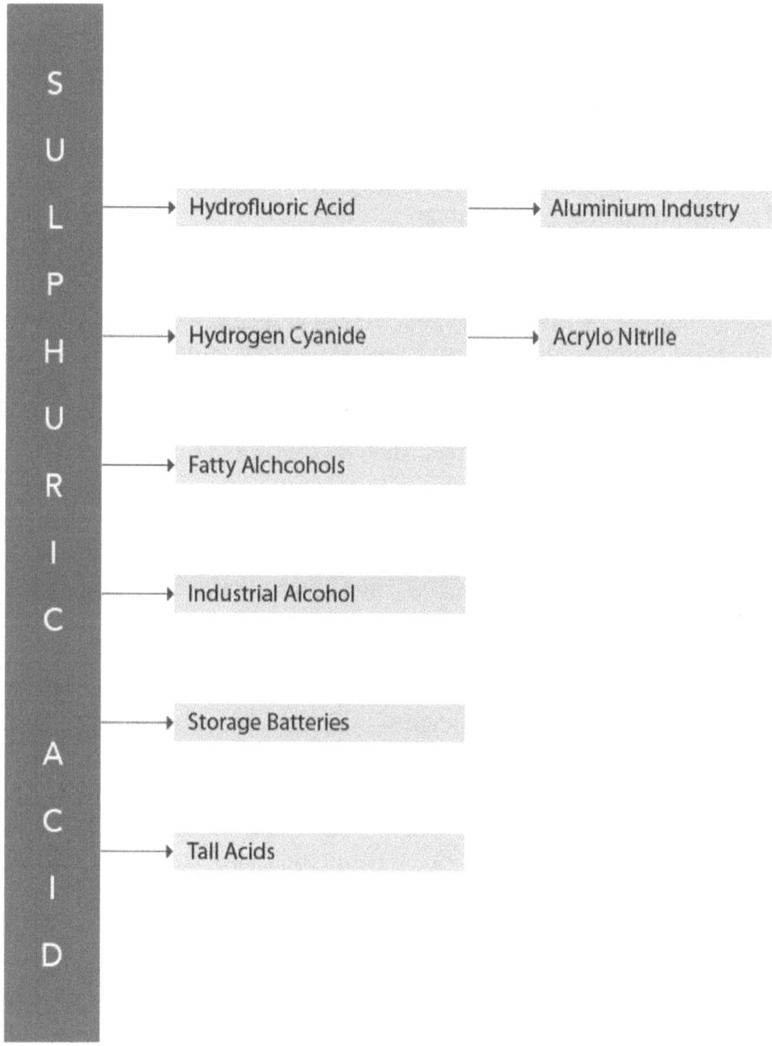

65. Hydrofluoric Acid

Hydrofluoric acid (HF), with the chemical formula HF, stands as a potent and versatile acid with a broad array of industrial and commercial applications. Its production often involves the reaction of calcium fluoride with sulfuric acid, resulting in a compound vital for various industries.

Sulfuric acid plays a pivotal role in the synthesis of hydrofluoric acid. Typically, the reaction entails digesting calcium fluoride, sourced mainly from the mineral fluorspar, with concentrated sulfuric acid. This process liberates hydrofluoric acid as a gas, with calcium sulfate as a byproduct.

Industrial and Commercial Applications:

a. **Manufacturing of Semiconductors, Glass, and Ceramics:** Hydrofluoric acid finds widespread use in the electronics industry, particularly in the manufacturing of semiconductors. It serves as a crucial etchant for removing oxides and contaminants from silicon wafers. Additionally, it aids in the production of specialized glasses and ceramics with precise properties.

b. **Petroleum Refining:** In the petroleum refining industry, hydrofluoric acid is employed to remove impurities, such as sulfur and nitrogen compounds, from gasoline and other fuels. Its ability to selectively react with these contaminants ensures the production of cleaner and more efficient fuels.

c. **Refrigerants and Herbicides:** Hydrofluoric acid serves as a key component in the production of refrigerants, contributing to the cooling systems essential for various applications. Moreover, it plays a role in the synthesis of herbicides, aiding in agricultural practices by controlling unwanted vegetation.

d. **Pharmaceuticals and Cleaning Products:** In the pharmaceutical industry, hydrofluoric acid finds application in the synthesis of certain pharmaceutical compounds. Additionally, it features in some cleaning products, where its corrosive properties help dissolve stubborn deposits and stains.

e. **Metal and Glass Etching:** Due to its corrosive nature, hydrofluoric acid serves as a potent etching agent for metals and glass. It facilitates precise engraving and marking on surfaces, making it indispensable in various manufacturing processes.

In essence, hydrofluoric acid, synthesized with the aid of sulfuric acid, emerges as a vital compound with diverse applications across industries. Its unique properties make it indispensable in processes ranging from semiconductor manufacturing to petroleum refining, underscoring its significance in modern industrial practices.

65a. Hydrofluoric Acid in the Aluminum Industry

Hydrofluoric acid (HF) plays a vital role in the aluminum industry, serving as an essential etching and cleaning agent. Its versatile properties find application in various stages of aluminum production and processing, enhancing efficiency and product quality.

a. **Surface Treatment and Cleaning:** In the production of aluminum, HF is employed to eliminate the oxide layer that naturally forms on the metal's surface. This oxide layer, if left untreated, can impede subsequent chemical reactions. HF effectively dissolves this layer, ensuring optimal surface conditions for further processing. Moreover, HF is utilized to clean aluminum surfaces before welding or other critical operations, ensuring proper adhesion and integrity.
b. **Manufacturing of Aluminum Fluoride:** HF also plays a pivotal role in the synthesis of aluminum fluoride, a crucial compound in aluminum production. Aluminum fluoride is added to molten aluminum to lower its melting point, facilitating the extraction and processing of the metal. Additionally, aluminum fluoride serves as a flux in the production of iron and steel, aiding in the removal of impurities and enhancing metallurgical processes.
c. **Energy Efficiency and Resource Optimization:** By enabling the removal of surface oxides and facilitating the production of aluminum fluoride, HF contributes to energy efficiency and resource optimization in the aluminum industry. The reduction in the metal's melting point due to aluminum fluoride leads to decreased energy consumption during aluminum processing, thereby enhancing overall sustainability and cost-effectiveness.

In summary, hydrofluoric acid plays a critical role in the aluminum industry, facilitating surface treatment, aluminum fluoride production, and energy-efficient processing. Despite its invaluable contributions, the utilization of HF demands meticulous attention to safety to ensure the well-being of personnel and the sustainability of operations.

66. Hydrogen Cyanide

Hydrogen cyanide (HCN) is a valuable chemical compound with diverse applications, synthesized via the Andrussow process, which employs sulphuric acid as a catalyst in the presence of other reactants.

In the Andrussow process, ammonia (NH_3) and methane (CH_4) undergo reaction with oxygen (O_2) in the presence of sulphuric acid (H_2SO_4) as a catalyst. The resulting chemical reaction yields hydrogen cyanide (HCN) along with water (H_2O).

Applications:

a. **Acrylic fibers and Plastics:** HCN serves as a precursor in the production of acrylic fibers and plastics, essential components in various industries including textiles, automotive, and consumer goods. Acrylic fibers find utility in clothing, carpets, and upholstery due to their durability and versatility.
b. **Fumigation:** As a potent fumigant, HCN is utilized to eradicate pests and insects in storage facilities, shipping containers, and aircraft. Its efficacy in pest control ensures the preservation of stored commodities and prevents infestation-related losses.
c. **Chemical Intermediates:** HCN acts as a crucial intermediate in the synthesis of diverse chemicals, including cyanuric chloride, sodium cyanide, and adiponitrile. These intermediates serve as building blocks for the manufacture of various industrial and consumer products, ranging from pesticides to pharmaceuticals.
d. **Pharmaceuticals:** HCN plays a pivotal role in pharmaceutical synthesis, particularly in the production of sodium cyanide. Sodium cyanide finds application in the extraction of precious metals such as gold and silver from ores, contributing to the mining industry's efficiency and productivity.

In conclusion, hydrogen cyanide, synthesized with the aid of sulphuric acid in the Andrussow process, finds widespread application across diverse industries, ranging from textiles to pharmaceuticals. While its versatility and efficacy make it indispensable in various processes, stringent safety precautions are essential to safeguard human health and environmental integrity.

66a. Acrylonitrile Production and Applications

Acrylonitrile (C_3H_3N) is a versatile chemical compound synthesized from hydrogen cyanide (HCN) in the Sohio process, where HCN undergoes reaction with propylene (C_3H_6) and oxygen (O_2). Acrylonitrile serves as a crucial chemical intermediate in various industries, facilitating the production of a wide array of products.

In the Sohio process, hydrogen cyanide (HCN) is combined with propylene (C_3H_6) and oxygen (O_2) to yield acrylonitrile (C_3H_3N) as the primary product. This reaction is fundamental to the production of acrylonitrile and forms the basis of its industrial synthesis.

Applications:

a. **Acrylic fibers:** Acrylonitrile acts as a pivotal monomer in the production of acrylic fibers, essential components in textiles such as clothing, carpets, and upholstery. The incorporation of acrylonitrile lends durability, colorfastness, and resilience to acrylic fibers, enhancing their suitability for diverse applications.

b. **Plastics:** Acrylonitrile serves as a key monomer in the production of various plastics, including acrylonitrile-butadiene-styrene (ABS), styrene-acrylonitrile (SAN), and acrylonitrile-styrene (AS) copolymers. These plastics find utility across numerous industries, from consumer goods to automotive components, owing to their strength, impact resistance, and versatility.
c. **Synthetic Rubber:** As a monomer, acrylonitrile contributes to the synthesis of nitrile rubber, a resilient elastomer employed in the fabrication of gaskets, seals, hoses, and other industrial components. Nitrile rubber's exceptional oil and chemical resistance make it indispensable in diverse engineering applications.
d. **Adhesives:** Acrylonitrile-based adhesives find widespread usage in laminating films, paper, textiles, and other materials. These adhesives offer strong bonding properties, moisture resistance, and durability, making them suitable for a broad spectrum of bonding applications across industries.
e. **Carbon fiber:** Acrylonitrile serves as a precursor in the production of carbon fiber, a lightweight and high-strength material extensively utilized in aerospace, automotive, and sporting goods industries. Carbon fibre's exceptional mechanical properties make it an ideal choice for applications requiring superior strength-to-weight ratios and structural integrity.

In summary, acrylonitrile, derived from hydrogen cyanide through the Sohio process, plays a pivotal role in diverse industries, including textiles, plastics, rubber, adhesives, and advanced materials. Its versatility and utility underscore its significance as a key component in the production of various products, while stringent safety measures are imperative to mitigate associated risks and ensure safe handling and utilization.

67. Fatty Alcohols

Fatty alcohols, essential components in numerous industries, are synthesized from Sulphuric Acid (H_2SO_4) via the Ziegler process. This intricate method involves the reaction of an olefin, such as ethylene (C_2H_4), with H_2SO_4 to yield an intermediate compound, subsequently hydrogenated to form fatty alcohols. The resulting fatty alcohols find extensive utility across various sectors:

The Ziegler process initiates with the reaction of an olefin, commonly ethylene (C_2H_4), with Sulphuric Acid (H_2SO_4), yielding an intermediate compound. This intermediate is then subjected to hydrogenation to yield fatty alcohols, crucial building blocks in diverse applications.

Applications:

a. **Personal Care Products:** Fatty alcohols serve as indispensable ingredients in personal care formulations, including shampoos, conditioners, and skin creams. Employed as emollients, thickeners, and emulsifiers, they enhance product texture, stability, and moisturizing properties, ensuring optimal performance and consumer satisfaction.

b. **Detergents:** Widely utilized as surfactants, fatty alcohols play a pivotal role in detergent formulations, encompassing laundry detergents, dishwashing detergents, and cleaning products. Their exceptional cleaning and foaming properties enable effective removal of dirt and stains, delivering superior cleaning performance.

c. **Textiles:** Fatty alcohols function as softeners and lubricants in textile production, enriching fabric texture and imparting desirable characteristics. From fabric softeners to spin finishes, they enhance fabric feel, drape, and durability, augmenting overall textile quality and appeal.

d. **Plastics:** Fatty alcohols serve as versatile additives in plastic production, acting as plasticizers and lubricants. Their incorporation into plastics, such as PVC and polypropylene, enhances flexibility, processability, and performance, ensuring the attainment of desired material properties.

e. **Chemical Intermediates:** Fatty alcohols serve as crucial intermediates in the synthesis of various chemicals, including fatty acids, esters, and amines. Their reactivity and structural versatility enable the production of a diverse array of value-added compounds, contributing to numerous industrial processes and applications.

In summary, fatty alcohols, derived from Sulphuric Acid through the Ziegler process, serve as multifunctional ingredients with wide-ranging applications in personal care, detergents, textiles, plastics, and chemical synthesis. Their versatility, coupled with their ability to enhance product performance and quality, underscores their significance as indispensable components in numerous industrial processes and consumer products.

68. Industrial Alcohol

Industrial alcohol, known as ethanol (C_2H_5OH), is synthesized from Sulphuric Acid (H_2SO_4) through the acid-catalyzed hydration of ethylene. This pivotal process involves the reaction of ethylene (C_2H_4) with water (H_2O) in the presence of H_2SO_4 as a catalyst, as depicted below:

Reaction: $C_2H_4 + H_2SO_4$ (as catalyst) $+ H_2O \rightarrow C_2H_5OH$

Applications:

a. **Fuel:** Industrial alcohol serves as a versatile fuel source, finding application as an additive in gasoline blends, commonly referred to as gasohol. Blending ethanol with gasoline enhances octane rating and reduces emissions, contributing to cleaner combustion and improved engine performance.

b. **Solvent:** Renowned for its solvent properties, industrial alcohol is extensively utilized in various industries as a solvent for coatings, inks, and cleaning products. Its ability to dissolve a wide range of substances makes it indispensable in manufacturing processes requiring efficient dissolution and dispersion.

c. **Personal Care Products:** In the realm of personal care, industrial alcohol serves as a vital component in the production of perfumes, deodorants, and hair sprays. Its solvent nature facilitates the dissolution and dispersion of fragrances and active ingredients, ensuring optimal product formulation and performance.
d. **Pharmaceuticals:** Industrial alcohol plays a crucial role in pharmaceutical manufacturing, serving as both a solvent and preservative in liquid medications, including cough syrups and oral solutions. Its compatibility with a variety of active ingredients and its antimicrobial properties contribute to its widespread use in pharmaceutical formulations.
e. **Beverages:** Integral to the production of alcoholic beverages, industrial alcohol serves as the primary fermentation product in the brewing and distillation processes of beer, wine, and spirits. Ethanol content determines the potency and flavor profile of the final beverage, making it a cornerstone of the beverage industry.

In conclusion, industrial alcohol, synthesized from Sulphuric Acid through the acid-catalyzed hydration of ethylene, serves as a versatile compound with diverse applications in fuel, solvents, personal care products, pharmaceuticals, and beverages. Its broad utility, coupled with its favorable properties and compatibility with various manufacturing processes, underscores its significance across industries and underscores its importance as a fundamental component in countless industrial and consumer applications.

69. Storage Batteries

Storage batteries, commonly referred to as lead-acid batteries, harness the power of sulphuric acid (H_2SO_4) as an essential electrolyte. In these batteries, sulphuric acid serves as a conductor, facilitating the flow of electrical current between the positive and negative plates. This process enables the battery to both store and release electrical energy efficiently. As electrical energy is discharged during use, the sulphuric acid within the battery undergoes depletion, necessitating periodic replenishment with fresh, ultra-pure sulphuric acid devoid of any iron or iron compounds.

Applications:

a. **Automotive:** Lead-acid batteries find widespread use in automotive applications, where they serve as the primary power source for starting engines and operating electrical accessories such as lights, radios, and air conditioning systems.

b. **Uninterruptible Power Supply (UPS) Systems:** These batteries are integral components of UPS systems, offering backup power solutions in instances of mains power failure. By seamlessly providing electrical energy during outages, lead-acid batteries safeguard critical equipment and infrastructure, ensuring uninterrupted operation in sensitive environments such as data centers and telecommunications facilities.
c. **Renewable Energy Systems:** Lead-acid batteries play a crucial role in renewable energy systems, such as solar and wind power installations, by storing surplus energy generated during periods of high production. This stored energy can then be utilized during peak demand or when renewable energy sources are unavailable, enhancing system reliability and grid stability.
d. **Marine Applications:** Lead-acid batteries are indispensable in marine environments, powering various onboard systems and equipment on boats and ships. From navigation aids to communication devices, these batteries ensure reliable electrical power, contributing to safe and efficient maritime operations.
e. **Telecommunications:** Lead-acid batteries serve as reliable backup power sources in telecommunications infrastructure, supporting critical communications networks during power outages or disruptions. By providing uninterrupted power supply, these batteries help maintain connectivity and ensure the continuity of essential services.

In summary, storage batteries powered by sulphuric acid play an indispensable role in a diverse array of applications, ranging from automotive and marine to renewable energy and telecommunications. By harnessing the unique properties of sulphuric acid as an electrolyte, these batteries enable reliable energy storage and delivery, ensuring continuous operation and resilience across various industries and sectors.

70. Tall Acids

Tall oil, a residue derived from the paper-making process, undergoes refinement into tall oil fatty acids (TOFA), also referred to as tall acids, utilizing a process employing sulphuric acid (H_2SO_4). This refining journey commences with the distillation of tall oil, followed by its interaction with H_2SO_4 to eliminate impurities and yield a more concentrated tall acid mixture.

Applications:

a. **Soaps and Detergents:** Tall acids serve as pivotal raw materials in the soap and detergent industry, functioning as emulsifiers and foam stabilizers. Their emulsifying properties aid in suspending oils and dirt, facilitating effective cleaning, while their foam-stabilizing attributes contribute to the formation of lather, enhancing product performance.

b. **Rubber:** Within the rubber manufacturing sector, tall acids act as processing aids, imparting vital functionalities such as dispersion and tackification. Their role as dispersing agents ensures uniform dispersion of fillers and additives, thereby enhancing rubber compound homogeneity and product consistency.
c. **Oilfield Chemicals:** Tall acids find application in the formulation of oilfield chemicals, where they serve as emulsifiers, dispersants, and surfactants. Their emulsifying properties facilitate the formation and stabilization of oil-water emulsions, while their dispersing capabilities aid in the uniform dispersion of additives and particulates.
d. **Coatings:** In the coatings industry, tall acids play a crucial role as a source of unsaturated fatty acids. These acids serve as precursors for reactive diluents and resins, contributing to the formation of cross-linked polymer networks, thereby enhancing coating adhesion, flexibility, and durability.
e. **Personal Care Products:** Tall acids feature prominently in the production of personal care items, including cosmetics and toiletries. Here, they serve as versatile ingredients, functioning as emollients to soften and moisturize the skin, thickeners to enhance product viscosity, and emulsifiers to stabilize oil-in-water or water-in-oil emulsions.

It is imperative to recognize that the properties and applications of tall acids are contingent upon their specific composition, influenced by factors such as the wood source utilized in tall oil production and the refining methodologies employed. Therefore, meticulous attention to detail in both sourcing and processing is essential to ensure the desired performance and functionality of tall acids across diverse industrial applications.

Summary

Sulfuric acid (H_2SO_4) is one of the most widely used and important industrial chemicals globally.

Its primary use can be summarized as follows:

a. **Manufacture of Fertilizers:** Sulfuric acid is a key component in the production of phosphate and ammonium sulfate fertilizers. It is used in the production of superphosphate, a widely used phosphate fertilizer, by reacting with phosphate rock. Additionally, sulfuric acid is used to produce ammonium sulfate fertilizer through the neutralization of ammonia.
b. **Chemical Manufacturing:** Sulfuric acid is a fundamental chemical used in the manufacturing of a wide range of chemical products. It is a vital reagent in various industrial processes, including the production of detergents, synthetic fibers (such as nylon), pigments, dyes, explosives (like TNT), pharmaceuticals, and plastics.

c. **Mineral Processing:** Sulfuric acid plays a crucial role in mining and metallurgical operations. It is used in leaching processes to extract metals from ores, such as copper, nickel, zinc, and uranium. Sulfuric acid is also employed in the production of metals like titanium and tantalum.

d. **Petroleum Refining:** Sulfuric acid is used in the refining of crude oil to produce gasoline, diesel, and other petroleum products. It's involved in processes such as alkylation and catalytic cracking, where it acts as a catalyst or as a medium for specific reactions.

e. **pH Control and Water Treatment:** Sulfuric acid is utilized for pH adjustment in various industries, including water treatment, wastewater treatment, and in the production of pulp and paper. It helps neutralize alkaline substances and aids in the removal of impurities.

f. **Battery Manufacturing:** Lead-acid batteries, commonly used in vehicles and uninterruptible power supplies, utilize sulfuric acid as the electrolyte. The acid facilitates the chemical reactions within the battery, allowing for the storage and release of electrical energy.

g. **Textile Industry:** Sulfuric acid is used in the production of synthetic fibers like rayon and viscose. It's involved in the chemical processing of textiles, including dyeing and finishing operations.

h. **Food Processing:** In the food industry, sulfuric acid is used for various purposes, such as in the production of food additives like citric acid, and in the processing of sugar, starch, and vegetable oils.

i. These applications demonstrate the versatility and indispensability of sulfuric acid across a wide range of industries, making it one of the most essential chemicals in modern manufacturing and industrial processes.

REFERENCES:

1. "Sulphuric acid safety data sheet" (PDF). arkema-inc.com.
2. Greenwood, Norman N.; Earnshaw, Alan (1997). Chemistry of the Elements (2nd ed.), Butterworth-Heinemann. ISBN 978-0-08-037941-8.
3. Merck Index, 11th Edition, 9625
4. Patnaik, Pradyot (2002), Handbook of Inorganic Chemicals. McGraw-Hill.
5. "Sulphuric Acid". The Columbia Encyclopedia (6th ed.). 2009. Retrieved 16 March 2010.
6. "Sulphuric acid". Encyclopædia Britannica. Vol. 26 (11th ed.). 1910–1911. pp. 65–69.
7. Hermann Müller "Sulphuric Acid and Sulphur Trioxide" in Ullmann's Encyclopedia of Industrial Chemistry, Wiley-VCH, Weinheim. 2000
8. Wildermuth, Egon; Stark, Hans; Friedrich, Gabriele; Ebenhöch, Franz Ludwig; Kühborth, Brigitte; Silver, Jack; Rituper, Rafael. "Iron

Compounds". Ullmann's Encyclopedia of Industrial Chemistry. Weinheim: Wiley-VCH.
9. "Acid or alkaline soil: Modifying pH - Sunset Magazine". www.sunset.com. 3 September 2004.
10. Austin, George T. (1984). Shreve's Chemical process industries (5th ed.). New York, NY: McGraw-Hill. p. 357
11. Schrödter, Klaus; Bettermann, Gerhard; Staffel, Thomas; Wahl, Friedrich; Klein, Thomas; Hofmann, Thomas (2008). "Phosphoric Acid and Phosphates". Ullmann's Encyclopedia of Industrial Chemistry. Weinheim: Wiley-VCH. doi:10.1002/14356007.a19_465.pub3.
12. ^ "Current EU approved additives and their E Numbers". Foods Standards Agency. 14 March 2012.
13. George, C.W.; Susott, R.A. (April 1971). "Effects of Ammonium Phosphate and sulphate on the Pyrolysis and Combustion of Cellulose". Research Paper INT-90. Intermountain Forest and Range Experiment Station: USDA Forest Service.
14. Kongshaug, Gunnar; Brentnall, Bernard A.; Chaney, Keith; Gregersen, Jan-Helge; Stokka, Per; Persson, Bjørn; Kolmeijer, Nick W.; Conradsen, Arne; Legard (2014). "Phosphate Fertilizers". Ullmann's Encyclopedia of Industrial Chemistry. Weinheim: Wiley-VCH. pp. 1–49
15. Zapp, Karl-Heinz (2012). "Ammonium Compounds". Ullmann's Encyclopedia of Industrial Chemistry. Weinheim: Wiley-VCH.
16. "Where Does Air Pollution Come From?". www.purakamasks.com. 2019-02-15.
17. "Citric acid". International Chemical Safety Cards. NIOSH. September 18, 2018.
18. Tang, Sheau-Chung; Yang, Jen-Hung (April 10, 2018). "Dual Effects of Alpha-Hydroxy Acids on the Skin". Molecules. 23 (4): 863
19. Science Lab.com. "Material Safety Data Sheet: Bromine MSDS". sciencelab.com. Archived from the original on 15 November 2017. Retrieved 27 October 2016.
20. ^ NIOSH Pocket Guide to Chemical Hazards. "#0064". National Institute for Occupational Safety and Health (NIOSH).
21. Dagani, M. J.; Barda, H. J.; Benya, T. J.; Sanders, D. C. "Bromine Compounds". Ullmann's Encyclopedia of Industrial Chemistry. Weinheim: Wiley-VCH.
22. "Preparation and purification of 1,2-dibromoethane" (PDF). Synlett. 28: 49–51. 2017.
23. Rossberg, M.; et al. "Chlorinated Hydrocarbons". Ullmann's Encyclopedia of Industrial Chemistry. Weinheim: Wiley-VCH.

24. Shao A, Hathcock JN (April 2008). "Risk assessment for the amino acids taurine, L-glutamine and L-arginine". Regulatory Toxicology and Pharmacology. 50 (3): 376–399.
25. Kinoshita, Shukuo; Udaka, Shigezo; Shimamoto, Masakazu (1957). "Studies on amino acid fermentation. Part I. Production of L-glutamic acid by various microorganisms". J Gen Appl Microbiol. 3 (3): 193–205.
26. Cheung, Hosea; Tanke, Robin S.; Torrence, G. Paul (2000). "Acetic Acid". Ullmann's Encyclopedia of Industrial Chemistry. Weinheim: Wiley-VCH.
27. Clarke H. T.;. Davis, A. W. (1941). "Oxalic acid (anhydrous)". Organic Syntheses: 421.; Collective Volume, vol. 1
28. Barclay, John (1866). "Injection of Acetic Acid in Cancer". Br Med J. 2 (305): 512.
29. H. Benninga (1990): "A History of Lactic Acid Making: A Chapter in the History of Biotechnology". Volume 11 of Chemists and Chemistry. Springer
30. Westhoff, Gerrit; Starr, John N. (2012). "Lactic Acids". Ullmann's Encyclopedia of Industrial Chemistry. Weinheim: Wiley-VCH. Westhoff, Gerrit; Starr, John N. (2012).
31. Freeman, F. "Chromic Acid" Encyclopedia of Reagents for Organic Synthesis (2001) John Wiley & Sons
32. Franz Wirsching "Calcium sulphate" in Ullmann's Encyclopedia of Industrial Chemistry, 2012 Wiley-VCH, Weinheim.
33. "Uses of Copper Compounds: Copper Sulphate". copper.org. Copper Development Association Inc
34. "Ethanol". PubChem. National Library of Medicine. Retrieved 28 September 2021.
35. Hillis O. Folkins (2005). "Benzene". Ullmann's Encyclopedia of Industrial Chemistry. Weinheim: Wiley-VCH.
36. Jörg F, Ulrich G, Simo TA (2005). "Toluene". Ullmann's Encyclopedia of Industrial Chemistry.
37. McKeown NJ (Feb 1, 2015). Tarabar A (ed.). "Toluene Toxicity, Background, Pathophysiology, Epidemiology"
38. Cannella, William J. (2000). "Xylenes and Ethylbenzene". Kirk-Othmer Encyclopedia of Chemical Technology.
39. Tremblay, Jean-François (2011). "Making Aromatics in Singapore". Chemical & Engineering News Archive.
40. Krauskopf LG (September 1993). "Plasticizer structure/performance relationships". Journal of Vinyl and Additive Technology. 15 (3): 140–147.
41. Hubinger JC, Havery DC (March 2006). "Analysis of consumer cosmetic products for phthalate esters". Journal of Cosmetic Science. 57 (2): 127–137
42. Gerd Collin; Hartmut Höke; Helmut Greim (2003). "Naphthalene and Hydronaphthalenes". Ullmann's Encyclopedia of Industrial Chemistry.

43. Tamimi, Faleh; Sheikh, Zeeshan; Barralet, Jake (2012). "Dicalcium phosphate cements: Brushite and monetite". Acta Biomaterialia. 8 (2): 474–487.
44. Vogel, A. "Anthraquinone". Ullmann's Encyclopedia of Industrial Chemistry
45. Boone, C.; Bond, C.; Stone, D. (2012). "Boric Acid General Fact Sheet". National Pesticide Information Center, Oregon State University Extension Services.
46. Harvey, S. C. (1980). "Antiseptics and Disinfectants; Fungicides; Ectoparasiticides". In Gilman, A. G.; Goodman, L. S.; Gilman, A. (eds.). Goodman & Gillman's The Pharmacological Basis of Therapeutics (6th ed.)
47. Luknitskii, F. I. (1975). "Chemistry of Chloral". Chemical Reviews. 75 (3): 259–289.
48. Jira, Reinhard; Kopp, Erwin; McKusick, Blaine C.; Röderer, Gerhard; Bosch, Axel; Fleischmann, Gerald. "Chloroacetaldehydes". Ullmann's Encyclopedia of Industrial Chemistry.
49. DDT and Its Derivatives: Environmental Aspects, Environmental Health Criteria monograph No. 83, Geneva: World Health Organization, 198
50. Anneken, David J.; Both, Sabine; Christoph, Ralf; Fieg, Georg; Steinberner, Udo; Westfechtel, Alfred (2006). "Fatty Acids". Ullmann's Encyclopedia of Industrial Chemistry. Weinheim: Wiley-VCH.
51. "Stearic acid". PubChem, US National Library of Medicine.
52. Duarte, A.M.; Caixeirinho, D.; Miguel, M.G.; Sustelo, V.; Nunes, C.; Fernandes, M.M.; Marreiros, A. (2012). "Organic Acids Concentration in Citrus Juice from Conventional Versus Organic Farming". Acta Horticulturae. 933 (933): 601–606.
53. Thomas, Alfred (2000). "Fats and Fatty Oils". Ullmann's Encyclopedia of Industrial Chemistry
54. Morris, Peter John Turnbull (1989). The American Synthetic Rubber Research Program. University of Pennsylvania Press. p. 258. ISBN 978-0-8122-8207-8.
55. Fabric Information: Acetate & Viscose, NY Fashion Center Fabrics.
56. Booth G (2007). "Nitro Compounds, Aromatic". Ullmann's Encyclopedia of Industrial Chemistry (6th ed.). Weinheim: Wiley-VCH.
57. "Nitrobenzene". Immediately Dangerous to Life or Health Concentrations (IDLH). National Institute for Occupational Safety and Health (NIOSH)
58. Kahl, Thomas; Schröder, K. W.; Lawrence, F. R.; Elvers, Barbara; Höke, Hartmut; Pfefferkorn, R.; Marshall, W. J. (2007). "Aniline". In Ullmann, Fritz (ed.). Ullmann's encyclopedia of industrial chemistry. Wiley: New York.
59. "Aniline". Immediately Dangerous to Life or Health Concentrations (IDLH). National Institute for Occupational Safety and Health (NIOSH).
60. Riemenschneider, Wilhelm; Bolt, Hermann M. "Esters, Organic". Ullmann's Encyclopedia of Industrial Chemistry. Weinheim: Wiley-VCH.

61. Favre, Henri A.; Powell, Warren H. (2014). Nomenclature of Organic Chemistry : IUPAC
62. S. N. Bizzari; M. Blagoev (June 2010). "CEH Marketing Research Report: FORMIC ACID". Chemical Economics Handbook.
63. Gerd Anger, Jost Halstenberg, Klaus Hochgeschwender, Christoph Scherhag, Ulrich Korallus, Herbert Knopf, Peter Schmidt, Manfred Ohlinger, "Chromium Compounds" in Ullmann's Encyclopedia of Industrial Chemistry, Wiley-VCH, Weinheim, 2005. doi:10.1002/14356007.a07_067
64. Jump up to:[a] [b] Freeman, F. "Sodium Dichromate" in Encyclopedia of Reagents for Organic Synthesis (Ed: L. Paquette) 2004, J. Wiley & Sons, New York
65. IARC (2012) [17-24 March 2009]. Volume 100C: Arsenic, Metals, Fibres, and Dusts. Lyon: International Agency for Research on Cancer.
66. "Salicylic acid". PubChem, US National Library of Medicine.
67. Boullard O, Leblanc H, Besson B (2000). "Salicylic Acid". Ullmann's Encyclopedia of Industrial Chemistry.
68. March, Jerry (1992), Advanced Organic Chemistry: Reactions, Mechanisms, and Structure (4th ed.)
69. Kosswig, Kurt (2000), "Sulfonic Acids, Aliphatic". Ullmann's Encyclopedia of Industrial Chemistry.
70. Völz, Hans G.; et al. (2006). "Pigments, Inorganic". Ullmann's Encyclopedia of Industrial Chemistry.
71. Odochian, Lucia (1995). "Study of the nature of the crystallization water in some magnesium hydrates by thermal methods". Journal of Thermal Analysis and Calorimetry. 45 (6): 1437–1448.
72. Dieter M. M. Rohe; Hans Uwe Wolf (2005). "Zinc Compounds". Ullmann's Encyclopedia of Industrial Chemistry.
73. Helmold Plessen (2000). "Sodium sulphates". Ullmann's Encyclopedia of Industrial Chemistry.
74. L. Lange, W. Triebel, "Sulfides, Polysulfides, and Sulfanes" in Ullmann's Encyclopedia of Industrial Chemistry 2000, Wiley-VCH, Weinheim.
75. Ulrich Wietelmann; Richard J. Bauer (2005). "Lithium and Lithium Compounds". Ullmann's Encyclopedia of Industrial Chemistry. Ulrich Wietelmann; Richard J. Bauer (2005). "Lithium and Lithium Compounds". Ullmann's Encyclopedia of Industrial Chemistry.
76. Logsden, John E.; Loke, Richard A. (1999). "Propyl Alcohols". In Kroschwitz, Jacqueline I. (ed.). Kirk-Othmer Concise Encyclopedia of Chemical Technology (4th ed.).
77. Siegfried Rebsdat; Dieter Mayer. "Ethylene Glycol". Ullmann's Encyclopedia of Industrial Chemistry.
78. Ethylene glycol. National Institute for Occupational Safety and Health. Emergency Response Database.

79. Anneken, David J.; Both, Sabine; Christoph, Ralf; Fieg, Georg; Steinberner, Udo; Westfechtel, Alfred (2006). "Fatty Acids". Ullmann's Encyclopedia of Industrial Chemistry.
80. Global Health and Education Foundation (2007). "Conventional Coagulation-Flocculation-Sedimentation". Safe Drinking Water is Essential. National Academy of Sciences.
81. Shimizu, S.; Watanabe, N.; Kataoka, T.; Shoji, T.; Abe, N.; Morishita, S.; Ichimura, H. "Pyridine and Pyridine Derivatives". Ullmann's Encyclopedia of Industrial Chemistry.
82. Sinha, A. K.; Sharma, U. K.; Sharma, N. (2008). "A comprehensive review on vanilla flavor: Extraction, isolation and quantification of vanillin and other constituents". International Journal of Food Sciences and Nutrition. 59 (4): 299–326.
83. "Pickling and Passivation". Australian Stainless Steel Development Association.
84. Helmut Fiegein "Cresols and Xylenols" in Ullmann's Encyclopedia of Industrial Chemistry" 2007; Wiley-VCH, Weinheim.
85. Weber, Manfred; Weber, Markus; Kleine-Boymann, Michael (2004). "Phenol". Ullmann's Encyclopedia of Industrial Chemistry. Weinheim: Wiley-VCH.
86. Di Carlo, F. J. (1975). "Nitroglycerin Revisited: Chemistry, Biochemistry, Interactions". Drug Metabolism Reviews. 4 (1): 1–38.
87. Fritz Ullmann "Anthraquinone". Ullmann's Encyclopedia of Industrial Chemistry. Weinheim: Wiley-VCH.

Chapter 5 : Exploring the Origins and Properties of Oleum

HISTORY:

Oleum, also referred to as fuming sulfuric acid, stands as a pivotal compound in various industrial realms such as dye production, detergent manufacturing, and pharmaceuticals. Its historical narrative unfolds back to the early 19th century.

In 1839, the notable French chemist Georges-Eugene Chevreul unearthed a groundbreaking revelation: the dissolution of sulfur trioxide in sulfuric acid engendered a markedly concentrated form of sulfuric acid. This innovation bore the name "oleum," derived from the Latin term for "oil," owing to its notably viscous and oily texture.

Chevreul's discovery heralded a new era in the industrial landscape, propelling the production of sulfuric acid into a central role in the creation of a diverse array of goods. Oleum, particularly instrumental in dye synthesis, showcased its prowess in engaging with organic compounds to yield an array of vibrant hues and pigments.

Over the ensuing years, the production and utilization of oleum underwent continuous evolution. Presently, oleum is predominantly synthesized by introducing sulfur trioxide to concentrated sulfuric acid, finding application across various industrial sectors, including pharmaceuticals, plastics, and pesticides.
Despite its utility, oleum remains a hazardous substance necessitating vigilant handling and storage practices. Its high reactivity and corrosiveness mandate specialized containment measures to forestall potential leaks or spills.

PHYSIO-CHEMICAL PROPERTIES:

Oleum, alternately known as fuming sulfuric acid or Nordhausen acid, represents a profoundly reactive and corrosive liquid variant of sulfuric acid. Formed by the addition of excess sulfur trioxide to concentrated sulfuric acid, this mixture may contain up to 25% SO_3, rendering it exceptionally acidic.

This colorless or faintly yellowish liquid exudes a potent odor reminiscent of sulfur dioxide gas, boasting a density of approximately 1.9 g/cm3 and a boiling point of around 340°C.

Oleum finds diverse utility in industrial settings, serving as a catalyst in chemical reactions, a desiccant in specific material production, and a reagent in compound synthesis. However, its highly corrosive nature demands stringent safety protocols and suitable protective gear for handling. Only individuals with requisite training and access to appropriate facilities should engage with oleum to mitigate risks of severe burns or respiratory complications.

SOURCES OF OLEUM

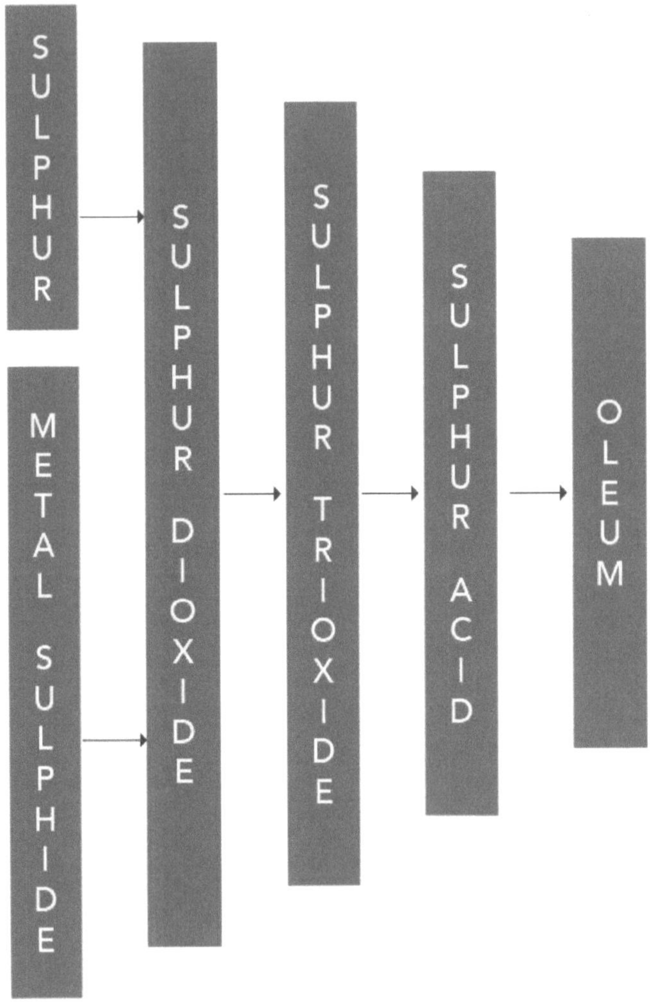

Oleum, recognized interchangeably as fuming sulfuric acid or pyrosulfuric acid, constitutes a blend of sulfur trioxide (SO_3) within concentrated sulfuric acid (H_2SO_4). This compound emerges from a process termed the contact process or chamber process.

Here's an insightful glimpse into the typical production pathway of oleum from sulfur:

a. **Sulfur Combustion:** Elemental sulfur (S_8) undergoes combustion to yield sulfur dioxide (SO_2). This reaction commonly unfolds through the incineration of molten sulfur in the presence of air: $S_8 + 8O_2 \rightarrow 8SO_2$

b. **Conversion of Sulfur Dioxide to Sulfur Trioxide:** The next phase involves the oxidation of sulfur dioxide to sulfur trioxide. Typically, this transformation leverages a vanadium(V) oxide catalyst: $2SO_2 + O_2\ (V_2O_5) \rightarrow 2SO_3$

c. **Absorption of Sulfur Trioxide:** The sulfur trioxide generated in the preceding step integrates with concentrated sulfuric acid to engender oleum:
$SO_3 + H_2SO_4 \rightarrow H_2S_2O_7$

d. **Dilution of Oleum:** To align with specific application requirements, the resultant oleum solution may undergo dilution with additional sulfuric acid or water, modulating its concentration accordingly.

d. **Purification and Storage:** The refined oleum solution may undergo purification protocols aimed at eliminating impurities before its subsequent storage or integration into diverse industrial processes.

The oleum derived through this systematic process serves as a vital component across a spectrum of industrial applications. It finds utility in the manufacture of dyes, explosives, and detergents, alongside contributing to the production of fertilizers, plastics, and pharmaceuticals.

APPLICATION OF OLEUM

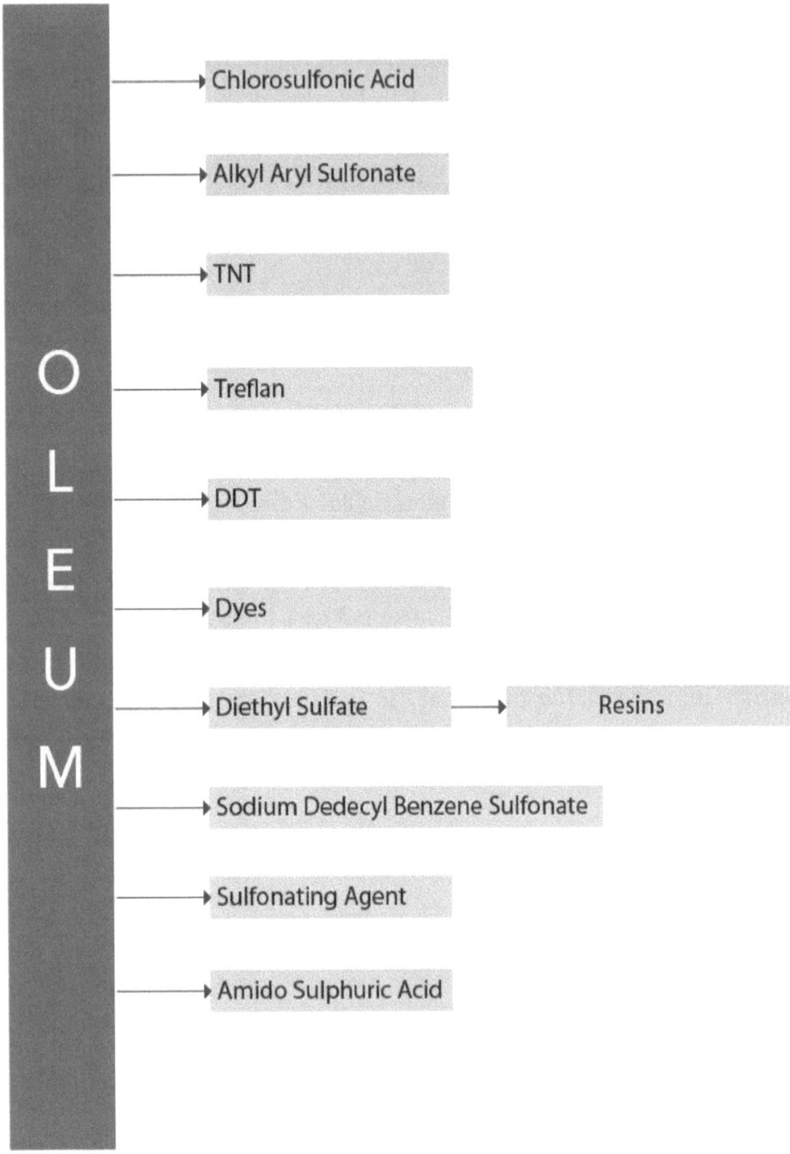

1. Chlorosulfonic Acid

Chlorosulfonic acid ($ClSO_3H$) is synthesized through the reaction between oleum—a mixture of sulfur trioxide dissolved in concentrated sulfuric acid—and chlorine gas. This reaction can be depicted as:

$SO_3Cl + H_2SO_4 \rightarrow ClSO_3H + H_2SO_4$

Chlorosulfonic acid is a highly reactive and corrosive substance widely utilized in various industrial processes, including:

a. **Sulfonation Reactions:** In organic synthesis, chlorosulfonic acid acts as a sulfonating agent, adding sulfonic acid groups to aromatic compounds. These groups are crucial in the production of dyes, surfactants, and pharmaceuticals.
b. **Esterification:** It facilitates the conversion of carboxylic acids into acid chlorides in esterification reactions, enabling the synthesis of diverse organic compounds.
c. **Dehydration Reactions:** Chlorosulfonic acid serves as a dehydrating agent, promoting the elimination of water from reactions to enhance product formation.
d. **Catalysis:** It acts as a catalyst in certain chemical reactions, particularly esterifications, etherifications, and polymerizations.
e. **Desulfonation:** Chlorosulfonic acid can remove sulfonic acid groups from organic compounds in desulfonation processes.

Given its corrosive and reactive nature, handling chlorosulfonic acid requires caution. Proper safety precautions, including the use of fume hoods and protective gear, are essential when working with this compound.

2. Alkyl Aryl Sulfonates

Alkyl aryl sulfonates (AAS) are derived from oleum through a process called sulfonation. This method involves the reaction of an alkyl or aryl compound with oleum to yield the corresponding alkyl aryl sulfonic acid. Subsequently, the sulfonic acid can be neutralized with a base, like sodium hydroxide, to form the desired alkyl aryl sulfonate. The reaction can be represented as follows:
$Ar\text{-}H + R\text{-}X + H_2SO_4 \rightarrow Ar\text{-}SO_3H + R\text{-}SO_3H + HX$

Here, Ar denotes an aryl group, R represents an alkyl group, X signifies a halogen, and HX denotes a hydrogen halide.

The resulting alkyl aryl sulfonates serve as anionic surfactants with versatile applications, including:

a. **Personal Care Products:** AAS function as emulsifiers and foaming agents in a myriad of personal care items, ranging from shampoos and bath soaps to toothpastes.
b. **Textile Industry:** In the textile sector, AAS are employed as detergents and emulsifiers to eliminate dirt and impurities from fibers, enhancing the tactile experience of textiles.
c. **Oilfield Applications:** AAS find utility in the oilfield realm to augment oil flow and regulate the precipitation of wax and asphaltenes.
d. **Detergents:** AAS serve as surfactants and emulsifiers in detergents, enhancing their cleaning efficacy and deterring scum formation.
e. **Agricultural Applications:** In agriculture, AAS act as surfactants to improve the dispersion and effectiveness of pesticides and herbicides.

It is imperative to exercise caution when handling oleum during the sulfonation process due to its highly corrosive and reactive nature. Adhering to appropriate safety protocols is essential to mitigate the risk of accidents and ensure the safe handling of oleum and the resultant alkyl aryl sulfonates.

3. Nitrocellulose

Nitrocellulose, also referred to as cellulose nitrate, is a highly flammable compound produced by nitrating cellulose using a mixture of nitric acid and sulfuric acid. Initially employed as guncotton, replacing conventional gunpowder, it found applications as a low-order explosive in mining and photography.

Nitrocellulose can be synthesized from oleum through a process known as nitration. This entails reacting cellulose, a natural polymer, with oleum and nitric acid to yield nitrocellulose.

The chemical equation for this reaction is:

Cellulose ($C_6H_{10}O5$) + HNO_3 + H_2SO_4 → Nitrocellulose + H_2O

Uses of Nitrocellulose

Nitrocellulose serves various purposes due to its highly flammable and versatile nature:

a. **Explosives:** It is a vital component in manufacturing smokeless gunpowder and other high-performance explosives.

b. **Coatings:** Nitrocellulose serves as a rapid-drying base for coatings like lacquers, varnishes, and enamels, facilitating application and improving finish quality.
c. **Printing Inks:** Used as a binder in printing inks, it enables quick-drying, glossy, and durable prints.
d. **Personal Care Products:** Nitrocellulose acts as a film-forming agent in nail polish and other personal care products.

4. TNT

Trinitrotoluene (TNT) is widely recognized for its explosive properties and is commonly used in military and industrial applications. TNT can be synthesized from oleum (sulfuric acid) and other chemicals through a process known as Nitration.

Nitration Process:

a. **Nitration of Toluene:** Toluene, a volatile organic compound, is first nitrated with a mixture of nitric acid and sulfuric acid to form mononitrotoluene (MNT).
   C_7H_8 (toluene)+$HNO_3 \rightarrow C_7H_7NO_2$ (mononitrotoluene) + H_2O
b. **Further Nitration:** Mononitrotoluene is then subjected to further nitration with the nitric acid-sulfuric acid mixture to produce dinitrotoluene (DNT).
   $C_7H_7NO_2$ (mononitrotoluene)+$HNO_3 \rightarrow C_7H_6N_2O_4$ (dinitrotoluene) + H_2O
c. **Final Nitration:** Dinitrotoluene undergoes another nitration with the same mixture to yield trinitrotoluene (TNT).
   $C_7H_6N_2O_4$ (dinitrotoluene) + $HNO_3 \rightarrow C_7H_5N_3O_6$ (trinitrotoluene) + H_2O

Uses of TNT

TNT finds extensive applications due to its explosive properties and stability:

a. **Military Applications:** TNT serves as a primary explosive material in military ordnance, including bombs, shells, and grenades.
b. **Construction Industry:** In the construction industry, TNT is used for blasting and excavation purposes, owing to its high explosive power.
c. **Chemical Synthesis:** TNT is occasionally utilized as a reagent in chemical synthesis to generate charge transfer salts.

5. Treflan

Treflan is a widely recognized preemergence herbicide containing trifluralin as its active ingredient. While not directly synthesized from oleum (sulphuric acid), trifluralin can be derived from various chemicals, including aliphatic dicarboxylic acids obtained initially from oleum.
Trifluralin, the key component of Treflan, is synthesized from oleum-derived materials through a series of chemical reactions.

Treflan serves as an effective tool in weed management across diverse agricultural crops:

a. **Weed Control:** Treflan efficiently controls a broad spectrum of weeds, including those commonly found in corn, soybeans, cotton, and cereals. By inhibiting weed seed germination, it prevents their emergence and growth, thereby safeguarding crop yields.
b. **Soil Residuality:** With a prolonged residual effect, Treflan persists in the soil for several months, ensuring sustained weed suppression. This characteristic minimizes the necessity for frequent reapplications, contributing to cost savings and operational efficiency.
c. **Low Toxicity:** Treflan exhibits low toxicity towards humans and mammals, making it a preferred choice for weed control in proximity to food crops. Its safety profile underscores its suitability for integrated pest management strategies.
d. **Minimal Soil Residue:** Following application, Treflan undergoes rapid degradation in the soil, leaving behind negligible residues. This feature mitigates the risk of long-term environmental impact and facilitates crop rotation practices.

6. DDT

Dichlorodiphenyltrichloroethane (DDT), once hailed as an effective insecticide, holds a complex legacy due to its environmental impacts and persistent nature.
While DDT is not directly synthesized from oleum, its precursor chemicals, such as chlorobenzene and chloral, are initially derived from oleum through chemical processes.

Originally developed as a pesticide and insecticide, DDT found widespread application in various sectors:

a. **Public Health Programs:** DDT played a pivotal role in controlling insect-borne diseases like malaria, yellow fever, and typhus, contributing significantly to public health initiatives globally.
b. **Agricultural Pest Control:** DDT was extensively used in agriculture to combat insect pests, including mosquitoes, ticks, and fleas, thereby enhancing crop yields and protecting agricultural produce.

Despite its controversial reputation, DDT offered certain advantages:

c. **Effective Pest Control:** DDT exhibited remarkable efficacy against a broad spectrum of insect pests, making it a preferred choice for pest management in various settings.
d. **Cost-Effectiveness:** DDT was relatively inexpensive to produce and widely accessible, enabling its widespread adoption in pest control programs.

e. However, the use of DDT posed significant challenges and drawbacks:
f. **Persistence:** DDT's persistence in the environment is a major concern, as it breaks down slowly and accumulates in soil and water bodies, posing long-term risks to ecosystems.
g. **Toxicity:** DDT is highly toxic to a wide range of non-target organisms, including birds, fish, and wildlife, raising ecological and health concerns.
h. **Bioaccumulation:** DDT and its metabolites bioaccumulate in the food chain, leading to elevated levels of contamination in top predators and posing risks to ecosystem stability.

In response to environmental and health concerns, many countries, including the United States, banned the majority of DDT uses in the 1970s. Presently, DDT is primarily utilized for disease vector control in select regions, subject to stringent regulations aimed at mitigating its adverse effects on human health and the environment.

7. Dyes

Dyes play a crucial role in coloring various materials and products, and they can be synthesized from oleum through a process called sulfonation. This section outlines the production process and applications of dyes derived from oleum.

The production of dyes from oleum involves several intricate chemical reactions:

a. **Sulfonation:** Oleum, a concentrated form of sulfuric acid, is reacted with organic compounds like benzene or toluene to introduce a sulfonic acid group ($-SO_3H$) onto the aromatic ring. Careful control is essential to prevent over-sulfonation due to the highly exothermic nature of this reaction.

b. **Nitration:** The resulting sulfonic acid undergoes nitration with nitric acid to add a nitro group (-NO_2) to the aromatic ring, another highly exothermic reaction requiring precise control.
c. **Reduction:** Certain dyes necessitate reduction to replace the nitro group with an amino group (-NH_2), achieved using reducing agents such as iron or zinc.
d. **Coupling:** The final step involves a coupling reaction, catalyzed by agents like copper or diazonium salts, where two molecules combine to form a larger chromophore-containing molecule.

Sulfonated dyes find applications across various industries:

a. **Textile Industry:** Widely employed for coloring cotton, wool, and synthetic fibers, sulfonated dyes enhance the aesthetic appeal of textiles.
b. **Leather Industry:** Sulfonated dyes are utilized to color leather goods such as shoes, bags, and accessories, contributing to their visual attractiveness.
c. **Paper Industry:** Sulfonated dyes are instrumental in producing colored papers and paper products, adding vibrancy to packaging and print materials.
d. **Food and Beverage Industry:** In the food and beverage sector, sulfonated dyes are used to color products like candy, soft drinks, and fruit juices, enhancing their visual appeal.

While dyes, including sulfonated ones, offer diverse coloring solutions, their use is regulated to safeguard human health and the environment. Some dyes have been associated with adverse effects, leading to restrictions or bans in many countries. Thus, it's imperative to assess the safety and environmental impact of dyes before their application in any industry.

8. Diethyl Sulphate

Diethyl sulphate (DES) is a chemical compound that can be synthesized from oleum (sulphuric acid) through esterification. This section provides an overview of the production process and the diverse applications of diethyl sulphate.
Diethyl sulphate is produced through the esterification reaction between ethanol and oleum. The balanced chemical equation for this reaction is as follows:

$$SO_3 + \frac{2C_2H_5OH}{\text{Ethanol}} \rightarrow \frac{(C_2H_5)2SO_4}{DES} + \frac{H_2O}{Water}$$

During this exothermic reaction, oleum, which is concentrated sulphuric acid containing dissolved sulphur trioxide (SO_3), reacts with ethanol to form diethyl sulphate and water.
Diethyl sulphate serves as a crucial intermediate in the production of various chemicals, finding applications in:

a. **Pharmaceuticals:** DES acts as a starting material in the synthesis of numerous pharmaceuticals, including anti-inflammatory drugs, antidepressants, and antihistamines. It contributes to the production of essential medications that improve human health.
b. **Pesticides:** As an intermediate, DES plays a role in the synthesis of certain pesticides, including herbicides and insecticides. These chemicals aid in agricultural practices by controlling pests and weeds, thereby enhancing crop yield.
c. **Detergents:** DES serves as a precursor in the synthesis of detergents and surfactants. These products are integral to cleaning formulations, providing effective removal of dirt and grease from various surfaces.

8a. Resins Production & Uses from Diethyl Sulphate (DES)

Resins are essential materials derived from diethyl sulphate (DES) through reactions with suitable polyalcohols, such as glycerol or pentaerythritol, forming ester bonds, the fundamental components of resins.
The production of resins involves chemical reactions between diethyl sulphate and various hydrocarbons, yielding resins with diverse properties based on the type of hydrocarbon utilized.

Here are some reactions for resin production:

a. **Reaction with Polyisobutene:** Oleum reacts with polyisobutene to yield a viscous resin. $(C_4H_8)n + H_2SO_4 \rightarrow (C_4H_8)nSO_3 + H_2O \rightarrow$ Viscous Resin
b. **Reaction with Styrene:** Oleum reacts with styrene to produce polystyrene resin. $C_8H_8 + H_2SO_4 \rightarrow C_8H_8SO_3 + H_2O \rightarrow$ Polystyrene Resin
c. **Reaction with Terpene:** Oleum reacts with terpene to generate a terpene resin.
d. **Reaction with Alpha-Pinene:** Oleum reacts with alpha-pinene to form a terpene resin. $nC_{10}H_{16}$ (alpha-pinene) $\rightarrow$ (Terpene Resin)

Resins find extensive applications across various industries, including:

a. **Coatings:** Serving as binders, resins are utilized in paints, varnishes, and adhesives, offering protective and decorative properties.
b. **Composites:** Resins play a vital role in the production of composites like fibreglass-reinforced plastics and carbon-fibre composites, providing strength and durability.
c. **Adhesives:** Resins are the primary components in adhesives, including hot-melt, pressure-sensitive, and structural adhesives, ensuring strong bonds in diverse materials.

d. **Inks:** Acting as binders, resins are employed in various inks such as offset, flexographic, and gravure inks, contributing to print quality and durability.
e. **Plastics:** Resins serve as the main constituents in numerous plastics like polyethylene, polypropylene, and polyvinyl chloride (PVC), offering versatility and applicability in various products.

9. Sodium Dodecyl Benzene Sulfonate

Sodium dodecyl benzene sulfonate (SDBS) is a surfactant derived from oleum through a process called sulfonation, offering versatile applications across various industries. The production of Sodium Dodecyl Benzene Sulfonate from Oleum involves the reaction between oleum (a mixture of sulphur trioxide and sulphuric acid) and dodecyl benzene (an alkyl benzene).

The reaction proceeds as follows:

a. **Sulfonation Reaction:** Dodecyl benzene reacts with sulphuric acid to form dodecyl benzene sulfonic acid.

1. $C_{12}H_{25}C_6H_5 + H_2SO_4 \rightarrow C_{12}H_{25}C_6H_4SO_3H + H_2O$

b. **Neutralization Reaction:** Dodecyl benzene sulfonic acid is then neutralized with sodium hydroxide to produce Sodium Dodecyl Benzene Sulfonate.

1. $C_{12}H_{25}C_6H_4SO_3H + NaOH \rightarrow C_{12}H_{25}C_6H_4SO_3Na + H_2O$

c. Overall Reaction:
$C_{12}H_{25}C_6H_5 + 2NaOH + SO_3 \rightarrow C_{12}H_{25}C_6H_4SO_3Na + H_2O + Na_2SO_4$

SDBS finds extensive usage as a surfactant due to its ability to reduce surface tension, enhancing the spreadability and wetting of liquids. Some common applications include:

a. **Detergents:** SDBS is a key ingredient in various detergents, including household cleaners, dishwashing liquids, and laundry detergents, ensuring effective cleaning performance.
b. **Personal Care Products:** Widely used in personal care items like shampoos, body washes, and bubble baths, SDBS acts as a surfactant, providing foaming and cleansing properties.
c. **Industrial Applications:** SDBS serves as a crucial surfactant in industrial processes such as emulsion polymerization, oilfield operations, and textile processing, facilitating efficient operations.

While SDBS offers numerous benefits, its usage may have environmental implications, including water pollution and adverse effects on aquatic ecosystems. Regulatory controls are in place to mitigate these risks, with restrictions or bans on SDBS application in some regions. Additionally, individuals may experience skin irritation or allergies upon exposure to SDBS, necessitating careful handling and usage precautions.

10. Sulfonating Agents

Sulfonating agents, vital in organic chemistry, facilitate the addition of sulfonic acid groups ($-SO_3H$) to diverse compounds. Oleum, comprising sulfur trioxide (SO_3) and sulfuric acid (H_2SO_4), serves as a primary precursor due to its ability to donate sulfonic acid groups effectively.

Two main reactions illustrate the conversion of oleum into sulfonating agents:

a. **Production of Fuming Sulfuric Acid (Oleum with >100% SO_3)**
   - Reaction: $SO_3(g) + H_2SO_4 \rightarrow H_2S_2O_7$
   - Resulting in fuming sulfuric acid, this reaction yields a potent sulfonating agent, capable of introducing sulfonic acid groups into various organic compounds.

b. **Production of Sulfuric Acid Monoesters**
   - Reaction: $R\text{-}OH + H_2SO_4 \rightarrow R\text{-}SO_3H + H_2O$
   - Utilizing oleum, this reaction generates sulfuric acid monoesters, possessing sulfonic acid and alkyl groups. These serve as milder sulfonating agents for a wide array of organic compounds.

Sulfonating agents derived from oleum find extensive use across industries:

c. **Synthesis of Surfactants:** Key in producing surfactants like sodium dodecylbenzene sulfonate (SDBS) and alkyl aryl sulfonates.

d. **Synthesis of Dyes:** Integral to the production of various dyes, including azo dyes and anthraquinone dyes.

e. **Synthesis of Pharmaceuticals:** Employed in manufacturing sulfonamide antibiotics and sulfonylurea hypoglycemic agents.

f. **Synthesis of Other Chemicals:** Utilized in the production of plasticizers, resins, fragrances, and various specialty chemicals.

The production and application of sulfonating agents are subject to stringent regulatory controls to safeguard human health and the environment. Certain sulfonating agents, especially those derived from oleum, may pose risks, leading to restrictions or bans in some regions. Hence, thorough evaluation of safety and environmental impact is imperative before employing sulfonating agents in any application.

11. Amidosulphuric Acid

Amidosulphuric acid (ASA), also known as sulfamic acid, is a significant compound synthesized from oleum and an amine compound through a process called amidation. This article discusses the production process of ASA from oleum and its diverse applications across industries.

ASA is commonly synthesized by reacting oleum (concentrated sulfuric acid containing excess sulfur trioxide) with urea. The chemical equation for this reaction is:

$2\ SO_3 + 2\ NH_2CONH_2 \rightarrow 4\ H_2NSO_3H + CO_2$

This reaction is exothermic and requires careful handling in a well-ventilated environment with appropriate safety precautions. Several reactions may occur during the production process:

a. **Acid-Base Reaction:** Oleum, a strong acid, reacts with urea, a weak base, forming ammonium ions and sulfate ions while releasing heat.
b. **Dehydration Reaction:** Excess sulfur trioxide in oleum removes water from urea, leading to the formation of biuret.
c. **Decomposition Reaction:** At high temperatures, ASA may decompose into sulfur trioxide and water, necessitating controlled temperature conditions to prevent this reaction.

ASA serves as an intermediate in the production of various chemicals, including:

a. **Surfactants:** Used in the synthesis of surfactants like alkyl aryl sulfonates and sodium lauryl sulfate.
b. **Pharmaceuticals:** Functions as a starting material for sulfonamide antibiotics, sulfonylurea hypoglycemic agents, and other pharmaceutical compounds.
c. **Therapeutic Agents:** Sulfamates derived from ASA are utilized in the design of therapeutic agents such as antibiotics, HIV inhibitors, anticancer drugs, anti-epileptic drugs, and weight loss drugs.
d. **Agricultural Chemicals:** Forms the basis for the production of herbicides and insecticides.
e. **Sweet-Tasting Compounds:** ASA derivatives are precursor compounds for sweet-tasting substances.

The production and utilization of ASA, like other chemicals derived from oleum, are subject to stringent regulatory controls to ensure human health and environmental safety. Some ASA derivatives, especially those used in pharmaceuticals and agricultural chemicals, may pose risks, leading to restrictions or bans in certain regions. Therefore, a comprehensive evaluation of safety and environmental impact is imperative before employing ASA-derived compounds in any application.

Summary

Oleum, also referred to as fuming sulfuric acid, stands as a highly reactive and corrosive compound pivotal as an intermediary in the synthesis of various chemicals. Below outlines its diverse applications:

a. **Sulfonating Agent:** Oleum plays a crucial role in the production of sulfonated compounds like detergents, dyes, and pharmaceuticals, functioning as a sulfonating agent.
b. **Chemical Synthesis:** Its utility extends to the synthesis of an array of chemicals, encompassing sulfuric acid, nitric acid, and derivatives such as chlorosulfonic acid and sulfanilic acid.
c. **Acid Catalyst:** Employed as an acid catalyst in numerous chemical reactions, notably aiding in polymer and ester production.
d. **Purification:** Oleum finds application in purifying oils and fats, effectively eliminating impurities from these substances.
e. **Laboratory Reagent:** Widely utilized as a laboratory reagent for chemical analysis and research purposes due to its reactive properties.
f. **Petroleum Refining:** Integral in petroleum refining processes, oleum contributes to the removal of impurities and enhancement of the final product's quality.
g. **Battery Manufacturing:** In the production of lead-acid batteries, oleum serves as a vital component within the electrolyte, facilitating battery functionality.

In essence, oleum emerges as a cornerstone compound within the chemical industry, pivotal not only in its direct applications but also as a precursor facilitating the synthesis of a myriad of other chemicals.

REFERENCES:

1. Greenwood, Norman N.; Earnshaw, Alan (1997). Chemistry of the Elements (2nd ed.). Butterworth-Heinemann. pp. 703–704. ISBN 978-0-08-037941-8.
2. Hermann Müller "sulphuric Acid and sulphur Trioxide" in Ullmann's Encyclopedia of Industrial Chemistry, Wiley-VCH, Weinheim. 2000
3. Merck Index, 11th Edition, 9625
4. Saunders, C. W.; Taylor, L. T. (1990). "A review of the synthesis, chemistry and analysis of nitrocellulose". Journal of Energetic Materials. 8 (3): 149–203.
5. "Trinitrotoluene". Merriam-Webster.com Dictionary
6. https://www.missouribotanicalgarden.org/gardens-gardening
7. DDT and Its Derivatives: Environmental Aspects, Environmental Health Criteria monograph No. 83, Geneva: World Health Organization, 1989
8. Science Daily (May 9, 2009). "Unprecedented Use of DDT Concerns Experts". ScienceDaily.com. Archived from the original on May 14, 2009. Retrieved May 30, 2009
9. DDT and Its Derivatives: Environmental Aspects, Environmental Health Criteria monograph No. 83, Geneva: World Health Organization, 1989
10. Dow (June 24, 2006). "Diethyl sulphate" (PDF). Product Safety Assessment. Dow Chemical Company. Retrieved 2016-03-05
11. Kurt Kosswig,"Surfactants" in Ullmann's Encyclopedia of Industrial Chemistry, Wiley-VCH, 2005, Weinheim
12. Bipin V. Vora; Joseph A. Kocal; Paul T. Barger; Robert J. Schmidt; James A. Johnson (2003). "Alkylation". Kirk-Othmer Encyclopedia of Chemical Technology.
13. Winum, J. Y.; Scozzafava, A.; Montero, J. L.; Supuran, C. T. (2005). "Sulfamates and their therapeutic potential". Medicinal Research Reviews. 25 (2): 186–228.

Chapter 6: Hydrogen Sulphide: Extraction, Risks & Utilization

Hydrogen sulphide (H_2S) emerges from microbial breakdowns of organic matter in oxygen-deprived environments like swamps and sewers, notably through anaerobic digestion facilitated by sulphate-reducing microorganisms. Its presence extends to volcanic gases, natural gas reserves, and occasional occurrences in well-drawn water.

HISTORY

Dating back to 1700 A.D., H2S's recognition as an environmental toxin traces to Bernardino Ramazzini's observations, detailing sewer gas-induced eye irritations among workers. Over the centuries, its role sparked research intrigue, from its chemical identification by H.C. van Swieten in the 17th century to toxicological studies in the 19th century. Tragically, it found application in World War II's gas chambers. Today, its primary uses revolve around sulphuric acid production and as a reducing agent in various industries.

PROPERTIES

Hydrogen sulphide is a colourless, poisonous, corrosive, and flammable gas, notable for its characteristic rotten egg odour. With a boiling point of -60.3°C and a melting point of -85.5°C, it boasts a slightly soluble nature in water and various solvents like ethanol and methanol. Its toxic, corrosive, and flammable nature necessitates careful handling in industrial settings, alongside stringent safety measures.

OCCURRENCE

This gas naturally occurs in volcanic gases, geothermal springs, swamps, and natural gas reserves, often stemming from organic matter decay. Its low odour threshold makes it detectable even at low concentrations, presenting potential hazards to human and animal health.

INDUSTRIAL APPLICATIONS

Hydrogen sulphide finds multifaceted utility in industry, contributing to sulphur, sulphuric acid, and organic chemical production. Its role spans metal ore processing, oil and gas refining, and as a precursor for compounds like methionine, crucial in animal feed.

SAFETY PRECAUTIONS

Due to its toxicity, meticulous handling and ventilation are essential in workplaces where H_2S exposure is plausible. Personal protective equipment (PPE) and monitoring devices aid in ensuring worker safety.

CHEMICAL REACTIONS

H2S engages in various chemical reactions, serving as a reducing agent and reacting with oxidising agents like chlorine and oxygen to form sulphur compounds. It can also react with metals to yield metal sulphides.

PHYSICAL PROPERTIES

- Molecular weight: 34.08 g/mol
- Density: 1.189 g/L at STP
- Boiling point: -60.3 °C
- Melting point: -85.5 °C
- Solubility in water: 7.6 mg/L at 25 °C
- Flammable range: 4.3% to 46% in air
- Vapour pressure: 4.7 kPa at 20 °C
- Viscosity: 0.32 cP at 20 °C
- Specific heat: 1.46 J/g·K
- Autoignition temperature: 572 °C
- Explosion limits: 4.3-46%

CONCLUSION

Hydrogen sulphide's diverse properties and industrial applications underscore its significance, albeit necessitating stringent safety measures to mitigate its inherent hazards. Understanding its properties, occurrences, and reactions is imperative for safe handling and utilization across various sectors.

SOURCES OF HYDROGEN SULPHIDE

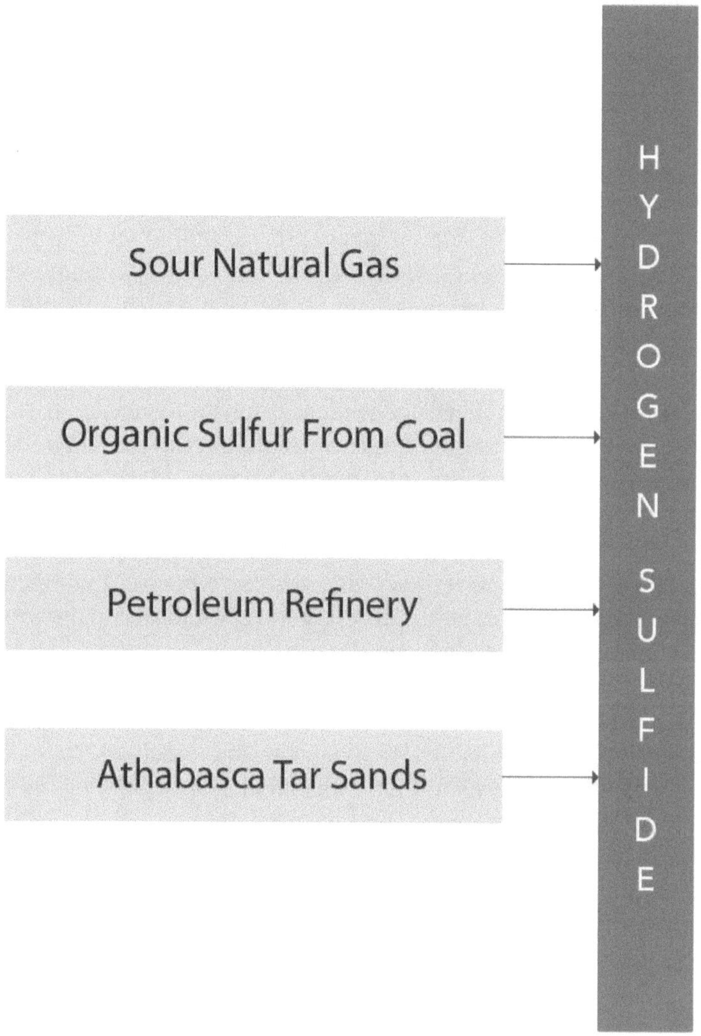

1. Hydrogen Sulphide Removal from Sour Natural Gas

In oil refineries and natural gas processing plants, the elimination of hydrogen sulphide and other organosulphur compounds is termed "sweetening," transforming the product into a more palatable form devoid of the sour, unpleasant odours associated with mercaptans and hydrogen sulphide.

a. **Sweetening Process:** To prepare raw natural gas containing hydrogen sulphide for use, it undergoes treatment to reduce impurities to acceptable levels, typically employing an amine treatment process. The extracted H_2S is commonly converted into elemental sulphur as a by-product through the Claus process or treated in a wet sulphuric acid process unit to yield sulphuric acid.
b. **Presence in Sour Natural Gas:** Hydrogen sulphide (H_2S) frequently resides in natural gas deposits, particularly in "sour" natural gas, distinguished by significant H2S content. The sour designation stems from the foul odour akin to rotten eggs, rendering such gas challenging to transport and process.
c. **Origins and Variation:** Thermochemical sulphate reduction in sedimentary basins contributes to the presence of H_2S in natural gas, wherein buried organic matter undergoes conversion into hydrocarbons and sulphur compounds, including hydrogen sulphide. The concentration of H_2S in natural gas can range widely, from trace levels to concentrations as high as 20%.
d. **Removal Process:** Due to its toxic and corrosive nature, hydrogen sulphide must be eradicated from natural gas before transportation and processing. Sour gas sweetening achieves this, employing chemical agents like amines or physical methods such as membrane separation to extract H_2S.
e. **Utilization and Transport:** Upon removal of hydrogen sulphide, the treated natural gas becomes suitable for transport via pipelines to end-users or processing facilities. Here, further refinement and separation yield diverse products like methane, propane, and butane, meeting various industrial and consumer demands.

In summary, the removal of hydrogen sulphide from sour natural gas via sweetening processes ensures the safety and quality of the end product, facilitating its efficient transportation and utilization across multiple sectors.

2. Hydrogen Sulphide from Organic Sulphur Coal

The industrial process of coking, which involves the production of coke from coal, occurs in airless kilns or coke ovens at extreme temperatures of up to 2,000 °C (3,600 °F). This process leads to the vaporization or decomposition of organic substances within the coal, releasing volatile and liquid byproducts such as coal gas and coal tar. The residual product, coke, consists of carbon and mineral residues fused together into

a hard, glassy solid. Alongside coke, additional byproducts of coking include coal tar pitch, ammonia (NH_3), and hydrogen sulphide (H_2S).
Hydrogen sulphide (H_2S) can be generated from coal containing organic sulphur compounds. When coal undergoes combustion, these sulphur compounds are converted into sulphur dioxide (SO_2) and sulphur trioxide (SO_3), which subsequently react with water vapor in the atmosphere to form sulphuric acid (H_2SO_4). This sulphuric acid can further react with atmospheric hydrogen ions ($H+$) to produce hydrogen sulphide (H_2S). This process occurs during the combustion of coal in power plants or other industrial settings where coal serves as a primary fuel source.

3. Hydrogen Sulphide from Petroleum Refinery Streams

Hydrogen sulphide (H_2S) emerges as a byproduct from petroleum refinery streams. Throughout the refining process, crude oil undergoes separation to isolate various hydrocarbon compounds. One resulting stream, known as "sour gas" or "sour crude," comprises a blend of hydrocarbons enriched with sulphur-containing compounds, including hydrogen sulphide (H_2S) and mercaptans. Although this stream is typically subjected to further processing to eliminate sulphur compounds, the removal process concurrently releases H_2S as a byproduct. Moreover, certain refining techniques like catalytic reforming also yield H_2S as a byproduct.

4. Hydrogen Sulphide Generation in Athabasca, Canada

Athabasca, situated in Alberta, Canada, boasts extensive oil sands deposits renowned for their abundance. Within these oil sands resides heavy crude oil known as bitumen, comprising hydrocarbons alongside various compounds, including sulphur. Throughout the extraction and processing of bitumen, hydrogen sulphide (H_2S) emerges as a byproduct. This H_2S can be found in tailings ponds and process streams within oil sands operations.

To mitigate the presence of sulphur-containing compounds, the industry employs an "upgrading" process. This method entails heating the bitumen and employing diverse chemical processes to isolate sulphur compounds. Despite these efforts, some H_2S may still be released as a byproduct during these processes.

It's crucial to emphasize that H_2S is a toxic gas demanding careful handling. The industry is obligated to implement necessary safety precautions and adhere to regulations to prevent the release of H_2S gas into the environment.

Summary

Various methods exist for the production of hydrogen sulphide, each offering distinct advantages depending on factors like desired purity, availability of raw materials, and cost considerations.

The primary methods include:

a. **Natural Gas Sweetening:** This widely utilized method involves the removal of hydrogen sulphide from natural gas through a process known as sweetening. Given the prevalence of hydrogen sulphide as a common impurity in natural gas, natural gas sweetening is a cost-effective and readily available option.
b. **Chemical Production:** Hydrogen sulphide can also be generated through chemical reactions, such as the reaction of sulphur with iron or the reaction of sodium polysulphide with an acid. This method offers flexibility and can be tailored to specific industrial requirements.
c. **Biological Production**: Certain microorganisms possess the capability to produce hydrogen sulphide through metabolic processes. Biological production provides an eco-friendly and sustainable approach to hydrogen sulphide generation, albeit with limitations in scalability.
d. **Thermal Decomposition:** Elemental sulphur can undergo thermal decomposition to yield hydrogen sulphide. While less commonly employed, this method offers a straightforward means of hydrogen sulphide production from elemental sources.
e. **Electrolysis:** Hydrogen sulphide production can also be achieved via the electrolysis of sulphuric acid or sulphates. This method offers precise control over production parameters but may involve higher energy costs.

The selection of the most suitable production method hinges on a comprehensive assessment of factors such as purity requirements, raw material availability, cost-effectiveness, and environmental considerations. Each method presents unique advantages and challenges, underscoring the importance of informed decision-making in hydrogen sulphide production processes.

APPLICATION OF HYDROGEN SULPHIDE

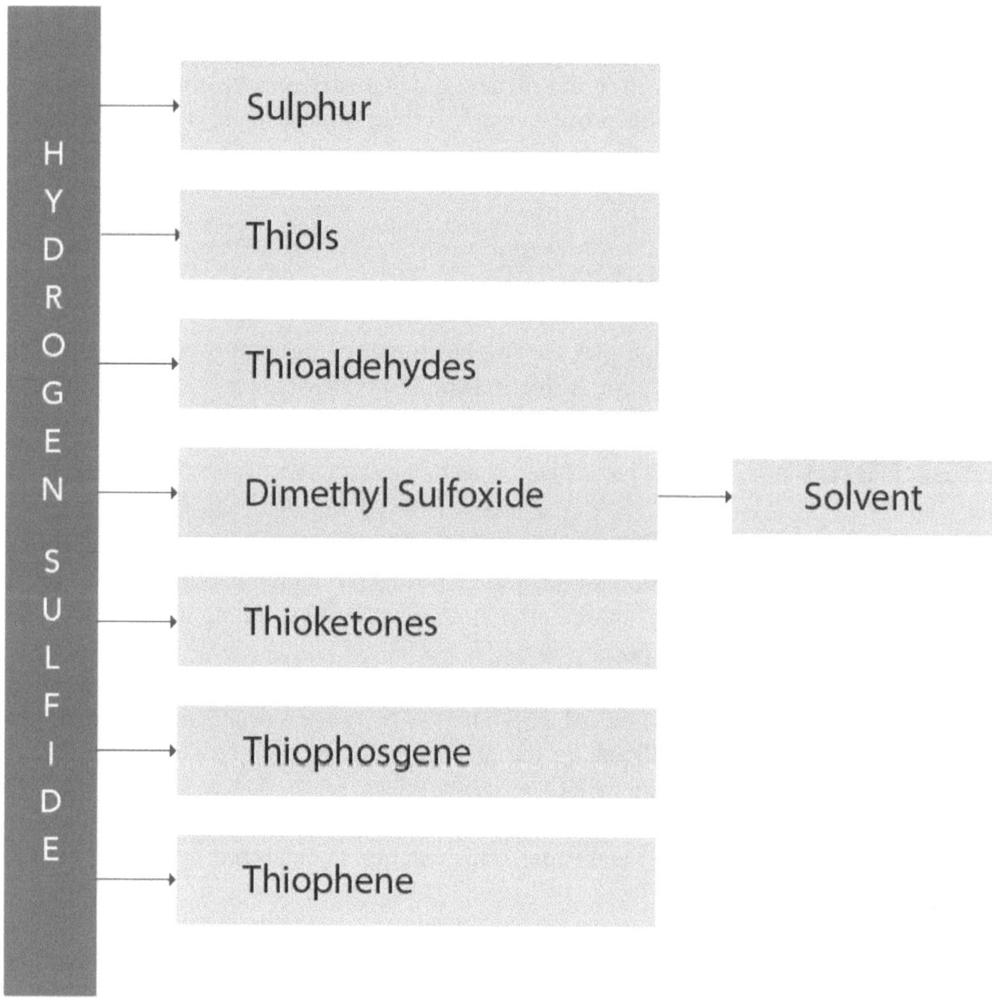

1. Sulphur

Sulphur extraction from hydrogen sulphide (H_2S) is accomplished through a method known as "scrubbing." During this process, H_2S gas is passed through a solution containing a suitable chemical reagent, such as iron chloride ($FeCl_3$). This reagent reacts with H_2S, yielding elemental sulphur (S) alongside other byproducts. The separated sulphur can then be collected as a solid.

Another prevalent technique for sulphur retrieval from hydrogen sulphide is the "Claus process," extensively utilized in the oil and gas industry and, to some extent, in the Viscose Rayon industry. This process involves the combustion of H_2S with an excess of oxygen in the presence of an oxidizing catalyst to produce water vapor (H_2O) and molten sulphur. The catalyst, typically enriched 'Bauxite mineral' or Alumina (Al_2O_3), facilitates the reaction:

$$2H_2S + O_2 \text{ catalyst} \frac{S_2}{\text{Molten form}} + \frac{2H_2O}{Water\ Vapour}$$

It's essential to acknowledge that the sulphur obtained through these processes may contain impurities, necessitating further refining to remove them.

2. Thiol

Thiols, also known as mercaptans, are organic compounds characterized by a sulphur atom bonded to a hydrocarbon group, often emitting strong odors reminiscent of garlic or rotten eggs. Remarkably, thiols serve as crucial odorants aiding in the detection of otherwise odorless natural gas.

The production of thiols from hydrogen sulphide (H_2S) is achieved through a process termed "thiolation." In this method, H_2S reacts with an appropriate organic compound, such as an alcohol or an alkane, in the presence of a suitable catalyst. This reaction yields a thiol along with water (H_2O) as a byproduct.

For instance, the reaction of H_2S with methanol (CH_3OH) results in the formation of methyl mercaptan (CH_3SH) and water. This reaction is catalyzed by an acid catalyst, typically sulphuric acid.

Thiols find extensive applications in various industries, notably in fragrance and flavor production. Additionally, they play a pivotal role in the oil and gas sector, where they are employed as odorants added to natural gas and propane to facilitate leak detection. However, it's imperative to acknowledge the flammability, toxicity, and potential environmental hazards associated with thiols. Hence, meticulous handling and adherence to safety regulations are paramount when working with these compounds.

3. Thioaldehyde

Thioaldehydes, a class of organic compounds featuring a sulphur atom bonded to an aldehyde group, play crucial roles in various applications, from fragrances to organic synthesis. The synthesis of thioaldehydes from hydrogen sulphide (H_2S) is achieved through a process known as "thioaldehyde synthesis."

In this process, H_2S reacts with an aldehyde or ketone in the presence of a suitable catalyst, leading to the formation of a thioaldehyde and water (H_2O) as byproducts. For instance, the reaction of H_2S with formaldehyde (HCHO) yields thiomalic aldehyde (CH_2SCHO) and water. Typically, a base catalyst like sodium hydroxide catalyzes this reaction.

While thioaldehydes offer diverse applications, including fragrance and flavor production, it's crucial to acknowledge their potential hazards. Thioaldehydes are flammable, toxic, and can pose environmental risks. Therefore, meticulous handling and adherence to safety regulations are essential to mitigate associated risks effectively.

4. Dimethyl Sulfoxide

Dimethyl sulfoxide (DMSO), an organic compound containing sulphur, is synthesized from hydrogen sulphide (H_2S) in a process known as dimethyl sulfoxide synthesis. This method involves the reaction of H_2S with methanol in the presence of an acid catalyst, such as sulphuric acid or hydrochloric acid.

During this reaction, DMSO and water are the primary products, with minor by-products like methanesulfonic acid. The reaction typically occurs under high temperatures (approximately 150-200 °C) and pressures (around 50-100 psi). The choice of catalyst and reaction conditions significantly influence the yield and purity of the DMSO obtained.

It's noteworthy that DMSO finds extensive applications in various industries, including pharmaceuticals, electronics, and chemical synthesis. However, due to its potential hazards, including toxicity and environmental concerns, careful handling and adherence to safety protocols are imperative throughout the production process.

4A. Solvent Production from Dimethyl Sulfoxide (DMSO)

Dimethyl sulfoxide (DMSO) serves as a highly polar, aprotic solvent with extensive applications across various industries. Its utility ranges from direct usage to further purification and distillation to attain a high-purity solvent.

The purification procedure typically involves distillation aimed at removing impurities like water, methanol, and other organic compounds. DMSO is subjected to heating until reaching its boiling point (189.2°C) under reduced pressure, after which the resulting vapor is condensed and collected. This distillation process may be repeated multiple times to achieve a solvent of high purity with minimal water content.

Moreover, DMSO acts as a precursor for crafting other solvents. Through chemical reactions, DMSO can generate a diverse array of solvents. For instance, its reaction with chlorine yields dimethyl sulfochloride (DMSC), a solvent adept at dissolving fats, oils, and waxes. Similarly, combining DMSO with acetone yields dimethyl sulfoxide-d6, a solvent prized in NMR spectroscopy.

In essence, DMSO presents itself as a versatile solvent, usable in its original state, or subject to further purification and distillation for high-purity applications. Furthermore, its reactivity enables the creation of an assortment of novel solvents.

Beyond its solvent properties, DMSO finds utility as a cryoprotectant in cell preservation. When added to cell media, it curtails ice formation, thereby safeguarding cells against damage during freezing. Typically, a concentration of approximately 10% is utilized with a gradual freezing method, allowing cells to be safely stored at temperatures as low as −80 °C (−112 °F) or even in liquid nitrogen.

5. Thioketones

Thioketones represent compounds housing a sulphur atom within a ketone functional group. Their synthesis from hydrogen sulphide (H_2S) is achieved through a process aptly termed "Thioketone Synthesis." This synthetic route entails the addition of H_2S to an alkene or alkyne, succeeded by an oxidation step. The resultant thioketone may undergo further functionalization or serve as a pivotal building block in organic synthesis endeavors.

6. Thiophosgene

Thiophosgene, also known as carbonyl sulphide (COS), emerges as a chemical compound obtainable from hydrogen sulphide (H_2S) via a method dubbed the "phosgene method." This process entails the reaction of H2S with chlorine in the presence of a catalyst, typically a metal chloride, to yield thiophosgene and water. Operating at elevated temperatures (approximately 150-200 °C) and pressures (around 50-100 psi), this reaction follows a precise equation:

$$H2S+Cl2+COCl2 \rightarrow CSCl2+2HCl+SOCl2$$

The efficacy and purity of the resulting thiophosgene hinge on several factors, including reaction conditions, catalyst type, and initial material impurities.

Thiophosgene serves as a pivotal intermediate in the synthesis of various chemicals, including thiocarbamates, thioureas, and thiols, among others. Additionally, it finds application in the production of pesticides and fungicides. However, due to its highly toxic nature, stringent safety precautions must be observed during handling. Inhalation of the substance may lead to respiratory irritation, burns, delayed pulmonary edema, and potentially fatal outcomes.

7. Thiophene

Thiophene, a colorless liquid with a scent reminiscent of benzene, is manufactured on a modest scale globally, totaling approximately 2,000 metric tons annually. The production process involves a vapor-phase reaction utilizing a sulphur source, typically carbon disulfide, and a C-4 source, commonly butanol. These compounds are subjected to an oxide catalyst at temperatures ranging from 500 to 550°C.

While thiophene synthesis from hydrogen sulphide (H_2S) is feasible through a process known as the "Thiophene Synthesis," it is not widely adopted in industrial settings due to its inefficiency. This method entails the reaction of H2S with an organic compound, such as an alkyne or an alkyl halide, in the presence of a palladium catalyst. The resultant product is a derivative of thiophene, which can undergo further functionalization or purification for diverse applications.

However, alternative routes are preferred in industrial practice owing to their higher efficiency compared to thiophene synthesis from H_2S.

Summary

Hydrogen sulfide (H_2S) has several applications across different industries, primarily due to its unique properties and chemical reactivity.

Some of the key applications of hydrogen sulfide include:

a. **Oil and Gas Industry:** Hydrogen sulfide is commonly encountered in natural gas and crude oil production. It is often present in reservoirs as a byproduct of microbial activity or as a result of the decomposition of organic matter. In the oil and gas industry, hydrogen sulfide must be removed from natural gas streams through processes such as amine gas treating, scavenging with iron-based compounds, or catalytic oxidation.

b. **Chemical Manufacturing:** Hydrogen sulfide is used in various chemical processes as a precursor or reagent. It is utilized in the production of sulphuric acid, elemental sulphur, and sulphur-containing compounds like thioethers and mercaptans. Additionally, hydrogen sulfide is employed in the synthesis of metal sulfides and organosulphur compounds.

c. **Mining and Metallurgy:** In the mining and metallurgical industries, hydrogen sulfide is often encountered in the extraction and processing of metal ores, particularly those containing sulphur-bearing minerals. Hydrogen sulfide is used in hydrometallurgical processes for the leaching and precipitation of metal sulfides, such as copper, zinc, and nickel.

d. **Analytical Chemistry:** Hydrogen sulfide is utilized as a reagent in analytical chemistry for the qualitative and quantitative detection of metal ions. It forms insoluble metal sulfides with many metal ions, which can be used for precipitation and gravimetric analysis. Hydrogen sulfide is also employed in gas chromatography as a sulphur-specific detector.

e. **Water Treatment:** In water treatment processes, hydrogen sulfide may be present as a result of microbial activity or the decomposition of organic matter. It is removed from water supplies using chemical oxidation methods, such as chlorination or ozonation, or through aeration and filtration processes. Hydrogen sulfide removal helps improve water quality and reduces unpleasant odors.

f. **Healthcare and Biomedical Research:** Hydrogen sulfide has garnered attention in healthcare and biomedical research due to its role as a signaling molecule in physiological processes. It has vasodilatory, anti-inflammatory, and cytoprotective properties, and research is ongoing to explore its potential therapeutic applications in conditions such as cardiovascular disease, inflammation, and neurodegenerative disorders.

g. **Wastewater Treatment:** In wastewater treatment facilities, hydrogen sulfide may be generated as a byproduct of anaerobic digestion or microbial decomposition. It is removed from wastewater using chemical oxidation methods or through biological processes such as aerobic digestion or sulphur-oxidizing bacteria.

These applications demonstrate the diverse roles that hydrogen sulfide plays in various industries, from industrial processes to healthcare and environmental management. However, it's important to note that hydrogen sulfide can be toxic and highly flammable, requiring careful handling and control measures to ensure safety and environmental protection.

REFERENCES:

1. Greenwood, Norman N.; Earnshaw, Alan (1997). Chemistry of the Elements (2nd ed.), Butterworth-Heinemann. ISBN 978-0-08-037941-8.
2. Merck Index, 11th Edition, 9625
3. Patnaik, Pradyot (2002), Handbook of Inorganic Chemicals. McGraw-Hill.
4. Espagno, Lucien; Hemmer, Henri; Perrot, Jacques (2000). "Hydrogen sulphide". Ullmann's Encyclopedia of Industrial Chemistry.
5. William M. Haynes (2016), CRC Handbook of Chemistry and Physics (97th ed.). Boca Raton: CRC Press. pp. 4–87.
6. Lindenmann, J.; Matzi, V.; Neuboeck, N.; Ratzenhofer-Komenda, B.; Maier, A; Smolle-Juettner, F. M. (December 2010). "Severe hydrogen sulphide

poisoning treated with 4-dimethylaminophenol and hyperbaric oxygen". Diving and Hyperbaric Medicine. 40 (4): 213–217.

7. NaturalGas.org website page Processing Natural Gas
8. Tiwari, H. P.; Sharma, R.; Kumar, Rajesh; Mishra, Prakhar; Roy, Abhijit; Haldar, S. K. (December 2014). "A review of coke making by-products". Coke and Chemistry. 57 (12): 477–484.
9. James G, Speight (2006). The Chemistry and Technology of Petroleum (Fourth ed.).
10. A. Capperucci; A. Degl'Innocenti; P. Scafato; P. Spagnolo (1995). "Synthetic Applications of Bis(trimethylsilyl) sulphide: Part II. Synthesis of Aromatic and Heteroaromatic o-Azido-Thioaldehydes".
11. Pegg, DE (2007). "Principles of Cryopreservation". In Day JG, Stacey GN (eds.). Cryopreservation and Freeze-Drying Protocols. Methods in Molecular Biology. Vol. 368. Humana Press. pp. 39–57.
12. Swanston, Jonathan (2006). "Thiophene". Ullmann's Encyclopedia of Industrial Chemistry. Weinheim: Wiley-VCH.

Chapter 7: Evolution of the Chemical Industry

HISTORICAL FOUNDATIONS:

The chemical industry boasts a millennia-spanning legacy rooted in the ingenuity of ancient civilizations like the Indian, Egyptians, Greeks and Chinese. Millennia ago, these cultures harnessed natural resources such as minerals, plants, and animal byproducts to craft medicines, dyes, and assorted materials.

INDUSTRIAL REVOLUTION CATALYST

With the advent of the Industrial Revolution in the 18th and 19th centuries, the chemical industry underwent a transformative shift, emerging as a distinct scientific and technological domain. Pioneering breakthroughs in chemistry and engineering birthed synthetic marvels like sulphuric acid, ammonia, and chlorine, paving the way for revolutionary products like synthetic dyes, fertilizers, and pharmaceuticals.

20TH CENTURY ADVANCEMENTS

The 20th century marked a period of monumental growth and innovation for the chemical industry. Propelled by cutting-edge technologies, groundbreaking scientific discoveries, and burgeoning economic demand, the sector propelled the development of modern marvels such as plastics, synthetic rubber, pesticides, and electronic components.

CONTEMPORARY LANDSCAPE

Today, the chemical industry stands as a global powerhouse, encompassing a vast array of products, technologies, and applications. Continuously evolving, the industry prioritizes sustainability, safety, and efficiency while spearheading the creation of novel materials and products tailored to meet the ever-evolving demands of society.

Diverse Scope of the Chemical Industry

Versatile Applications: The chemical industry is a multifaceted realm encompassing companies engaged in the production, manufacturing, and distribution of chemicals and chemical products. These commodities find utility across diverse sectors, including agriculture, construction, healthcare, consumer goods, transportation, and energy.

Industry Ecosystem: Comprising a spectrum ranging from multinational giants to specialized niche players, the chemical industry leverages an array of technologies and processes to innovate and produce goods. An extensive network of suppliers, distributors, and service providers supports its operations, facilitating research and development endeavors.

Sustainable Paradigm Shift: In response to contemporary imperatives, the chemical industry is spearheading initiatives geared towards sustainable and eco-friendly product development. These endeavors encompass the creation of materials designed to minimize waste, conserve resources, and curtail the use of hazardous chemicals, in addition to supporting renewable energy technologies and energy-efficient building materials.

Chemistry's Integral Role

Elemental Foundations: At its core, the chemical industry harnesses an array of elemental building blocks, including carbon, nitrogen, oxygen, hydrogen, chlorine, sulfur, phosphorus, and sodium. These elements serve as vital ingredients in a plethora of chemical reactions and processes, underpinning the creation of diverse products and materials.

Chemical Ingenuity: Fueled by a profound understanding of chemistry, the industry leverages the unique properties of these elemental constituents to drive innovation, develop novel materials, and engineer cutting-edge technologies. From plastics and fuels to fertilizers and disinfectants, chemistry's versatile toolkit enables the creation of solutions that address society's evolving needs.

APPLICATION OF SULPHUR & SULPHUR COMPOUNDS IN CHEMICAL INDUSTRY

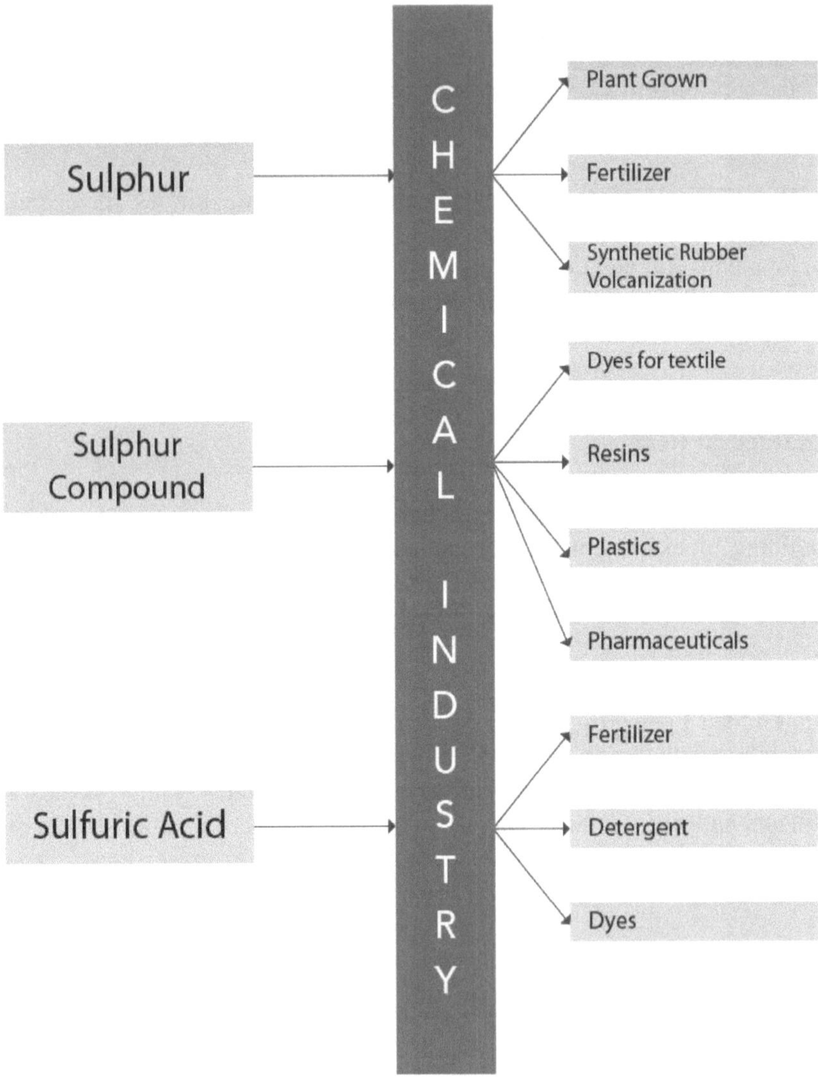

1. Sulphur-Based Solutions for Plant Growth

a. **Nurturing Plant Vitality**: Sulphur stands as an indispensable nutrient in fostering robust plant growth, pivotal for synthesizing proteins, enzymes, and vital plant compounds. Several sulphur-based products cater to plant nutrition, each serving distinct roles in bolstering agricultural success.
b. **Sulphur Fertilisers**: Sulphur fertilisers serve as vital supplements, furnishing plants with essential sulphur for optimal growth. Crucial for protein, vitamin, and hormone production, sulphur fertilisers also contribute to soil pH modulation, enhancing nutrient accessibility for plants.
c. **Sulphur-Infused Pesticides:** Combatting plant diseases is streamlined with sulphur-containing pesticides, efficacious against maladies like powdery mildew, rust, and scab. These pesticides thwart fungal spores and modify plant surface pH, curbing disease proliferation effectively.
d. **Elemental Sulphur:** A versatile ally in soil management, elemental sulphur is enlisted to ameliorate soil pH, particularly in alkaline soils, fostering heightened nutrient uptake by plants. Furthermore, it doubles as a potent agent against soil-borne pests and diseases.
e. **Sulphur Dioxide**: In the realm of pest control, sulphur dioxide emerges as a formidable fumigant, safeguarding stored grains and food products against infestations. Its application extends to the wine industry, where it preserves wine quality and stability during storage.

Cultivating Agricultural Excellence: Collectively, sulphur-based products represent indispensable assets in agricultural and horticultural endeavors. By enhancing soil nutrient profiles, curbing pests and diseases, and elevating crop yield and quality, these solutions epitomize agricultural innovation at its finest.

2. Synthetic Rubber Vulcanization with Sulphur

a. Sulphur stands as a linchpin in the vulcanization of synthetic rubber, a process pivotal for fortifying rubber's strength, resilience, and longevity. Various products stemming from sulphur-driven vulcanization contribute to an array of industrial applications.
b. Sulphur emerges as the quintessential vulcanizing agent for synthetic rubber, fostering intermolecular cross-linkages that imbue rubber with enhanced physical and mechanical attributes. Through vulcanization, sulphur orchestrates the formation of a robust three-dimensional network within rubber, rendering it more robust, elastic, and resistant to wear, tear, and environmental stressors.

c. Augmenting the vulcanization kinetics, accelerators collaborate with sulphur to expedite the process, reducing time and temperature requirements. Zinc oxide, stearic acid, and sulphur donors like thiazoles and sulfenamides are among the accelerators commonly deployed alongside sulphur, enhancing productivity and efficiency in rubber processing.
d. Facilitating sulphur and accelerator efficacy, activators optimize their distribution and accessibility within the rubber compound. By enhancing cross-linking efficiency, activators such as magnesium oxide and zinc oxide bolster the quality and performance of vulcanized rubber products.
e. The indispensability of sulphur in synthetic rubber vulcanization catalyzes the production of diverse rubber components vital across industries. From automotive to aerospace, vulcanized rubber finds application in tires, hoses, belts, gaskets, and seals, underscoring its pivotal role in machinery and equipment fabrication.

3. Elevating Textile Harnessing Sulphur Compounds for Dyeing

a. Embracing a spectrum of vibrant hues, sulphur dyes, revered for their water-insoluble nature, grace both natural and synthetic fibers like cotton, wool, silk, and polyester. Particularly favored for denim and casual wear, these dyes undergo specialized processes to adhere seamlessly to fibers, enriching textiles with enduring color.
b. Cementing their presence through chemical bonds, reactive dyes forge covalent ties with fibers, enhancing dye permanence and vibrancy. Among them, sulphur-containing reactive dyes emerge as stalwarts for cotton and cellulosic fibers, showcasing radiant hues and exceptional colorfastness.
c. Imbuing textiles with timeless allure, vat dyes, resistant to water solubility until activated by reducing agents, grace cellulose fibers with brilliance. Indigo and Anthraquinone derivatives stand out in this category, with sulphur-laced variants lending their steadfastness and brilliance to cotton, silk, and other fibers, ideal for enduring outdoor textiles.
d. Pioneering synthetic fiber aesthetics, disperse dyes adorn polyester and nylon with captivating hues, finding their niche in sportswear and outdoor gear. Among them, sulphur-infused disperse dyes reign supreme, elevating polyester textiles with enduring color and resilience.
e. Sulphur compounds, revered for their versatility and resilience, emerge as quintessential components in textile dyeing. Their diverse array of dyes, spanning natural and synthetic fibers, epitomize vibrancy, longevity, and resistance to fading, championing a kaleidoscope of textile applications with enduring elegance.

4. Sulphur Compounds Role in Resin Production

Resins, pivotal in myriad applications, epitomize versatility in both natural and synthetic realms.

a. **Natural Resin:** Harvested from select trees, natural resin, with its adhesive properties, finds its place in varnishes, pharmaceuticals, and beyond.
b. **Synthetic Resin:** Advancements in chemical synthesis ushered in the era of synthetic resin, a polymer blend of carbon, hydrogen, sulphur, and traces of oxygen.
c. These high molecular weight compounds redefine industrial applications.
d. **Sulphur's Impact on Resins:** Sulphur compounds, instrumental in resin formulation, catalyze innovation across various resin types:
e. **Phenolic Resins:** Born from the marriage of phenol, formaldehyde, and sulphur catalysts, phenolic resins adorn industries with their utility in adhesives, coatings, and molding compounds.
f. **Epoxy Resins:** Crafted from bisphenol-A, epichlorohydrin, and sulphur-based curing agents, epoxy resins reign as adhesives, coatings, and composites, epitomizing versatility.
g. **Sulphur-Modified Resins:** Infused with sulphur, these resins enhance adhesion and chemical resistance, finding homes in industrial coatings and adhesives, fortifying applications with durability and reliability.
h. **Polysulfide Resins:** Forged from diols and sulphur-based curing agents, polysulfide resins serve as sealants and adhesives, sealing industries with their reliability and versatility.
i. **Sulphurized Oils:** Emerging from the fusion of oils and sulphur, sulphurized oils lubricate industries, fostering efficiency and resilience in metalworking fluids.

In summation, sulphur compounds emerge as indispensable agents in resin production, empowering industries with resilience, versatility, and reliability across a multitude of applications.

5. Sulphur Compounds Impact on Plastic Production

Sulphur compounds emerge as key components in the manufacturing process of diverse plastic products, contributing to their durability, resilience, and versatility.

a. **Polyvinyl Chloride (PVC):** Sulphur takes center stage as a thermal stabilizer in PVC production. Renowned for its versatility, PVC finds its way into an array of applications, spanning from pipes and flooring to electrical cable insulation and medical devices.
b. **Thermoplastic Elastomers (TPEs):** Harnessing both rubber-like elasticity and thermoplastic processing, TPEs owe their resilience to sulphur-based curatives. Serving as seals, gaskets, and automotive parts, TPEs stand as testaments to sulphur's role in enhancing plastic functionality.
c. **Styrene-Butadiene Rubber (SBR):** Sulphur-based crosslinkers form the backbone of SBR production. Widely utilized in tires, conveyor belts, and footwear, SBR epitomizes sulphur's contribution to the automotive and footwear industries.
d. **Polysulfone (PSU):** The marriage of bisphenol-A with a sulphur-containing diamine gives birth to PSU, a thermoplastic hailed for its versatility. From medical devices to aircraft parts, PSU showcases sulphur's versatility across a myriad of applications.
e. **Sulphur-Containing Additives:** Thiurams and dithiocarbamates, serving as accelerators, propel the production of various rubber types. From natural rubber to synthetic counterparts like polybutadiene and nitrile rubber, sulphur-containing additives pave the way for enhanced rubber performance across industries.

In essence, Sulphur compounds stand as indispensable constituents in plastic production, weaving resilience, versatility, and functionality into plastic products that shape industries worldwide.

6. Sulphur Compounds Vital Role in Pharmaceuticals

Sulphur compounds stand as stalwarts in the pharmaceutical realm, boasting a rich history of contributions across various drug formulations and therapeutic applications.

a. **Antibiotics:** Integral to numerous antibiotics, sulphur either forms part of the molecule or augments its activity as a functional group. Esteemed examples encompass sulfamethoxazole, trimethoprim, and penicillin, showcasing sulphur's prowess in combating bacterial infections.
b. **Antifungals:** Sulphur-laden antifungal medications combat diverse fungal infections, ranging from toenail fungus to ringworm. Terbinafine and sulconazole exemplify the efficacy of sulphur in addressing fungal ailments with precision.
c. **Anticoagulants:** Heparin, a cornerstone anticoagulant, relies on sulphur for its therapeutic efficacy. Widely utilized in preventing blood clots across various

medical contexts, heparin underscores sulphur's significance in cardiovascular health.

d. **Anti-inflammatory Drugs:** Sulfasalazine, a potent anti-inflammatory agent, owes its efficacy to its sulphur moiety. Renowned for its utility in managing inflammatory bowel disease, sulfasalazine underscores sulphur's role in modulating immune responses.
e. **Radiopharmaceuticals:** Harnessing the power of radioactive sulphur, radiopharmaceuticals emerge as pivotal tools in diagnostic imaging and cancer therapy. Sulphur's unique properties play a pivotal role in the synthesis of compounds vital for precise disease diagnosis and treatment.

In summation, Sulphur compounds stand as indispensable components in pharmaceutical development and production, ushering in an era of therapeutic breakthroughs and improved patient outcomes.

7. Sulphuric Acid's Role in Fertilizer Production

Sulphuric acid stands as a cornerstone in fertilizer manufacturing, facilitating the creation of essential products vital for agricultural productivity. Here are some fertilizer examples, crafted with the aid of sulphuric acid, along with their applications:

a. **Superphosphate Fertilizers:** Derived by treating rock phosphate with sulphuric acid, superphosphate fertilizers are phosphorus-rich formulations crucial for robust plant growth. Widely employed in agriculture, these fertilizers bolster crop yields by providing an essential nutrient foundation.
b. **Ammonium Sulphate Fertilizers:** Formulated through the reaction of sulphuric acid with ammonia, ammonium sulphate fertilizers boast a balanced composition of nitrogen and sulphur. These nutrients play pivotal roles in plant development, making these fertilizers instrumental in fostering healthy growth and maximizing crop productivity.
c. **Potassium Sulphate Fertilizers:** Crafted by combining sulphuric acid with potassium chloride, potassium sulphate fertilizers offer a potent blend of potassium and sulphur essential for plant vitality. These fertilizers serve as vital components in agricultural practices, enhancing crop yields and fortifying plant resilience.
d. **Nitrogen-Based Fertilizers:** Sulphuric acid's versatility extends to the production of nitrogen-based fertilizers, including ammonium nitrate and urea. These fertilizers, rich in nitrogen, serve as cornerstone nutrients for robust plant growth, supporting agriculture by ensuring optimal crop development.

In essence, Sulphuric acid's pivotal role in fertilizer production underscores its significance in agricultural practices, driving enhanced crop yields and promoting sustainable food production to meet the demands of a growing population.

8. Sulphuric Acid's Contribution to Detergent Production

Detergents, synthetic cleansing compounds, owe their effectiveness to surfactants, and sulphuric acid plays a key role in their production. From laundry to dishwashing and industrial applications, sulphuric acid aids in creating potent cleaning solutions.

Here's a glimpse into the diverse array of detergent products crafted with its assistance:

a. **Laundry Detergents:** In the realm of laundry care, sulphuric acid contributes to various formulations, be it liquid or powder detergents. Its role lies in neutralizing the alkali used during production, enhancing the cleaning efficacy of the final product. With sulphuric acid's assistance, laundry detergents tackle stains and grime effectively.

b. **Dishwashing Detergents:** For sparkling dishes, both liquid and powder dishwashing detergents utilize sulphuric acid in their formulation. By neutralizing alkali components, sulphuric acid enhances the cleaning prowess of these detergents, ensuring grease-free, spotless results.

c. **Industrial Detergents:** In manufacturing settings, industrial detergents, designed for heavy-duty cleaning and degreasing, benefit from sulphuric acid's involvement. Its inclusion aids in breaking down stubborn stains and soils, ensuring thorough cleanliness in industrial environments.

d. **Surfactants:** Sulphuric acid also serves in the production of surfactants, crucial ingredients in various detergents. Surfactants facilitate the breakdown and removal of dirt and stains by reducing water's surface tension. This enables deeper penetration into fabrics and surfaces, ensuring comprehensive cleaning.

In essence, sulphuric acid's pivotal role in detergent production underscores its significance in fostering cleanliness, both in household and industrial settings. Its contribution enhances the efficacy of cleaning products, ensuring optimal performance in tackling diverse cleaning challenges.

9. Sulphuric Acid Role in Dye Production

In the realm of dye manufacturing, sulphuric acid emerges as a crucial component, facilitating the creation of vibrant hues across various substrates.

Here's a glimpse into the diverse array of dye products crafted with its assistance:

a. **Synthetic Dyes:** Sulphuric acid serves as a catalyst in the production of synthetic dyes, notably azo dyes. Widely utilized in the textile industry for dyeing cotton, wool, and silk, azo dyes are synthesized by reacting an amine with a diazonium salt in the presence of sulphuric acid.
b. **Acid Dyes:** For dyeing protein fibers like wool, silk, and synthetic fibers such as nylon, acid dyes reign supreme. Sulphuric acid's role in the production of acid dyes is pivotal, creating the acidic conditions necessary for the dye to adhere effectively to the fibers.
c. **Natural Dyes:** Even in the realm of natural dyes like madder, indigo, and cochineal, sulphuric acid plays a vital role. It helps create the acidic environment required for these natural dyes to form strong bonds with the fibers, ensuring lasting and vibrant colors.
d. **Pigment Dyes:** Sulphuric acid's significance extends to pigment dye production, integral to paints, inks, plastics, and various applications. By facilitating the acidic conditions necessary for pigment binding, sulphuric acid ensures the vivid and enduring nature of these dyes.

In summary, sulphuric acid's multifaceted role in dye production underscores its significance in the realm of coloration. From synthetic to natural dyes and pigment formulations, its contribution ensures the creation of vibrant colors and lasting bonds across a diverse range of substrates.

Summary

Sulfur and sulfur compounds are extensively used in the chemical industry due to their diverse properties and reactivity.

Here are some of the key applications of sulfur and sulfur compounds in the chemical industry:

a. **Sulfuric Acid Production:** Sulfuric acid (H_2SO_4) is one of the most widely used chemicals in the world and has numerous industrial applications. The Contact process is the most common method for sulfuric acid production, which involves the oxidation of sulfur dioxide (SO_2) to sulfur trioxide (SO_3) using oxygen over a vanadium pentoxide catalyst, followed by the absorption of SO_3 in water to form sulfuric acid.
b. **Fertilizer Production:** Sulfur is an essential nutrient for plant growth, and sulfur compounds are used in the production of sulfur-containing fertilizers, such as ammonium sulfate and sulfur-coated urea. These fertilizers provide plants with sulfur in a form that is readily available for uptake, helping to improve crop yields and soil fertility.

c. **Petroleum Refining**: Sulfur compounds play crucial roles in various processes within the petroleum refining industry. For example, sulfur compounds present in crude oil are removed through hydrodesulfurization (HDS), a process that converts sulfur-containing compounds into hydrogen sulfide (H_2S) and removes them from petroleum fractions.

d. **Polymerization:** Sulfur compounds are used as modifiers and cross-linking agents in polymerization processes. For example, sulfur is used in the vulcanization of rubber, where it forms cross-links between polymer chains, imparting elasticity, strength, and durability to rubber products.

e. **Mining and Ore Processing:** Sulfur compounds are used in mining and ore processing as flotation agents to separate valuable minerals from gangue materials. Compounds such as sodium sulfide (Na_2S) and potassium ethyl xanthate (KEX) are commonly used as flotation reagents to facilitate the separation of minerals through selective hydrophobicity.

f. **Chemical Synthesis:** Sulfur compounds serve as intermediates and reagents in various chemical synthesis processes. For example, sulfur dioxide (SO_2) is used in the synthesis of sulfuric acid, sulfites, and sulfur dioxide derivatives. Organic sulfur compounds, such as thiols, thioketones, and thioethers, are used in the synthesis of pharmaceuticals, agrochemicals, and fine chemicals.

g. **Water Treatment:** Sulfur compounds are used in water treatment processes to remove contaminants and improve water quality. For example, sodium thiosulfate ($Na_2S_2O_3$) is used as a dechlorination agent to neutralize chlorine in water, while sulfur dioxide is used in water treatment plants to control pH levels and remove dissolved metals.

h. **Textile Industry:** Sulfur compounds are used in the textile industry as sulfur dyes, which are widely used for dyeing cotton, wool, and other natural fibers. Sulfur dyes have excellent color fastness properties and are resistant to fading and washing.

These are just a few examples of the wide-ranging applications of sulfur and sulfur compounds in the chemical industry. Their versatility and reactivity make them indispensable for various industrial processes and product

REFERENCES:

1. Austin, George T. (1984). Shreve's Chemical process industries (5th ed.). New York, NY: McGraw-Hill. p. 357
2. Greenwood, Norman N.; Earnshaw, Alan (1997). Chemistry of the Elements (2nd ed.), Butterworth-Heinemann. ISBN 978-0-08-037941-8.
3. Merck Index, 11th Edition, 9625
4. Patnaik, Pradyot (2002), Handbook of Inorganic Chemicals. McGraw-Hill

Chapter 8: Fueling Progress: Exploring the Petroleum Industry

HISTORY:

The origins of the petroleum industry trace back to ancient civilizations utilizing natural oil seepages for lighting, heating, and medicinal purposes. However, its formal establishment as a pivotal economic force began in the 19th century.
In 1859, Edwin Drake drilled the first commercial oil well in Titusville, Pennsylvania, heralding the inception of the modern petroleum sector. The subsequent discovery of oil reserves in the United States spurred the rapid growth of oil refineries and pipeline networks for efficient market transportation.
Through the late 19th and early 20th centuries, global oil and natural gas discoveries propelled the industry's expansion, giving rise to major multinational entities like Standard Oil, known for their influential market presence and aggressive business strategies.
During World War II, the petroleum industry assumed a vital role in fueling military operations, providing essential resources for tanks, aircraft, and naval vessels. Post-war, technological advancements fueled further growth, enhancing exploration, drilling, and refining processes.

Yet, the industry grappled with myriad challenges and controversies, including environmental impacts like oil spills, pollution, and climate change concerns. In response, there has been a shift towards renewable energy initiatives to reduce dependency on fossil fuels.

DESCRIPTION:
The petroleum industry constitutes a sprawling and intricate domain, encompassing the exploration, extraction, refinement, transportation, and distribution of crude oil and natural gas. Petroleum, a non-renewable fossil fuel, originates from the ancient organic remains of plants and animals.

Comprising major corporations, independent producers, and service providers, the industry employs a diverse workforce ranging from scientists and engineers to technicians and administrative personnel.

Despite its economic significance and pivotal role in global energy supply, the petroleum sector faces critical environmental and safety challenges. Efforts to address these issues include the adoption of sustainable practices such as carbon capture, renewable energy integration, and enhanced energy efficiency measures.

SUMMARY:

The petroleum industry encompasses the multifaceted processes of exploring, extracting, refining, and distributing crude oil and natural gas, vital yet finite fossil fuels. This complex sector involves a diverse array of companies and a sizable workforce, spanning roles from scientists and engineers to technicians, operators, and administrative personnel.

Exploration and production efforts employ advanced technologies like seismic imaging and drilling to locate and extract oil and gas reserves from beneath the earth's surface. Following extraction, these resources are transported through pipelines, tankers, and other means to refineries. There, intricate refining processes transform crude oil into a spectrum of essential products, including gasoline, diesel, jet fuel, and lubricants.

While the petroleum industry has been pivotal in driving global economic growth and meeting energy demands for decades, it confronts significant challenges concerning environmental sustainability and safety. Issues such as oil spills, pollution, and climate change necessitate a shift towards more sustainable practices and technologies. Initiatives like carbon capture and storage, renewable energy integration, and enhanced energy efficiency are emerging as key strategies to address these concerns and shape the industry's future.

APPLICATION OF SULPHUR & SULPHUR COMPOUNDS IN PETROLEUM INDUSTRY

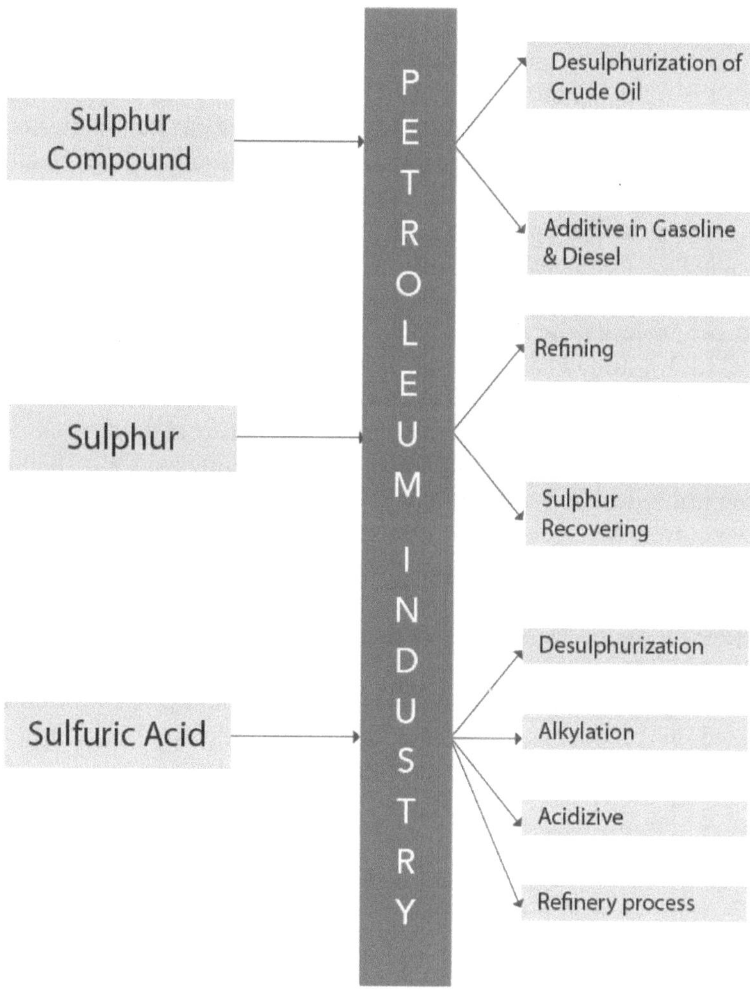

1. Sulphur Compounds as Crude Oil Desulphurization

a. Desulphurization is a vital process in the petroleum industry aimed at eliminating sulphur compounds from crude oil and its derivatives to mitigate environmental pollution, health risks, and equipment corrosion. Various desulphurization methods are employed, including hydrodesulphurization, catalytic cracking, and extractive desulphurization.

b. **Hydrodesulphurization (HDS)** stands as the most prevalent method for sulphur removal from crude oil and its derivatives. This process involves reacting the oil or its derivatives with hydrogen gas at elevated temperature and pressure in the presence of a catalyst. This catalytic action breaks down sulphur compounds into hydrogen sulfide (H_2S) and other gases, facilitating their separation and removal. HDS effectively eliminates sulphur compounds from various petroleum products such as gasoline, diesel, and jet fuel.

c. The sulphur compounds extracted during desulphurization present opportunities for conversion into valuable resources. For instance, hydrogen sulfide can undergo the Claus process, where it is burned to yield sulphur dioxide (SO_2), followed by reaction with additional hydrogen sulfide to produce elemental sulphur. Elemental sulphur finds utility in various applications, including fertilizer, detergent, and industrial production.

d. Additionally, sulphur can be transformed into sulphuric acid, a cornerstone of the chemical industry. Sulphuric acid is indispensable in the manufacturing of fertilizers, dyes, detergents, and other industrial products. Moreover, sulphur can be utilized to generate sulphur dioxide, crucial in food preservation, winemaking, and other processes.

e. In essence, desulphurization serves as a pivotal process within the petroleum industry, ensuring the elimination of sulphur compounds from crude oil and its derivatives. The sulphur compounds recovered during this process can be harnessed to produce elemental sulphur and sulphuric acid, offering a spectrum of valuable industrial applications.

2. Sulphur Compounds as Additives in Gasoline and Diesel Fuels

Sulphur compounds serve as valuable additives in gasoline and diesel fuels, enhancing performance and mitigating emissions.

Two prominent sulphur-containing additives utilized in fuel are:

a. **Methylcyclopentadienyl Manganese Tricarbonyl (MMT)**
   MMT, an organic compound containing manganese and sulphur, is incorporated into gasoline to elevate its octane rating, a measure of resistance to knocking or pinging. Besides augmenting fuel efficiency, MMT aids in diminishing engine deposits. However, its usage is contentious due to concerns regarding emissions and vehicle performance.

b. **Dimethyl Disulfide (DMDS)**
   DMDS, another sulphur-containing compound, is introduced into diesel fuel to bolster its lubricity, thereby minimizing wear and tear on engine components. Furthermore, DMDS enhances combustion efficiency and curtails particulate matter emissions, contributing to reduced environmental impact.

c. **Ultra-Low Sulphur Diesel (ULSD)**
   In addition to these additives, ultra-low sulphur diesel (ULSD) represents a variant of diesel fuel with a sulphur content of less than 15 parts per million (ppm). Compared to conventional diesel fuel with sulphur levels as high as 5000 ppm, ULSD significantly diminishes emissions, particularly in urban locales. By curtailing sulphur levels, ULSD effectively reduces particulate matter and nitrogen oxides emissions, thereby ameliorating air quality

3. Sulphur and Sulphur Compounds in Crude Oil Refining

Apart from desulphurization, the refining process of crude oil yields various sulphur-containing byproducts that find applications across diverse industries.
Sulphur stands as a key component in the synthesis of sulphuric acid, a versatile chemical utilized in the manufacturing of fertilizers, detergents, dyes, and numerous industrial goods.

a. **Elemental Sulphur**
   Derived during refining, elemental sulphur serves as a pivotal raw material for producing sulphuric acid, fertilizers, and other industrial products.
b. **Carbon Disulfide**
   Carbon disulfide, a sulphur-containing compound, finds utility in the production of rayon, cellophane, and other cellulose-based products. Additionally, it serves as a solvent for rubber, plastics, and various materials.
c. **Sulphur Dioxide**
   Generated during refining operations, sulphur dioxide plays a crucial role in food preservation, winemaking, and diverse product manufacturing processes.
d. **Hydrogen Sulfide**
   As a byproduct of refining, hydrogen sulfide can be transformed into elemental sulphur using the Claus process, contributing to the production of essential industrial materials.
e. **Sulphur Compounds as Fuel Additives**
   Sulphur compounds serve as additives in gasoline and diesel fuels, enhancing lubricity and reducing emissions, thereby improving overall performance.

4. Recovery of Sulphur During Petroleum Refining

Sulphur recovery encompasses the conversion of sulphur-containing compounds, primarily hydrogen sulfide (H_2S), encountered during petroleum refining into elemental sulphur.

The reclaimed sulphur finds versatile applications across various industries:

b. **Fertilizers Production:** Elemental sulphur plays a vital role in fertilizer manufacturing. It serves as a precursor to sulphuric acid, a key component in producing phosphoric acid, which, in turn, is utilized in formulating fertilizers like ammonium phosphate and triple superphosphate.
c. **Chemical Industry:** Sulphur finds extensive use in chemical synthesis, particularly in the production of sulphuric acid. This acid serves as a foundational material in battery manufacturing, detergent formulation, and pigment production. Additionally, sulphur is a crucial raw material in generating sulphur dioxide, a commonly employed preservative in the food industry.
d. **Petroleum Refining:** Due to its prevalence as an impurity in crude oil, sulphur must be extracted before refining crude into gasoline and other petroleum products. The reclaimed sulphur can subsequently be sold for various industrial applications.

e. **Metal Production:** In metal manufacturing processes, sulphur acts as a potent reducing agent, facilitating the removal of oxygen from metal ores. It finds application in the production of metals such as copper, lead, and zinc.

Various techniques are employed for sulphur recovery, with the Claus process emerging as the most prevalent method. This process entails the conversion of hydrogen sulfide into sulphur dioxide through oxidation, followed by its further reaction with hydrogen sulfide to yield elemental sulphur. Alternative methods include the Stretford process and the Thiopaq process, each offering unique advantages in specific refining contexts.

5. Desulphurization from Sulphuric Acid

Desulphurization of sulphuric acid is a vital process aimed at reducing sulphur content to enhance its quality and minimize environmental repercussions.

The primary desulphurization products find diverse applications:

a. **Gypsum:** Gypsum stands as the predominant desulphurization byproduct, formed through the reaction of calcium carbonate or calcium oxide with sulphuric acid. With versatile utility, gypsum finds its way into construction materials, soil conditioning agents, and serves as a foundational component in plaster and cement production.

b. **Elemental Sulphur:** Elemental sulphur, attainable through the Claus process by hydrogen sulfide reacting with sulphur dioxide, holds significant industrial value. It serves as a fundamental ingredient in the production of sulphuric acid, fertilizers, and various chemical formulations.

c. **Ammonium Sulphate:** Resulting from the reaction of ammonia with sulphuric acid, ammonium sulphate emerges as a valuable product. Widely utilized as a fertilizer and food additive, it contributes to agricultural productivity and food processing industries.

d. **Magnesium Sulphate**: Magnesium sulphate, derived from the interaction of magnesium oxide or magnesium carbonate with sulphuric acid, finds multifaceted application. It serves as a potent laxative, aids in fertilizer production, and addresses magnesium deficiency in plant cultivation.

Overall, desulphurization products derived from sulphuric acid present a spectrum of industrial and commercial uses, offering avenues for recycling and repurposing to mitigate waste and environmental impact.

6. Alkylation from Sulphuric Acid

Alkylation, a chemical process merging alkyl groups with organic compounds, is facilitated by sulphuric acid, often acting as a catalyst.

Here are key products and their applications:

a. **Gasoline**
   Sulphuric acid catalyzes the alkylation of isobutane with olefins, yielding high-octane gasoline termed "alkylate." This product enhances gasoline's octane rating, serving as a vital blending component, thus optimizing fuel performance.
b. **Linear Alkylbenzene (LAB)**
   LAB, a prevalent surfactant in household and industrial cleaners, is produced by alkylating benzene with linear alpha olefins, a process enabled by sulphuric acid. LAB's versatile applications span diverse cleaning formulations, facilitating effective dirt removal in various settings.
c. **Alkylated Aromatics**
   In the production of alkylated aromatics, sulphuric acid acts as a catalyst, yielding compounds vital as solvents, additives, and intermediates in manufacturing plastics, resins, and synthetic rubbers. These substances play integral roles in diverse chemical processes and end-product formulations.
d. **Cumene**
   Sulphuric acid contributes to cumene production, a valuable solvent and precursor in phenol and acetone manufacturing. This process, enabled by sulphuric acid, supports the synthesis of key chemicals pivotal in a range of industrial applications.

Overall, sulphuric acid's catalytic prowess in alkylation processes underpins the production of essential compounds across chemical and petroleum industries. These products, integral to industrial and commercial spheres, find applications in a myriad of consumer goods and industrial processes.

7. Acidizing with Sulphuric Acid

Acidizing, a pivotal process in the oil and gas industry, entails the injection of acid into a wellbore to dissolve obstructions hindering the flow of oil or gas. Sulphuric acid, a prominent acidizing agent, facilitates the creation of conduits for fluid movement.

Here are the primary acidizing products and their applications:

a. **Calcium Sulphate**
   Upon reacting with calcium carbonate in the rock formations, sulphuric acid generates calcium sulphate. This byproduct finds utility in an array of domains, including construction materials, soil enhancement, and as a retarder in cement production.
b. **Hydrogen Sulfide**
   Sulphuric acid's reaction with sulphur compounds yields hydrogen sulfide. This versatile product serves in sulphur and sulphuric acid production, as well as a reagent in various chemical processes.
c. **Iron Sulphate**
   Interaction with iron in rock formations results in the formation of iron sulphate. This compound finds application in dye and pigment manufacturing, alongside water treatment chemical production.
d. **Aluminum Sulphate**
   Sulphuric acid's reaction with aluminum generates aluminum sulphate, valuable in water treatment, paper production, and as a mordant in textile dyeing.

In essence, sulphuric acid-driven acidizing enhances oil and gas flow in wellbores, yielding products with diverse industrial applications. However, meticulous management is imperative to mitigate potential environmental impacts, safeguarding groundwater and natural resources.

8. Role of Sulphuric Acid in the Refining Process

Sulphuric acid serves as a cornerstone in the petroleum refining industry, playing a pivotal role in numerous processes.

Here are some key refinery process products and their applications derived from sulphuric acid:

a. **Catalytic Cracking**: In the fluid catalytic cracking (FCC) process, sulphuric acid acts as a catalyst, transforming heavy crude oil into lighter hydrocarbons. This process yields valuable products such as gasoline, diesel, and jet fuel.
b. **Alkylation**: Utilized as a catalyst in the alkylation process, sulphuric acid enables the combination of isobutane with olefins to produce high-octane gasoline components known as alkylate. Alkylate serves as a prized blending component in gasoline.
c. **Desulphurization**

Sulphuric acid plays a crucial role in desulphurization, removing sulphur from petroleum products like diesel and gasoline. The resultant low-sulphur product minimizes environmental impact.

d. **Oil Refining**: In the refining process, sulphuric acid is instrumental in purging impurities from crude oil, including sulphur, nitrogen, and metals. This enhances the quality of the end product, facilitating easier processing into various refined products.
e. **Corrosion Control**: Sulphuric acid also serves in corrosion control within refineries, safeguarding pipelines, tanks, and other equipment. This maintenance practice ensures the safety, reliability, and longevity of refining operations.

In essence, sulphuric acid stands as an indispensable component in petroleum refining, optimizing efficiency and product quality across diverse processes. The resulting products find extensive industrial and commercial applications, serving as vital constituents in numerous consumer goods.

Summary

Sulfur compounds play significant roles in various processes within the petroleum industry, spanning from exploration and refining to end-product applications. Here are some of the primary uses of sulfur compounds in the petroleum industry:

a. **Hydrodesulfurization (HDS):** One of the major applications of sulfur compounds in the petroleum industry is in the hydrodesulfurization process. Hydrodesulfurization is a crucial step in petroleum refining that involves the removal of sulfur compounds from crude oil, petroleum fractions, and other hydrocarbon streams. Sulfur compounds in crude oil, such as thiols, sulfides, and thiophenes, are converted into hydrogen sulfide (H_2S) and removed by reaction with hydrogen over catalysts under high temperature and pressure conditions.
b. **Catalysts in Refining:** Sulfur compounds are used as catalysts in various refining processes. For example, transition metal sulfides, such as molybdenum sulfide (MoS2) and cobalt-molybdenum sulfide (Co-Mo-S), are commonly used as catalysts in hydrodesulfurization and hydrocracking reactions to improve the efficiency of sulfur removal and the production of cleaner fuels.
c. **Sulfur Recovery:** Hydrogen sulfide produced during refining processes, such as hydrodesulfurization, is converted into elemental sulfur through the Claus process or other sulfur recovery methods. Elemental sulfur is a valuable byproduct of the petroleum industry and is used in various industrial applications, including the production of sulfuric acid and fertilizers.

d. **Corrosion Inhibition:** Sulfur compounds are used as corrosion inhibitors in petroleum refining and transportation systems to protect equipment and pipelines from corrosion caused by hydrogen sulfide and other corrosive species present in crude oil and petroleum products. Organic sulfur-containing compounds, such as mercaptans and thiophenes, are often added to crude oil or petroleum products as corrosion inhibitors.
e. **Fuel Additives**: Sulfur compounds are used as fuel additives in gasoline and diesel to improve fuel lubricity, reduce emissions, and enhance combustion efficiency. However, strict regulations on sulfur content in fuels have led to the development of sulfur-free or low-sulfur fuels to reduce environmental pollution, particularly sulfur dioxide (SO_2) emissions, which contribute to air pollution and acid rain.
f. **Chemical Feedstock:** Sulfur compounds obtained from petroleum refining processes, such as elemental sulfur and sulfuric acid, are used as chemical feedstocks in various industries, including the production of fertilizers, detergents, pesticides, and sulfur-containing chemicals.

Overall, sulfur compounds play crucial roles in the petroleum industry, from facilitating refining processes to providing valuable byproducts and enabling the production of cleaner fuels and chemicals. However, efforts to minimize sulfur content in petroleum products are underway to address environmental concerns and comply with regulatory standards for air quality and emissions control.

REFERENCES:

1. Greenwood, Norman N.; Earnshaw, Alan (1997). Chemistry of the Elements (2nd ed.), Butterworth-Heinemann. ISBN 978-0-08-037941-8.
2. Merck Index, 11th Edition, 9625
3. Patnaik, Pradyot (2002), Handbook of Inorganic Chemicals. McGraw-Hill.
4. "sulphuric Acid". The Columbia Encyclopedia (6th ed.). 2009. Retrieved 16 March 2010.
5. "Sulphuric acid". Encyclopædia Britannica. Vol. 26 (11th ed.). 1910–1911. pp. 65–69.
6. Austin, George T. (1984). Shreve's Chemical process industries (5th ed.). New York, NY: McGraw-Hill. p. 357

Chapter 9: The Evolution of the Rubber Industry

HISTORY:

The roots of the rubber industry stretch back to the ancient civilizations of Mesoamerica, where indigenous communities utilized the latex from rubber trees to craft various items. However, it wasn't until Charles Goodyear's invention of vulcanization in 1839 that the industry witnessed a transformative leap forward. Vulcanization, achieved by heating natural rubber with sulphur, yielded a more robust material, catalyzing the production of durable and versatile rubber goods.

During the late 1800s, rubber production underwent a geographical shift from South America to Southeast Asia, particularly in regions like Malaya (now Malaysia) and Indonesia. These areas provided ideal conditions for cultivating rubber trees, establishing themselves as major hubs for natural rubber production.

The advent of synthetic rubber in the early 20th century, marked by DuPont's development of neoprene in 1931, further revolutionized the industry. Synthetic rubber, derived from petroleum, boasted properties akin to natural rubber but offered superior resistance to heat, chemicals, and aging. This innovation proved pivotal, especially during World War II, when rubber became indispensable for military equipment manufacturing.

RUBBER INDUSTRY OVERVIEW:

The rubber industry stands as a global juggernaut encompassing the production, processing, and marketing of both natural and synthetic rubber products. Natural rubber, sourced from rubber trees predominantly found in Southeast Asia, Africa, and South America, coexists with synthetic rubber derived from petroleum-based chemicals.

Rubber's versatility finds expression across diverse sectors, ranging from automotive and consumer goods to medical and construction industries. In automotive manufacturing, rubber finds application in tires, belts, hoses, and gaskets, while in consumer goods, it forms the foundation of footwear, gloves, balloons, and toys. Medical equipment, too, relies heavily on rubber for products like gloves and catheters.

Despite its pervasive utility, the rubber industry faces ongoing challenges, particularly concerning sustainability and environmental impact. While natural rubber production can have adverse ecological consequences if mismanaged, initiatives are underway to explore sustainable alternatives, such as biotechnology-driven synthetic rubber derived from renewable sources.

KEY PLAYERS AND CHALLENGES:

Tire manufacturing giants like Bridgestone, Michelin, and Indian companies dominate the landscape, alongside chemical conglomerates like Dow Chemical, BASF, and LG Chem, which spearhead synthetic rubber production. Supporting this ecosystem are a plethora of suppliers, distributors, and service providers, including raw material suppliers and logistics facilitators.

In conclusion, the rubber industry's trajectory mirrors a saga of innovation, from ancient civilizations to modern-day technological advancements. As it navigates challenges and embraces sustainability, the industry remains a linchpin of global commerce, driving progress across myriad sectors with its versatile and indispensable products.

APPLICATION OF SULPHUR & SULPHUR COMPOUNDS IN RUBBER INDUSTRIES

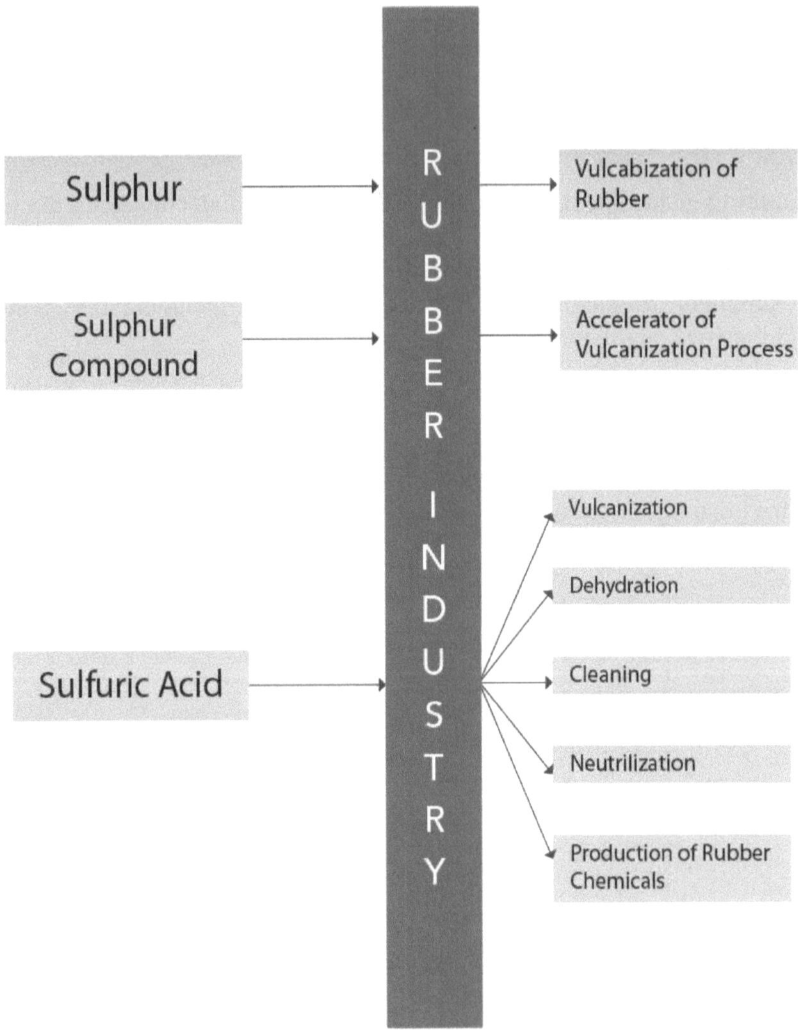

1. Sulphur in Rubber Vulcanization

Vulcanization, a chemical process pioneered by Charles Goodyear in the 1830s, enhances the strength, durability, and elasticity of natural rubber.

Here's an in-depth look at the role of sulphur in this crucial process:

a. **Vulcanization Process :** During vulcanization, sulphur molecules react with rubber polymer chains, creating cross-links that fortify the rubber, rendering it more resilient against heat, abrasion, and chemicals.
b. **Process Parameters :** Vulcanization involves heating the rubber with sulphur and various additives, including accelerators and activators, at specific temperatures and pressures tailored to the rubber compound and desired product properties.
c. **Chemical Additives**
   Several chemical products enhance the vulcanization process:
   - Activators: Zinc oxide and stearic acid boost sulphur reactivity, expediting cross-link formation.
   - Antioxidants: Phenols and amines shield rubber from degradation due to heat, oxygen, and ozone exposure.
   - Fillers: Carbon black and silica bolster rubber strength, stiffness, and wear resistance.
d. **Vulcanized rubber finds extensive use across industries:**
   - Tires: Vital for vehicles, vulcanized rubber tires offer robustness, durability, and heat resistance.
   - Industrial Products: Hoses, belts, seals, gaskets, and conveyor belts benefit from vulcanized rubber's performance and reliability.
   - Footwear: Boots, sneakers, and sandals owe their flexibility, durability, and wear resistance to vulcanized rubber.
   - Sporting Goods: Balls, grips, and protective gear made from vulcanized rubber boast strength and resilience.
   - Consumer Products: Gloves, balloons, and toys crafted from vulcanized rubber exhibit elasticity, durability, and damage resistance.

2. Sulphur Compounds as Vulcanization Accelerators

In rubber vulcanization, sulphur and sulphur-containing compounds like zinc dimethyldithiocarbamate (ZDMC), tetramethylthiuram disulfide (TMTD), and dibenzothiazyl disulfide (MBTS) play a crucial role. These compounds foster cross-linking among rubber polymer chains, bolstering the rubber's stiffness and tensile strength.

Accelerating Vulcanization

Sulphur compounds such as ZDMC, TMTD, and MBTS serve as accelerators in vulcanization, expediting the vulcanization rate and enhancing cross-linking uniformity. Additionally, they contribute to improving the rubber's heat resistance and aging properties.

Commonly Used Accelerators and Their Applications

a. **Mercaptobenzothiazole (MBT):** Widely applied in tire, hose, belt, and other rubber product manufacturing, MBT accelerates vulcanization, enhancing rubber's physical properties like tensile strength and abrasion resistance.
b. **Thiurams:** Utilized in footwear, belt, and hose production, thiurams excel at low temperatures, hastening vulcanization when sulphur is the primary curing agent.
c. **Dithiocarbamates:** Employed as secondary accelerators to augment vulcanization, dithiocarbamates find application in tire, conveyor belt, and other rubber product manufacturing.
d. **Thioureas:** Ideal for rubber products requiring high heat resistance, thioureas are used in seals, gaskets, and industrial rubber item production.
e. **Guanidines:** Effective at low temperatures, guanidines accelerate the curing speed of rubber compounds, improving production efficiency in glove, footwear, and sporting goods manufacturing.

The integration of accelerators in the vulcanization process significantly enhances rubber production efficiency, leading to reduced curing times, heightened production rates, and improved physical properties in the final product. However, selecting the appropriate accelerator type and dosage for each rubber compound is imperative to ensure optimal performance and safety.

3. Acid-Catalyzed Vulcanization in Rubber Production

While sulphuric acid isn't commonly employed as the primary curing agent in vulcanization, it finds application as an acid catalyst in a specialized process known as acid-catalyzed vulcanization. This method, although less common, serves specific purposes in the production of certain rubber products, particularly those requiring unique properties such as acid resistance.

Here's an overview of some components utilized in acid-catalyzed vulcanization:

a. **Sulphuric Acid:** Sulphuric acid acts as the primary acid catalyst in acid-catalyzed vulcanization. Its role is critical in facilitating the vulcanization process for specialized rubber items like acid-resistant seals and gaskets.
b. **Zinc Oxide:** In conjunction with sulphuric acid, zinc oxide serves as a promoter in acid-catalyzed vulcanization. It enhances the speed and efficiency of vulcanization, contributing to the production of high-quality rubber products.
c. **Stearic Acid:** Stearic acid is employed as a processing aid in acid-catalyzed vulcanization. Its function includes improving the flow characteristics of the rubber compound during processing and enhancing the dispersion of other additives, thereby optimizing the overall production process.

Acid-catalyzed vulcanization, though relatively slower and less efficient compared to sulphur-cured vulcanization with accelerators, offers distinct advantages for specialized applications requiring specific properties. However, it's imperative to handle sulphuric acid cautiously due to its corrosive and potentially hazardous nature.

4. Sulphuric Acid in Dehydration Processes

Sulphuric acid finds multifaceted applications in facilitating dehydration processes in the rubber industry, serving several crucial functions:

a. **Latex Concentration:** Natural rubber latex, sourced from rubber trees, typically contains a substantial water content. Sulphuric acid is instrumental in dehydrating latex, thereby concentrating it. Addition of sulphuric acid induces dehydration, effectively removing water molecules and resulting in a higher concentration of rubber solids. This process is indispensable for manufacturing various rubber products like gloves, tires, and rubber bands.
b. **Rubber Vulcanization:** Sulphuric acid contributes to the vulcanization process of rubber, wherein polymer chains are cross-linked to enhance strength, elasticity, and durability. Acting as a dehydrating agent during vulcanization, sulphuric acid aids in fostering sulphur bridges crucial for cross-linking. Dehydration of the rubber compound promotes the formation of these sulphur bridges, vital for reinforcing the rubber matrix.
c. **Rubber Chemical Modification:** Utilizing sulphuric acid, rubber undergoes chemical modification under controlled conditions. This treatment introduces diverse chemical alterations to rubber molecules. Dehydration reactions may ensue as part of these modifications, resulting in adjustments to the rubber's properties such as strength, flexibility, and resistance to heat or chemicals.
d. **Rubber Recycling:** In rubber recycling processes, sulphuric acid plays a pivotal role in breaking down rubber compounds and segregating them from

other materials. Through dehydration reactions, sulphuric acid facilitates the depolymerization of rubber, breaking it into smaller molecular components. These components can then be repurposed or further processed into new rubber products, contributing to sustainable resource utilization.

In conclusion, Sulphuric acid's efficacy in facilitating dehydration reactions renders it indispensable across various facets of the rubber industry, spanning from latex concentration to rubber vulcanization, chemical modification, and recycling. However, prudent handling and strict adherence to safety protocols are imperative due to its corrosive nature and associated hazards.

5. Utilizing Sulphuric Acid for Cleaning in the Rubber Industry

Sulphuric acid is a potent cleaning agent widely employed in the rubber industry owing to its robust acidic properties and efficacy in dissolving organic materials. Here's how sulphuric acid is harnessed for various cleaning tasks within rubber manufacturing:

a. **Equipment Cleaning:** Rubber manufacturing equipment, including mixing tanks, molds, and extrusion machinery, can accumulate resilient rubber residues, oils, and contaminants during production. Sulphuric acid serves as a powerful cleaning agent to eradicate these stubborn deposits. Diluted sulphuric acid solutions are often circulated through equipment or used in soak tanks to dissolve and dislodge the accumulated residues, facilitating their subsequent rinsing and removal.

b. **Decontamination:** Rubber processing entails the use of diverse chemicals, lubricants, and mold release agents, leaving behind residues on surfaces. Sulphuric acid effectively decontaminates surfaces by breaking down these residues and neutralizing any alkaline substances. This ensures thorough cleaning, eliminating contaminants before advancing to subsequent processing stages.

c. **Surface Preparation:** Preparation of rubber surfaces before treatments such as painting, coating, or bonding is crucial for optimal adhesion and performance. Sulphuric acid aids in surface preparation by eliminating oils, greases, and contaminants, fostering a clean and chemically activated surface conducive to enhanced adhesion of coatings or adhesives.

d. **Mold Cleaning:** Rubber molds used in manufacturing are prone to accumulating residues over time, impacting product quality and production efficiency. Sulphuric acid is instrumental in mold cleaning, dissolving rubber residues, release agents, and other contaminants adhering to mold surfaces. Proper mold cleaning ensures maintained mold quality, prolonged lifespan, and consistent product quality.

e. **Maintenance Cleaning:** Regular maintenance cleaning is imperative to prevent residue buildup and maintain hygienic conditions in rubber manufacturing facilities. Sulphuric acid-based cleaning solutions are deployed for periodic cleaning of floors, walls, and surfaces, effectively eliminating dirt, grime, and tenacious deposits, thereby upholding a clean and sanitized working environment.

In summary, Sulphuric acid's prowess in cleaning applications is indispensable across various facets of rubber manufacturing, ensuring equipment longevity, product quality, and workplace hygiene. However, prudent handling and adherence to safety protocols are paramount due to its corrosive nature.

6. Utilizing Sulphuric Acid for Neutralization

Sulphuric acid (H_2SO_4) finds application in neutralization processes within the rubber industry, serving various essential functions:

a. **Neutralizing Alkaline Substances:** During rubber processing, alkaline substances like sodium hydroxide (NaOH) or calcium hydroxide ($Ca(OH)_2$) may be utilized for curing or adjusting pH levels. Following these processes, neutralization of any residual alkalinity is imperative to meet desired rubber product specifications.

b. **pH Adjustment:** Rubber compounds often require precise pH levels to optimize processing conditions and achieve desired product properties. Sulphuric acid, being a potent acid, is adept at pH adjustment, effectively lowering pH levels when necessary to meet specific manufacturing requirements.

c. **Controlling Reactions:** In certain rubber manufacturing processes, maintaining optimal pH levels is crucial for controlling reactions and ensuring product quality. Sulphuric acid can be employed to regulate pH, thereby facilitating controlled reaction kinetics and desired product characteristics.

d. **Cleaning and Decontamination:** Additionally, sulphuric acid plays a role in cleaning rubber equipment and surfaces by neutralizing alkaline residues or contaminants, as highlighted in the previous response. Its ability to neutralize alkaline substances aids in thorough cleaning and decontamination, ensuring equipment and surfaces are free from unwanted residues.

In summary, Sulphuric acid's role in neutralization processes is pivotal within the rubber industry, encompassing tasks such as neutralizing alkaline substances, pH adjustment, reaction control, and cleaning. However, careful handling and adherence to safety protocols are essential due to its corrosive nature

7. Application of Sulphuric Acid in Production of Rubber Chemicals

Sulfuric acid (H_2SO_4) plays a crucial role in the production of rubber compounds, particularly in the vulcanization process, which is essential for improving the mechanical properties and durability of rubber products.

Here's how sulfuric acid is utilized in the production of rubber compounds:

a. **Vulcanization:** Vulcanization is a chemical process that involves the cross-linking of rubber molecules to improve their mechanical properties, such as elasticity, strength, and resistance to heat, wear, and aging. Sulfuric acid is not directly involved in vulcanization; instead, it is used in the preparation of sulfur-containing compounds, such as sulfuric acid esters or sulfates, which serve as vulcanizing agents during the vulcanization process.

b. **Sulfuric Acid Esters:** Sulfuric acid esters, also known as sulfur donors, are compounds that release sulfur during the vulcanization process. These compounds, such as sulfuric acid esters of phenols or alcohols, are added to rubber formulations along with sulfur to facilitate the cross-linking of rubber molecules. Sulfuric acid esters decompose at elevated temperatures, releasing sulfur atoms that form cross-links with the polymer chains of rubber.

c. **Accelerators and Activators:** Sulfuric acid may indirectly influence the vulcanization process by being used in the synthesis of accelerators and activators, which enhance the efficiency of sulfur vulcanization. Accelerators are chemicals that speed up the vulcanization process, while activators enhance the activity of accelerators. Sulfur-containing accelerators and activators, such as mercaptobenzothiazole derivatives or thiuram compounds, are commonly used in rubber compounding to improve the speed and efficiency of vulcanization.

d. **Neutralization:** In some rubber compounding processes, sulfuric acid may be used for neutralization purposes. Rubber formulations often contain acidic or basic ingredients that need to be adjusted to a specific pH range for optimal processing and performance. Sulfuric acid may be added to adjust the pH of rubber compounds or to neutralize alkaline substances present in the formulation.

e. **Reaction Medium:** Sulfuric acid can act as a reaction medium or catalyst in certain chemical reactions involved in rubber compounding. For example, it may be used in the synthesis of sulfur-containing additives or modifiers that enhance the properties of rubber compounds, such as antioxidants, antiozonants, or processing aids.

Overall, while sulfuric acid is not directly involved in the vulcanization process itself, it is used in various ways to facilitate the preparation of sulfur-containing compounds, accelerators, and activators, which are essential for the vulcanization of rubber compounds and the production of high-quality rubber products.

Summary:

Sulfur compounds play crucial roles in the rubber industry, particularly in the process known as vulcanization, which is essential for improving the mechanical properties and durability of rubber products. Here are the main applications of sulfur compounds in the rubber industry:

a. **Vulcanization:** Vulcanization is a chemical process that involves the cross-linking of rubber molecules to improve their mechanical properties, such as elasticity, strength, and resistance to heat, wear, and aging. Sulfur compounds, primarily elemental sulfur (S_8), are the most common vulcanizing agents used in rubber vulcanization. During vulcanization, sulfur forms cross-links with the polymer chains of rubber, creating a three-dimensional network structure that enhances the rubber's properties.

b. **Accelerators:** Accelerators are chemicals that speed up the vulcanization process, reducing the time and temperature required for vulcanization to occur. Various sulfur-containing compounds, such as mercaptobenzothiazole derivatives (e.g., MBT, MBTS), thiurams (e.g., TMTD, TMTM), and dithiocarbamates (e.g., ZDEC, ZDBC), are commonly used as accelerators in conjunction with sulfur. These accelerators enhance the efficiency of sulfur vulcanization and allow for better control over the vulcanization process.

c. **Activators:** Activators are substances that enhance the activity of accelerators in the vulcanization process. Zinc oxide (ZnO) and stearic acid are commonly used as activators in sulfur vulcanization. Zinc oxide acts as a co-agent in the formation of cross-links between rubber chains, while stearic acid serves as a lubricant and helps disperse other ingredients evenly in the rubber compound.

d. **Antioxidants and Antiozonants:** Sulfur-containing antioxidants and antiozonants are added to rubber formulations to protect rubber products from degradation caused by oxygen and ozone exposure, which can lead to cracking, aging, and loss of mechanical properties. These compounds scavenge free radicals and reactive oxygen species, preventing them from reacting with rubber molecules. Examples include sulfenamide-based antioxidants and phenolic antioxidants with sulfur functionalities.

e. **Processing Aids:** Certain sulfur-containing compounds, such as sulfur-containing fatty acids or their derivatives, can act as processing aids in rubber compounding. These compounds improve the processability of rubber compounds during mixing, milling, and extrusion, leading to better dispersion of ingredients and uniformity in the final rubber product.

Overall, sulfur compounds are indispensable in the rubber industry for vulcanization, accelerating the curing process, enhancing rubber properties, and protecting rubber products from degradation, contributing to the production of high-quality and durable rubber goods for various applications.

REFERENCES:

1. Dunstan, Wyndham Rowland (1911). "Rubber" . In Chisholm, Hugh (ed.). Encyclopædia Britannica. Vol. 23 (11th ed.). Cambridge University Press. p. 795.
2. Ducháček, Vratislav; Kuta, Antonín (October 1986). "Long-time sulfenamide-accelerated sulphur vulcanization of natural rubber/chlorobutyl rubber compounds". Journal of Applied Polymer Science. 32 (5): 4849–4855.
3. Greenwood, Norman N.; Earnshaw, Alan (1997). Chemistry of the Elements (2nd ed.), Butterworth-Heinemann. ISBN 978-0-08-037941-8.
4. Merck Index, 11th Edition, 9625
5. Austin, George T. (1984). Shreve's Chemical process industries (5th ed.). New York, NY: McGraw-Hill. p. 357

Chapter 10: Tracing the Path of Agriculture: An Evolutionary Journey

ANCIENT ROOTS:

The roots of agriculture intertwine with the dawn of human civilization, when our ancestors transitioned from nomadic lifestyles to settled communities, nurturing crops for sustenance. Around 12,000 years ago, the Fertile Crescent witnessed the birth of agriculture as people began cultivating plants and domesticating animals, heralding a new era in human history.

Advancements Through Ages:
Ancient civilizations such as the Egyptians, Greeks, and Romans pioneered sophisticated agricultural techniques, embracing innovations like irrigation, crop rotation, and animal husbandry. These advancements empowered societies to flourish, fostering surplus production and supporting burgeoning populations.

Medieval Renaissance:
During the Middle Ages, Europe witnessed the zenith of agricultural dominance, with the advent of transformative practices like the three-field system and heavy plows. Introduction of novel crops from the Americas, like potatoes and maize, fueled productivity, propelling medieval economies forward.

Industrial Revolution and Beyond:
The dawn of the Industrial Revolution in the 18th and 19th centuries sparked a mechanization revolution in agriculture. Innovations like the seed drill, combine harvester, and tractor ushered in unprecedented efficiency, albeit accompanied by the consolidation of land ownership and displacement of smaller farmers.

Modern Marvels:
Today, agriculture stands as a testament to technological prowess and global connectivity. Biotechnology, precision farming, and genetic advancements have revolutionized yields and quality. Yet, modern agriculture grapples with existential challenges like climate change, soil degradation, and biodiversity loss, underscoring the need for sustainable practices.

Unveiling the Agriculture Industry's Tapestry

Diverse Sectors:
The agriculture industry encompasses a broad spectrum of activities, spanning from crop and livestock production to food processing and distribution. Cultivation of diverse crops and the rearing of livestock are pivotal components, underpinning global food security.

Value-Added Processes:
Beyond primary production, the industry encompasses value-added processes like food processing, packaging, and distribution. From converting raw agricultural goods into packaged foods to delivering them to consumers' doorsteps, these activities ensure food reaches every corner of the globe.

Supporting Ecosystem:
A thriving ecosystem of research, equipment manufacturing, and marketing amplifies the industry's impact. Research and development drive innovation, while equipment manufacturing furnishes farmers with cutting-edge tools. Marketing and sales strategies amplify product reach, connecting producers with consumers seamlessly.

Navigating the Landscape of Agriculture: A Comprehensive Overview
The agriculture industry encompasses a diverse array of activities aimed at cultivating crops and raising livestock to meet humanity's sustenance needs.

Here's a breakdown of its fundamental components:

a. **Crop Cultivation:** Crop cultivation forms the backbone of agriculture, spanning a vast array of plants ranging from staple grains and fruits to specialty crops like flowers, cotton, jute, and herbs.
b. **Livestock Rearing:** Livestock rearing involves the responsible management of animals for the production of meat, milk, eggs, wool, leather, and other essential products, ensuring sustainable practices and animal welfare.
c. **Farming Techniques:** The adoption of farming techniques is pivotal, encompassing practices such as irrigation, fertilization, integrated pest

management, and animal husbandry to optimize yields while minimizing environmental impact.

d. **Agribusiness Support:** Agribusiness entities play a crucial role in bolstering the agriculture sector, encompassing seed and fertilizer suppliers, farm machinery manufacturers, and food processing and distribution firms, fostering a robust ecosystem.

e. **Government Intervention:** Government policies wield significant influence, ranging from agricultural subsidies and environmental regulations to trade policies, shaping the industry's trajectory and ensuring its socio-economic sustainability.

f. **Embracing Sustainability:** With the global population on the rise, the imperative for sustainable agricultural practices intensifies, driving initiatives aimed at mitigating environmental degradation while ensuring food security and resource conservation.

g. **Technological Advancements:** Technological innovation serves as a catalyst for agricultural progress, spanning from precision agriculture techniques and drone technology for crop monitoring to genetic engineering and biotechnology for crop enhancement and disease resistance.

In essence, the agriculture industry stands at the intersection of tradition and innovation, tasked with meeting the world's growing demand for food and resources while safeguarding the planet's ecological balance. Through collaborative efforts and adaptive strategies, it charts a course toward a sustainable and prosperous future.

APPLICATION OF SULPHUR & SULPHUR COMPOUNDS IN AGRICULTURE INDUSTRY

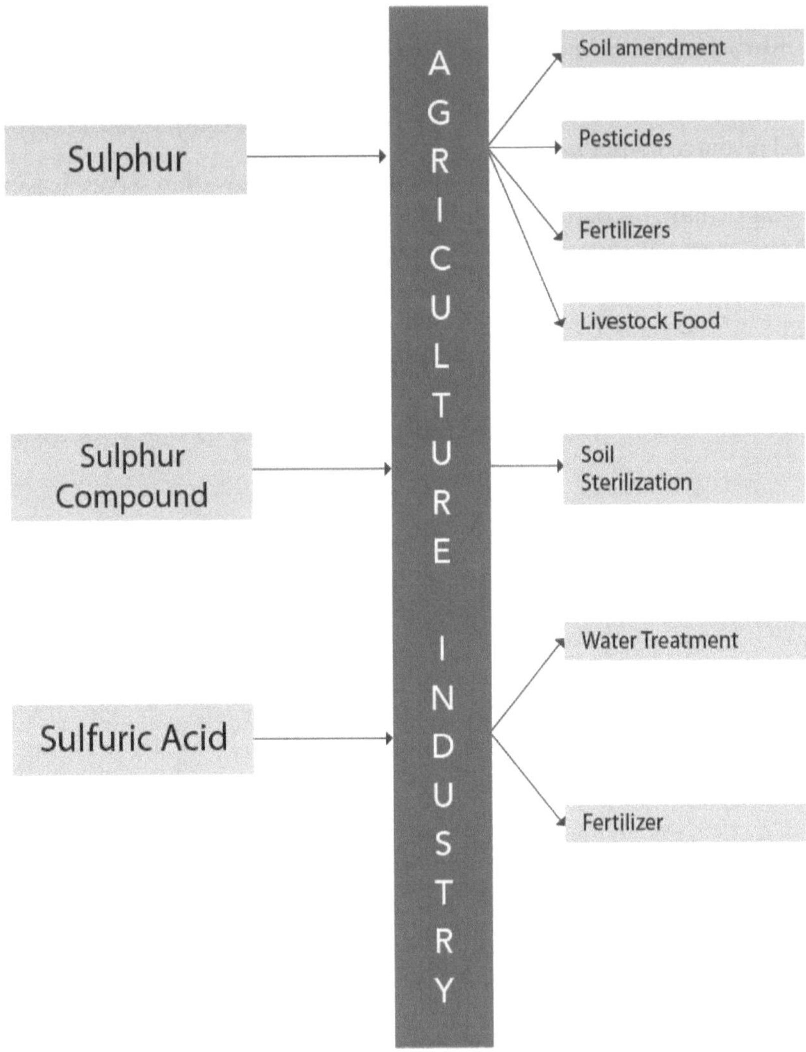

1. Soil Amendment by Sulphur

Sulphur plays a pivotal role in soil amendment, primarily by regulating soil pH. By lowering the pH of alkaline soils, sulphur creates a more acidic environment, facilitating the availability of vital nutrients like iron, zinc, and manganese to plants. Additionally, sulphur aids in synthesizing specific plant compounds that bolster crops against pest infestations and diseases.

a. **Fertilizer:** As an essential nutrient for plant growth, sulphur serves as a fertilizer when applied to soil, thereby enhancing crop yields. Its application contributes to augmenting the protein content in crops such as wheat and legumes, thus improving nutritional quality and overall crop productivity.
b. **Insecticide:** Sulphur stands as a potent insecticide, effectively controlling a wide array of pests in crops, including mites, psyllids, and thrips. By mitigating pest infestations, sulphur helps safeguard crop health and preserves yield potential.
c. **Fungicide:** Utilized as a fungicide, sulphur offers effective protection against various fungal diseases prevalent in crops, including powdery mildew, rust, and scab. Its fungicidal properties contribute to maintaining crop quality and yield, particularly in humid environments conducive to fungal growth.
d. **Livestock Feed Supplement:** In addition to its role in crop cultivation, sulphur serves as a valuable supplement in livestock feed, promoting animal health and performance. By incorporating sulphur into feed formulations, farmers can enhance overall livestock management practiccs and optimizc production outcomcs.

While sulphur offers multifaceted benefits as a soil amendment, its effectiveness hinges on several factors, including soil type, climate, and the specific crop being cultivated. To determine the optimal application rate and method, consulting with local agricultural experts or soil testing laboratories is advisable. By leveraging sulphur judiciously, farmers can enhance soil fertility, promote crop health, and foster sustainable agricultural practices.

2. Sulphur-Based Pesticides

Ground into a fine powder, elemental sulphur serves as an effective fungicide, combating fungal diseases like powdery mildew, rust, and scab in crops. Applied as a dust or spray, it provides reliable protection against harmful pathogens.

a. **Wettable Sulphur:** Formulated to mix seamlessly with water, wettable sulphur creates a sprayable suspension ideal for controlling pests such as mites, psyllids, and thrips in crops. Its versatility and efficacy make it a preferred choice for integrated pest management strategies.
b. **Lime-Sulphur:** A blend of sulphur and lime, lime-sulphur acts as both a fungicide and insecticide, targeting a wide range of fungal diseases and pests like scale insects and mites. Its dual-purpose nature enhances crop protection while minimizing the need for multiple pesticide applications.
c. **Sulphur Dioxide:** Deployed as a gas fumigant, sulphur dioxide effectively controls pests in stored grains and commodities, ensuring product quality and longevity. It also finds application in vineyards and other crops, offering targeted pest management solutions.

3. Harnessing Sulphur for Fertilizer Production

As a cornerstone of fertilizer manufacturing, sulphuric acid serves as a catalyst for producing various fertilizers like ammonium sulphate, superphosphate, and triple superphosphate.

These formulations enrich soil fertility by supplying sulphur alongside essential nutrients crucial for robust crop growth.

a. **Ammonium Sulphate:** Harnessing sulphur in the form of sulphate ions, ammonium sulphate fertilizer addresses nitrogen and sulphur deficiencies in crops, particularly in soils lacking both nutrients. Its dual-purpose composition supports balanced nutrition, fostering healthy plant development.
b. **Elemental Sulphur:** Offering a slow-release mechanism, elemental sulphur serves as a sustainable fertilizer, gradually providing sulphur to crops over an extended duration. Particularly valuable in alkaline soils, it not only enriches sulphur levels but also aids in pH regulation, enhancing nutrient uptake by plants.

c. **Gypsum:** Containing calcium sulphate, gypsum fertilizer improves soil structure and drainage, while concurrently supplying sulphur to crops. Its multifaceted benefits extend beyond sulphur provision, making it a versatile solution for optimizing soil health and crop productivity.
d. **Sulphur-Coated Urea:** Blending nitrogen-rich urea with sulphur, this innovative fertilizer formulation ensures a gradual nutrient release, offering a balanced supply of nitrogen and sulphur to crops. Its slow-release nature minimizes nutrient leaching and enhances nutrient utilization efficiency.
e. **Tailoring Sulphur Fertilization:** Effective utilization of sulphur-based fertilizers hinges on understanding soil characteristics, climatic conditions, and crop requirements. Collaborating with agricultural experts or conducting soil tests empowers farmers to tailor sulphur fertilization strategies, optimizing crop performance and yield potential. By leveraging sulphur's nutritional benefits judiciously, farmers can cultivate resilient, high-yielding crops while promoting sustainable agricultural practices.

4. Enhancing Livestock Feed with Sulphur Supplements

Sulphur plays a vital role in promoting the health and performance of livestock, serving as an essential nutrient in their diet.

Here are various sulphur supplements commonly incorporated into livestock feed:

a. **Organic Sulphur Compounds:** Derived from natural sources like plants, organic sulphur compounds are pivotal for animals' physiological functions, including protein synthesis and immune support. Supplementing livestock feed with these compounds ensures optimal health and vitality.
b. **Inorganic Sulphur Compounds:** Inorganic sulphur, such as sulphates and sulphites, serves as a crucial nutrient for rumen microbes in ruminant animals. These microbes aid in the digestion process, making inorganic sulphur supplementation essential for efficient feed utilization.

c. **Sulphur Amino Acids:** Containing sulphur, amino acids like methionine and cysteine are indispensable for protein synthesis, growth, and development in livestock. Adding sulphur amino acids to feed formulations enhances overall protein quality and supports healthy growth in animals.

d. **Sulphur Dioxide:** Occasionally utilized as a preservative, sulphur dioxide helps prevent spoilage and extends the shelf life of livestock feed. Its antimicrobial properties safeguard feed quality, ensuring that animals receive nutritious and safe feed.

5. Utilizing Sulphur Compounds for Soil Sterilization

Sulphur compounds offer effective solutions for soil sterilization, aiding in the management of soil-borne pathogens and pests detrimental to crop growth and yield. Here's a glimpse into the soil sterilization products and their applications:

a. **Soil Fumigants**

- Key Compounds: Methyl bromide, methyl isothiocyanate, and others.
- Functionality: These sulphur compounds serve as potent soil fumigants, penetrating the soil as gases to combat diverse soil-borne pests and diseases. By effectively targeting pathogens and pests, they contribute to soil sterilization and crop protection.
- Considerations: While highly effective, soil fumigants pose environmental risks due to their toxicity. Consequently, their usage is regulated in numerous countries to mitigate adverse impacts on the environment.

b. **Biopesticides**

- Key Compounds: Dithiocarbamates and others.
- Functionality: Certain sulphur compounds function as biopesticides, offering a natural approach to controlling soil-borne pests and diseases. By disrupting crucial cellular processes in pests and pathogens, they facilitate their elimination, promoting soil sterilization and crop health.
- Considerations: Biopesticides present a more environmentally friendly alternative to conventional soil fumigants. However,

prudent use and adherence to recommended application guidelines are essential to maximize efficacy while minimizing ecological impacts.

c. **Integrated Approach to Soil Sterilization**
  - Holistic Strategies: Soil sterilization should be part of a comprehensive pest and disease management plan, complementing practices like crop rotation and the cultivation of disease-resistant crop varieties.
  - Ecosystem Preservation: Overreliance on soil sterilization products can foster the development of resistant pests and pathogens, adversely affecting soil health and ecosystem balance. Therefore, judicious usage is paramount.
  - Expert Guidance: Consulting with agricultural specialists is advisable to tailor soil sterilization strategies to specific crops and soil conditions. Their insights can ensure optimal efficacy while safeguarding environmental sustainability.

By integrating sulphur compounds into soil sterilization practices responsibly, farmers can mitigate pest and disease pressures, fostering healthy soil ecosystems and robust crop yields.

6. Enhancing Water Quality with Sulphuric Acid

Sulphuric acid serves various roles in water treatment processes, contributing to the purification and safety of water resources.

Here's how it's utilized:

a. **pH Adjustment:** Sulphuric acid effectively adjusts the pH of water, making it more acidic. This adjustment inhibits the proliferation of bacteria and other microorganisms, promoting safer water conditions.

b. **Coagulation:** Utilized as a coagulating agent, sulphuric acid aids in the removal of suspended particles and impurities from water. By inducing the particles to clump together, it facilitates their filtration, enhancing water clarity.

c. **Disinfection:** While less common than chlorine, sulphuric acid can serve as a disinfectant in water treatment, effectively eliminating harmful bacteria and microorganisms, thus safeguarding public health.

d. **Scale and Corrosion Control:** Sulphuric acid's application helps mitigate the formation of scale and corrosion in water systems by reacting with minerals prone to causing such issues. This preventive measure preserves the integrity of pipes and equipment, prolonging their lifespan.

e. **Demineralization of Water**
    - Sulphuric Acid Dosing: In specific water treatment processes, sulphuric acid is employed for demineralization. Through a method called sulphuric acid dosing, the acid is added to water to lower its pH, facilitating the removal of minerals and impurities. This reaction causes the minerals to precipitate or form insoluble compounds, making them easier to separate from the water. Consequently, the mineral content and water hardness are reduced, yielding purified water.

It's crucial to handle sulphuric acid with care due to its highly corrosive nature. Proper protective equipment and handling protocols must be followed to mitigate risks. Additionally, neutralizing any residual acid is imperative before releasing treated water into the environment.

While sulphuric acid is effective in water treatment, it's essential to explore other methods and chemicals based on specific water composition and treatment requirements. Consulting water treatment professionals or experts ensures the adoption of the most suitable approach for each scenario, promoting efficient and safe water treatment practices.

7. Utilizing Sulphuric Acid in Fertilizer Production

Sulfuric acid (H_2SO_4) is extensively used in the production of fertilizers, particularly those containing sulfur as an essential nutrient for plant growth.

Here's how sulfuric acid is utilized in fertilizer production:

a. **Phosphoric Acid Production:** Sulfuric acid is commonly used in the production of phosphoric acid (H_3PO_4), which is a key component of phosphate fertilizers. In the wet process, phosphate rock is treated with sulfuric acid to produce phosphoric acid and calcium sulfate (gypsum) as a byproduct. The phosphoric acid produced is then used as a raw material for various phosphate fertilizers, such as diammonium phosphate (DAP) and triple superphosphate (TSP).

b. **Ammonium Sulfate Production:** Sulfuric acid is used in the production of ammonium sulfate ($(NH_4)2SO_4$), which is a nitrogen and sulfur-containing fertilizer. Ammonium sulfate is produced by reacting ammonia gas with sulfuric acid to form ammonium sulfate crystals. The resulting product is a soluble fertilizer that provides both nitrogen and sulfur nutrients to plants.
c. **Sulfur-Containing Fertilizers:** Sulfuric acid is utilized in the production of various sulfur-containing fertilizers, such as single superphosphate (SSP) and elemental sulfur-based fertilizers. In the production of single superphosphate, sulfuric acid is used to react with phosphate rock to produce a mixture of monocalcium phosphate and gypsum. Elemental sulfur-based fertilizers are produced by the oxidation of elemental sulfur using sulfuric acid to form sulfuric acid esters or sulfates, which are then incorporated into fertilizer formulations.
d. **Acidulation of Phosphate Rock:** Sulfuric acid is also used in the acidulation of phosphate rock to improve its solubility and availability of phosphorus for plant uptake. In this process, finely ground phosphate rock is mixed with sulfuric acid to produce phosphoric acid and calcium sulfate. The resulting phosphoric acid is then used in fertilizer formulations.
e. **pH Adjustment and Nutrient Uptake:** Sulfuric acid is added to fertilizer formulations to adjust the pH of the final product and to enhance the solubility and availability of other nutrients, such as phosphorus, potassium, and micronutrients. Proper pH levels are essential for maximizing nutrient uptake by plants and ensuring optimal growth and yield.

Overall, sulfuric acid plays a vital role in the production of various fertilizers, contributing to the supply of essential nutrients to plants and improving agricultural productivity. Its versatility and effectiveness make it a cornerstone in fertilizer manufacturing processes worldwide

Summary

Sulfur and sulfur compounds play several important roles in the agriculture industry, contributing to soil fertility, plant nutrition, and pest management. Here are some of the key applications:

a. **Soil Amendment**: Sulfur is an essential nutrient for plant growth and development. It is a component of various amino acids, vitamins, and enzymes involved in plant metabolism. Sulfur compounds, such as elemental sulfur and sulfur-containing fertilizers (e.g., ammonium sulfate, potassium sulfate), are used to replenish sulfur levels in soils deficient in this nutrient. Sulfur amendments help improve soil fertility, enhance crop yields, and optimize plant nutrition.
b. **Acidifying Agent:** Sulfur compounds, particularly elemental sulfur, are used as acidifying agents to lower soil pH in alkaline soils. Acidification of soils helps to correct pH imbalances and create optimal growing conditions for acid-loving crops such as blueberries, rhododendrons, and azaleas. Sulfur applications can also enhance the availability of certain nutrients, such as phosphorus and micronutrients, by reducing their fixation in alkaline soils.
c. **Pesticides and Fungicides:** Sulfur-based pesticides and fungicides are widely used in agriculture for the control of various pests, diseases, and fungal pathogens. Elemental sulfur and sulfur-containing compounds, such as sulfur dioxide and sulfur-based fungicides (e.g., sulfur dust, lime sulfur), are effective against a broad spectrum of pests and diseases, including powdery mildew, mites, rusts, and fungal infections. Sulfur compounds act by disrupting the cellular metabolism and growth of pathogens, preventing their proliferation and spread.
d. **Nutrient Synergism:** Sulfur interacts synergistically with other essential nutrients, such as nitrogen and phosphorus, in plant metabolism and nutrient uptake. Sulfur fertilization can enhance the efficiency of nitrogen fertilizers by promoting nitrogen assimilation and reducing nitrogen losses through leaching and volatilization. Additionally, sulfur aids in the utilization of phosphorus by plants, leading to improved phosphorus uptake and utilization efficiency.
e. **Biological Symbiosis:** Sulfur plays a crucial role in biological symbiotic relationships between plants and soil microorganisms, such as mycorrhizal fungi and rhizobia. Sulfur compounds are involved in

the formation of symbiotic associations between plant roots and mycorrhizal fungi, which facilitate nutrient uptake, especially phosphorus and micronutrients, and enhance plant growth and resilience to environmental stresses. Rhizobia bacteria, which form symbiotic relationships with leguminous plants, require sulfur for nitrogen fixation and nodulation.

f. **Seed Treatment:** Sulfur compounds are used in seed treatments to protect seeds from soil-borne diseases and pests and promote germination and early seedling growth. Sulfur-based seed treatments, such as sulfur dust or sulfur-coated seeds, provide protection against fungal pathogens and damping-off diseases, ensuring healthy seedling establishment and crop vigor.

Overall, sulfur and sulfur compounds play multifaceted roles in agriculture, contributing to soil health, plant nutrition, pest and disease management, and crop productivity. Integrating sulfur-based strategies into agricultural practices helps optimize nutrient management, enhance crop resilience, and promote sustainable agriculture.

REFERENCES:

1. Stevenson, G. C. (1971). "Plant Agriculture Selected and Introduced by Janick Jules and Others San Francisco: Freeman (1970), pp. 246". Experimental Agriculture. Cambridge University Press (CUP). 7 (4): 363.
2. Pimentel, D.; Berger, D.; Filberto, D.; Newton, M. (2004). "Water Resources: Agricultural and Environmental Issues". BioScience. 54 (10): 909–918.
3. Hobbs, Peter R; Sayre, Ken; Gupta, Raj (12 February 2008). "The role of conservation agriculture in sustainable agriculture". Philosophical Transactions of the Royal Society B: Biological Sciences. 363 (1491): 543–555.
4. Greenwood, Norman N.; Earnshaw, Alan (1997). Chemistry of the Elements (2nd ed.), Butterworth-Heinemann. ISBN 978-0-08-037941-8.
5. Merck Index, 11th Edition, 9625
6. Austin, George T. (1984). Shreve's Chemical process industries (5th ed.). New York, NY: McGraw-Hill. p. 357

Chapter 11: Evolution of the Pharmaceutical Industry

HISTORICAL PERSPECTIVE

The pharmaceutical industry, originating in the late 19th century, has seen significant growth propelled by the discovery of new medicines and advancements in the chemical sector.

a. **Early Development:** Medicines were initially derived from natural sources like plants and minerals, prepared by physicians or apothecaries.
b. **Chemical Synthesis:** The late 1800s saw the emergence of chemical synthesis in the laboratory, leading to breakthrough medicines such as aspirin in 1897.
c. **Establishment:** By the early 20th century, specialized pharmaceutical companies focused on drug development, production, and marketing, leading to rapid industry growth.
d. **Regulatory Measures:** Governments worldwide introduced regulations to oversee drug safety and efficacy, exemplified by the establishment of the FDA in 1906.
e. **Recent Trends:** Modern challenges include rising drug development costs, the advent of generic drugs, and addressing emerging diseases through technology, including biotechnology and personalized medicine.

Key Features of the Pharmaceutical Industry

a. **Research and Development:** Heavy investment in R&D for discovering new drugs and therapies, encompassing both basic and applied research.
b. **Drug Development:** Stringent testing through preclinical and clinical trials to ensure safety and efficacy before market approval.
c. **Manufacturing:** Complex processes and stringent quality control measures to produce drugs in large quantities.
d. **Distribution:** A network of wholesalers, pharmacies, and healthcare providers ensure safe and accurate drug distribution.

e. **Sales and Marketing:** Significant investment in promoting products and educating healthcare providers about their benefits.
f. **Regulation:** Extensive oversight by government agencies like the FDA and international bodies to ensure drug safety and efficacy.

Pharmaceutical Business Stages

a. **Research and Development:** Focus on discovering new drugs and therapies.
b. **Drug Development:** Rigorous testing through preclinical and clinical trials to ensure safety and efficacy.
c. **Manufacturing:** Complex processes and strict quality control to produce drugs in bulk.
d. **Distribution:** Network involving wholesalers, pharmacies, and healthcare providers for distribution.
e. **Sales and Marketing:** Heavy investment in promotional activities to educate about product benefits.
f. **Regulation:** Compliance with strict regulations to ensure drug safety and efficacy, overseen by government agencies.
g. **Innovation:** Driving innovation in medicine through the development of new drugs and therapies.
h. **Patents and Intellectual Property:** Reliance on patents and intellectual property protection to recoup investment in drug development.

Overall, the pharmaceutical industry is a critical component of healthcare, driving innovation while navigating complex regulatory landscapes and addressing evolving global health challenges.

APPLICATION OF SULPHUR & SULPHUR COMPOUNDS IN PHARMACEUTICAL INDUSTRY

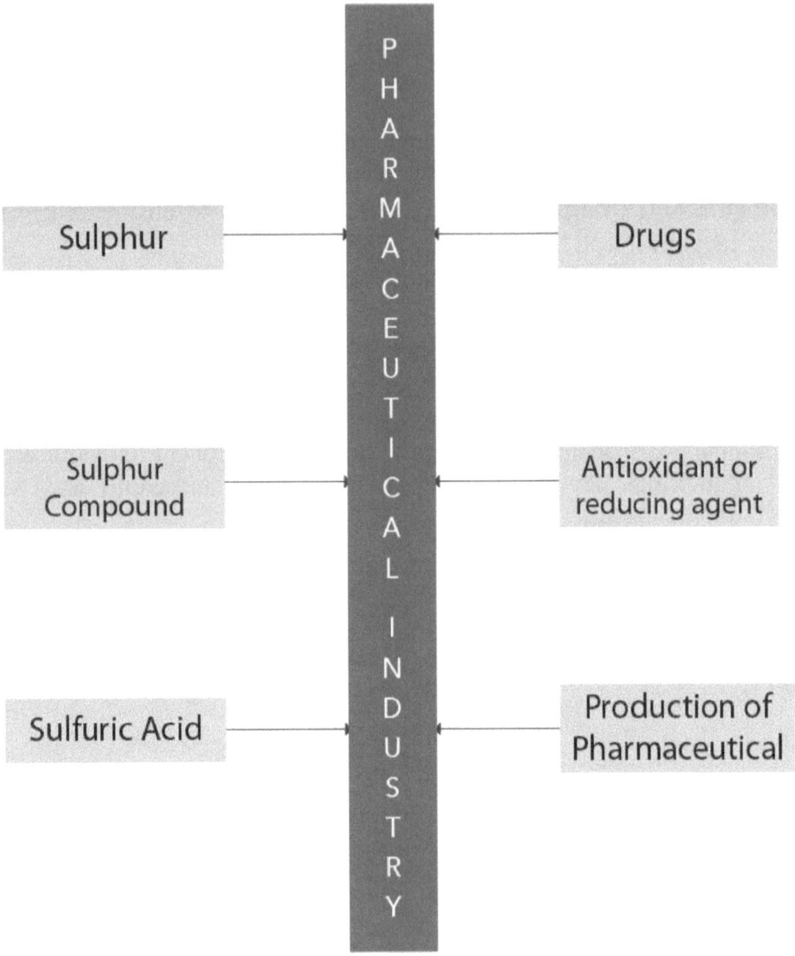

1. Drugs from Sulphur

a. **Sulfonamide Antibiotics:** Sulfonamide antibiotics are crucial in bacterial infection treatment, featuring a sulfonamide functional group derived from sulphur. Examples include sulfamethoxazole, sulfadiazine, and sulfacetamide.
b. **Sulfonylurea Drugs:** These drugs aid in type 2 diabetes management by stimulating insulin release from the pancreas, containing a sulfonylurea functional group sourced from sulphur. Examples of medications are glyburide, glimepiride, and glipizide.
c. **Topical Sulphur Preparations:** Utilized in addressing skin issues like acne, rosacea, and seborrheic dermatitis, these formulations incorporate sulphur to reduce inflammation and impede bacterial growth.
d. **Chemical Reactions:** Oxidation of sulphur in topical preparations leads to various sulphur compounds, including sulfides and sulphates. These compounds, in interaction with skin proteins and lipids, yield derivatives with antimicrobial, anti-inflammatory, and keratolytic properties, enhancing therapeutic effects.
e. **Sulphur-Containing Supplements**
    - Sulphur contributes to dietary supplements aimed at joint health and treating osteoarthritis.
    - Examples: Glucosamine sulphate and chondroitin sulphate are prominent supplements in this category.

Sulphur serves as a vital element in pharmaceuticals, facilitating the production of diverse drugs and supplements crucial for treating various conditions.

2. Antioxidant or Reducing Agent from Sulphur Compounds

a. **Glutathione:** A sulphur-containing compound, acts as a potent antioxidant in the body, shielding cells from oxidative harm by neutralizing free radicals and reactive molecules. Besides its intrinsic role as an antioxidant, glutathione is utilized as a dietary supplement with purported health benefits.
b. **Sodium Sulfite:** It functions as a reducing agent, commonly applied in the food industry to deter food product oxidation. Additionally, it serves as a preservative in select medications and as a reducing agent in photographic development. Widely utilized in food preservation and photographic processing due to its reducing properties.
c. **Thiols:** A sulphur-containing compound, exhibit dual roles as reducing agents or antioxidants, dependent on their chemical attributes. They find applications in pharmaceutical and industrial product manufacturing. Integral in

pharmaceutical and industrial processes, thiols contribute to various formulations and production procedures.

d. **Sulphur Dioxide:** Functioning as a reducing agent, sulphur dioxide features in diverse industrial settings, such as paper production and wastewater treatment. It also serves as a preservative in certain food and beverage items. Employed across multiple industries for its reducing properties, sulphur dioxide aids in industrial processes and food preservation.

Sulphur compounds, exhibiting versatile roles as both reducing agents and antioxidants, are indispensable in numerous industrial applications. Their capacity to mitigate oxidative damage and preserve product quality underscores their significance across various sectors.

3. Pharmaceutical Products from Sulphuric Acid

a. **Aspirin:** A widely used pain reliever, is synthesized by acetylating salicylic acid with acetic anhydride, a process catalyzed by sulphuric acid. The reaction involves salicylic acid ($C_7H_6O_3$) and acetic anhydride ($C_4H_6O_3$), yielding acetylsalicylic acid (aspirin) and acetic acid. Primarily employed to alleviate pain and inflammation, aspirin serves as a cornerstone medication for various conditions.

b. **Antibiotics:** Sulphuric acid facilitates fermentation processes crucial for the production of antibiotics, fostering optimal conditions for microorganism growth. Additionally, it aids in the purification and isolation of antibiotics. Antibiotics combat bacterial infections and are integral in medical treatment protocols worldwide.

c. **Vaccines:** In vaccine production, particularly for vaccines like the human papillomavirus (HPV) vaccine, sulphuric acid plays a role in recombinant DNA technology, aiding in the isolation and purification of proteins utilized in vaccine formulation. Vaccines confer immunity against infectious diseases, preventing illness and contributing to public health initiatives.

d. **Steroids:** Sulphuric acid is instrumental in the synthesis and purification of steroid drugs, including cortisone and prednisone, essential for managing inflammation, autoimmune disorders, and hormone imbalances. Steroids have diverse therapeutic applications, addressing a spectrum of medical conditions ranging from inflammatory diseases to hormonal imbalances.

Sulphuric acid's multifaceted role in pharmaceutical production encompasses catalyzing chemical reactions, facilitating fermentation, and aiding in purification processes. Its involvement underscores its significance in the pharmaceutical industry, contributing to the synthesis of crucial medications and vaccines.

Summary

Sulfur and sulfur compounds have various applications in the pharmaceutical industry, both as active pharmaceutical ingredients (APIs) and as intermediates in the synthesis of pharmaceutical compounds.

Here are some common applications:

a. **Antibiotics:** Several antibiotics contain sulfur in their chemical structure. For example, penicillins, cephalosporins, and carbapenems, which are widely used to treat bacterial infections, contain sulfur-containing functional groups, such as thiazolidine rings or β-lactam rings.

b. **Antifungals:** Sulfur-containing compounds are used as antifungal agents in pharmaceuticals. Examples include azoles, such as fluconazole and ketoconazole, which contain sulfur atoms in their structures and are effective against fungal infections.

c. **Antivirals:** Some antiviral drugs contain sulfur as part of their chemical structure. For instance, nucleoside analogs like zidovudine (AZT) and lamivudine (3TC), used to treat HIV/AIDS and hepatitis B, respectively, contain sulfur atoms.

d. **Sulfonamides:** Sulfonamides are a class of drugs that contain a sulfur-based functional group (SO_2NH_2). They are widely used as antimicrobial agents to treat bacterial infections, including urinary tract infections and respiratory tract infections.

e. **Chelating Agents:** Sulfur-containing chelating agents, such as dimercaprol (BAL) and dimercaptosuccinic acid (DMSA), are used in chelation therapy to treat heavy metal poisoning by forming stable complexes with toxic metals like mercury, lead, and arsenic.

f. **Anti-inflammatory Agents**: Some anti-inflammatory drugs, such as thiol-containing nonsteroidal anti-inflammatory drugs (NSAIDs) like tiopronin and mesalamine, contain sulfur atoms in their structures and are used to reduce inflammation and pain.

g. **Vitamins and Coenzymes:** Sulfur-containing compounds are essential components of vitamins and coenzymes involved in various biochemical processes. For example, biotin, a B vitamin, contains a sulfur atom in its structure and is involved in fatty acid synthesis and gluconeogenesis.

h. **Radiopharmaceuticals:** Sulfur-containing radiopharmaceuticals, such as sulfur colloid and technetium-99m sulfur colloid, are used in nuclear medicine imaging procedures, such as lymphoscintigraphy and liver imaging.

These examples highlight the diverse roles of sulfur and sulfur compounds in the pharmaceutical industry, from serving as active ingredients in drugs to playing crucial roles in biochemical processes and diagnostic imaging.

REFERENCES:

1. McGuire, John L.; Hasskarl, Horst; Bode, Gerd; Klingmann, Ingrid; Zahn, Manuel (2007). "Pharmaceuticals, General Survey". Ullmann's Encyclopedia of Industrial Chemistry. Wiley.
2. Zaffiri L, Gardner J, Toledo-Pereyra LH (April 2012). "History of antibiotics. From salvarsan to cephalosporins". J Invest Surg. 25 (2): 67–77.
3. "Emergence of Pharmaceutical Science and Industry: 1870-1930". Chem Eng News. 83 (25). 20 June 2005.
4. Sneader, Walter (31 October 2005). "13 Neurohormones". Drug Discovery: A History. John Wiley & Sons. pp. 155–156.

Chapter 12 : Purifying H2O: Advances in Water Treatment

HISTORICAL EVOLUTION:

Water treatment has ancient origins, with civilizations like the Indian, Egyptians, Greeks, and Romans employing rudimentary purification techniques like sedimentation, filtration, and boiling. In response to industrialization, the 19th century witnessed the establishment of water treatment facilities, notably in London amid a cholera epidemic. Subsequent centuries saw the development of innovative technologies, including chlorine disinfection and sand filtration, paving the way for modern water treatment.

Water Treatment Process:

a. **Coagulation and Flocculation:** Chemical additives promote the aggregation of impurities, forming larger particles known as floc for easier removal.
b. **Sedimentation:** Floc settles at the bottom of treatment tanks, allowing for its separation from the water.
c. **Filtration:** Water passes through various filters like sand or activated carbon to eliminate finer particles and contaminants.
d. **Disinfection:** Chemicals such as chlorine, ozone, or UV light are used to eradicate remaining bacteria and viruses.
e. **Distribution:** Treated water is stored and distributed through a network of pipes to consumers.

Additional processes like pH adjustment, softening, and advanced techniques such as Reverse Osmosis or nanofiltration may be employed based on water source and intended use.

Key Components in Water Treatment

a. **Chlorine:** Widely utilized as a disinfectant, chlorine effectively eradicates harmful microorganisms from water.
b. **Aluminum Sulphate (Alum):** As a coagulant, alum aids in the formation of floc, facilitating impurity removal.
c. **Activated Carbon:** Employed as a filter medium, activated carbon efficiently eliminates organic compounds, tastes, and odors from water.

d. **Sodium Hypochlorite:** Serving as an alternate disinfectant, sodium hypochlorite ensures continued microbial control.
e. **Calcium Carbonate:** Used for pH adjustment and hardness removal, calcium carbonate enhances water quality.
f. **Reverse Osmosis Membranes:** These membranes are instrumental in advanced filtration systems, removing dissolved solids and impurities.
g. **UV Light:** A potent disinfectant, UV light disrupts bacterial and viral DNA, rendering them incapable of replication.

Water treatment encompasses a spectrum of chemical elements and compounds aimed at purifying water for various uses. The utilization of chlorine, alum, activated carbon, and other components ensures the production of clean and safe drinking water that complies with regulatory standards, catering to diverse water sources and end-user requirements

APPLICATION OF SULPHUR COMPOUNDS & SULPHURIC ACID IN WATER TREATMENT

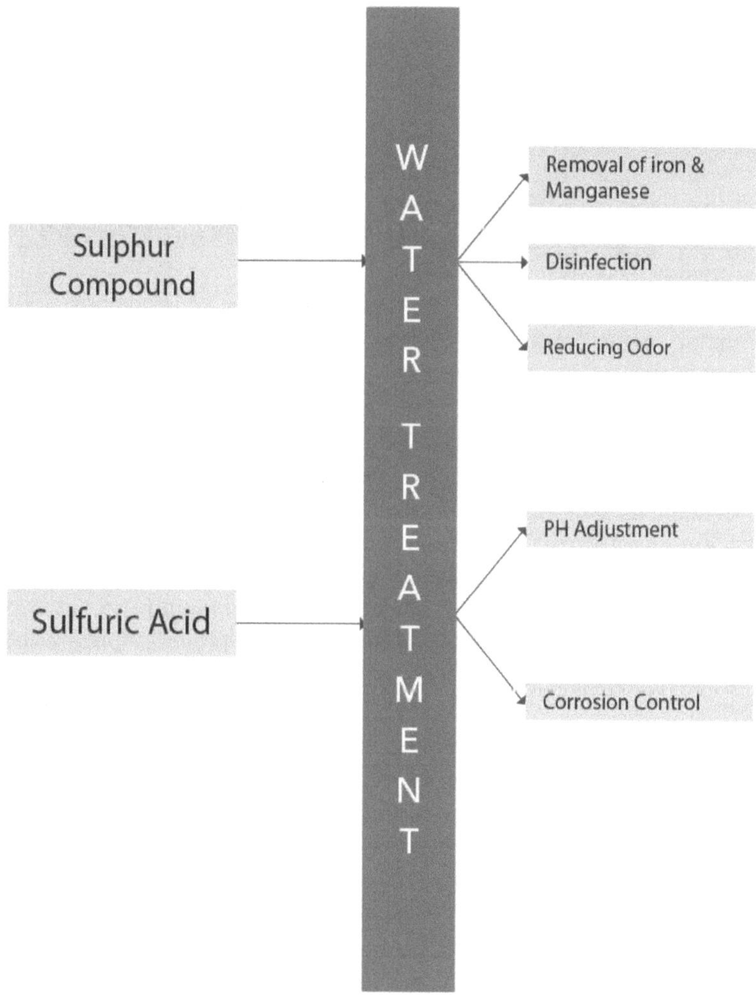

1. Removing Sulphur Compounds & Metallic Impurities from Water

Iron and manganese are commonly found in groundwater sources and can cause issues such as staining, unpleasant taste, and odor in drinking water. Sulfur compounds can be used effectively to remove iron and manganese from water through various treatment methods.

Here are some common approaches:

a. **Chemical Oxidation followed by Filtration:** Sulfur compounds, such as potassium permanganate ($KMnO_4$) or sodium chlorite ($NaClO_2$), can be added to the water to oxidize soluble forms of iron and manganese into their insoluble forms. Once oxidized, the precipitated iron and manganese can be removed from the water by filtration using a media filter, such as sand, anthracite, or manganese greensand. This process is known as chemical oxidation followed by filtration.

b. **Biological Oxidation and Filtration:** Sulfur compounds, particularly elemental sulfur or organic compounds containing sulfur, can be used to support biological oxidation processes for the removal of iron and manganese. In biological oxidation systems, sulfur compounds serve as electron donors for iron-oxidizing and manganese-oxidizing bacteria, which convert soluble forms of iron and manganese into insoluble precipitates. The precipitated iron and manganese can then be removed by filtration through a media filter.

c. **Aeration and Filtration:** Aeration involves exposing water to air or oxygen to facilitate the oxidation of soluble iron and manganese into their insoluble forms. Sulfur compounds, such as elemental sulfur or sulfur dioxide (SO_2), can be added to the water to enhance the oxidation process. Once oxidized, the precipitated iron and manganese can be removed by filtration using a media filter.

d. **Ion Exchange:** Sulfur compounds, such as sulfuric acid (H_2SO_4), can be used in ion exchange processes for the removal of iron and manganese ions from water. In this process, water is passed through an ion exchange resin bed containing sulfonic acid functional groups. The iron and manganese ions are exchanged with hydrogen ions (H^+) on the resin, effectively removing them from the water.

e. **Precipitation and Filtration:** Sulfur compounds can be used to precipitate iron and manganese ions as insoluble sulfide compounds. For example, hydrogen sulfide gas (H_2S) or sodium sulfide (Na_2S) can be added to the water to react with iron and manganese ions, forming insoluble iron sulfide (FeS) and manganese sulfide (MnS) precipitates. The precipitates can then be removed by filtration.

f. It's important to note that the specific treatment method used will depend on factors such as the concentration of iron and manganese in the water, the desired water quality standards, and the available treatment technology. Additionally, proper handling and disposal of sulfur compounds should be ensured to prevent environmental contamination and health risks.

2. Purification of Water by Disinfection

Water disinfection is a critical process aimed at ensuring the safety and cleanliness of water for consumption. Sulphur compounds play crucial roles in various disinfection processes.

Here's how they are utilized:

a. **Chlorine Dioxide (ClO_2)**
- Potent Disinfectant: Chlorine dioxide effectively eliminates a wide range of microorganisms, including bacteria, viruses, and protozoa.
- Common Usage: It is widely employed in water treatment plants for disinfecting drinking water and wastewater.
- Advantages: Chlorine dioxide is preferred over chlorine in some cases due to its higher efficacy, reduced formation of disinfection by-products (DBPs), and ability to penetrate biofilms.

b. **Sulfur Dioxide (SO_2)**
- Historical Use: Sulfur dioxide has a history of use as a disinfectant in water treatment, particularly for its effectiveness in eliminating bacteria and controlling odors.
- Pathogen Neutralization: It effectively neutralizes harmful pathogens and acts as a preservative, preventing microbial growth in water storage tanks and distribution systems.

c. **Sulphur Compounds in Disinfection By-Products (DBPs) Control**
- Indirect Role: Sulphur compounds indirectly contribute to water disinfection by controlling the formation of disinfection by-products (DBPs).
- DBP Formation Reduction: Chlorination and other disinfection processes produce DBPs like trihalomethanes (THMs) and haloacetic acids (HAAs), which can pose health risks.
- DBP Reduction: Sulphur compounds like sulfur dioxide help reduce DBP formation by scavenging reactive chlorine species, thus lowering their concentration in treated water.

d. Sulphuric Acid (H_2SO_4) for pH Adjustment

- pH Optimization: Sulphuric acid is utilized for adjusting pH levels in water treatment, crucial for enhancing the effectiveness of disinfection processes.
- Effective Disinfection: Maintaining the proper pH range ensures that disinfectants like chlorine or chlorine dioxide remain active and efficient in pathogen elimination.

In summary, sulphur compounds play pivotal roles in water disinfection, directly as disinfectants, and indirectly by controlling DBP formation and optimizing disinfection processes. These compounds are instrumental in safeguarding the safety and quality of drinking water supplies globally.

3. Eliminating Sulphur Compound Impurities for Odorless Water

Sulphur compounds in water often result in unpleasant odors reminiscent of "rotten eggs." To mitigate or eradicate these odors, various treatment methods are available:

a. Aeration

- Process: Aeration involves exposing water to air, facilitating the oxidation of sulphur compounds into odorless sulphate.
- Odor Reduction: This method effectively reduces the concentration of sulphur compounds in water, eliminating associated odors.

b. Activated Carbon Filtration

- Principle: Water passes through a bed of activated carbon, which absorbs and removes sulphur compounds and other contaminants.
- Effectiveness: Activated carbon filters, available in granular or powdered forms, efficiently eliminate sulphur odors, ensuring odorless water.

c. Chlorination

- Application: Chlorine treatment oxidizes sulphur compounds, aiding in odor elimination.
- Consideration: While effective, chlorination can produce byproducts necessitating careful management for water safety.

d. Biological Treatment

- Methods: Biofiltration and bioreactors utilize microorganisms to degrade and convert sulphur compounds into odorless forms.
- Efficacy: Biological treatment offers a natural and effective approach to reducing sulphur odors in water.

e. **Hydrogen Peroxide**
   - Oxidation: Hydrogen peroxide oxidizes sulphur compounds, transforming them into odorless derivatives.
   - Consideration: While effective, the cost and management of byproducts should be carefully evaluated.

Eliminating sulphur compound odors in water is crucial for ensuring safe and palatable drinking water. The selection of a treatment method depends on factors such as sulphur compound concentration, water chemistry, and desired treatment level. Each method offers unique advantages and considerations, contributing to the overall goal of providing odorless, high-quality water.

4. PH Adjustment of Water Using Sulphuric Acid

Sulphuric acid serves as a potent agent for pH adjustment in water treatment, crucial for achieving desired pH levels by neutralizing acidic or alkaline water.

Below are the products and applications of sulphuric acid in water treatment:

a. **Acid Injection**
   - Process: Sulphuric acid is injected into water through a system to decrease pH levels, commonly employed in industrial and municipal water treatment for tasks like wastewater treatment and potable water production.

b. **Neutralization**
   - Purpose: Sulphuric acid neutralizes alkaline water by lowering its pH level to a more neutral state.

c. **Precipitation**
   - Function: Sulphuric acid aids in precipitating dissolved metals, including iron, manganese, and aluminum, from water. By lowering the water's pH, metals precipitate out, facilitating their removal via filtration.

d. **pH Adjustment in Agriculture**
   - Usage: Agricultural applications involve using sulphuric acid to adjust the pH of irrigation water. Lowering pH levels enhances nutrient availability and uptake by plants, promoting healthier growth.

e. **pH Adjustment in Swimming Pools**
   - Importance: Sulphuric acid plays a role in maintaining optimal pH levels in swimming pool water, ensuring water quality and preventing equipment corrosion.

Utilizing sulphuric acid in water treatment necessitates diligent handling and control to uphold worker safety and environmental integrity. Employing appropriate equipment like dosing pumps and monitoring systems is essential for accurate acid addition control. Additionally, wearing personal protective equipment during acid handling is imperative.

5. Corrosion Control in Water Treatment with Sulphuric Acid

Sulphuric acid proves effective for corrosion control in water treatment, although its potent corrosive nature necessitates careful handling.

Below are the products and applications of sulphuric acid for corrosion control in water treatment:

a. **pH Adjustment**
   - Purpose: Sulphuric acid is utilized to adjust the pH of water, mitigating the corrosive impact of alkaline water on metal pipes and equipment.
   - Scenario: Elevated pH levels above 8 render water alkaline, posing a risk of corrosion to metal infrastructure. Sulphuric acid addition lowers pH levels, diminishing water alkalinity and minimizing corrosion risk.

b. **Passivation**
   - Process: Sulphuric acid facilitates passivation by forming a protective oxide layer on metal surfaces, shielding them from corrosion.
   - Action: By removing existing corrosion and promoting oxide layer formation, sulphuric acid ensures sustained protection against corrosion.

c. **Scale Removal**
   - Function: Sulphuric acid serves to dissolve scale deposits within pipes and equipment, preventing flow restriction, heat transfer inefficiency, and corrosion escalation.
   - Objective: Dissolving scale eases removal and prevents further accumulation, safeguarding water infrastructure integrity

Summary

Sulfur and sulfur compounds play various roles in water treatment processes, primarily in the removal of contaminants and the disinfection of water. Here are some common applications of sulfur and sulfur compounds in water treatment:

a. **Reduction of Heavy Metals:** Sulfur compounds, such as sodium sulfide (Na_2S) or hydrogen sulfide (H_2S), can be used to precipitate heavy metals from water. These compounds react with metal ions, forming insoluble metal sulfide precipitates that can be easily removed by filtration or sedimentation. This process is particularly effective for the removal of metals like lead, cadmium, mercury, and copper.
b. **Dechlorination:** Sulfur compounds, such as sodium sulfite (Na_2SO_3) or sodium bisulfite ($NaHSO_3$), are commonly used as dechlorinating agents in water treatment. Chlorine is often added to water as a disinfectant, but excess chlorine can be harmful. Sulfur compounds react with chlorine to form chloride ions and harmless sulfate ions, effectively removing chlorine from the water.
c. **pH Adjustment:** Sulfuric acid (H_2SO_4) is frequently used in water treatment for pH adjustment. It can be added to water to lower the pH and neutralize alkalinity. Proper pH control is essential for optimizing the efficiency of other treatment processes, such as coagulation, flocculation, and disinfection.
d. **Biological Treatment:** Elemental sulfur or sulfur compounds can be used to support biological treatment processes for the removal of contaminants in water. For example, sulfur can be added to denitrification systems to provide an electron donor for denitrifying bacteria, which convert nitrates into nitrogen gas. Sulfur compounds may also support the growth of sulfate-reducing bacteria, which can reduce sulfate to sulfide in anaerobic conditions.
e. **Odor Control:** Hydrogen sulfide (H_2S) is a common source of odor in water, often resulting from the decomposition of organic matter containing sulfur. Sulfur compounds, such as sodium sulfide (Na_2S) or potassium permanganate ($KMnO_4$), can be used to oxidize hydrogen sulfide into odorless sulfate ions. This process helps control foul odors and improve the aesthetic quality of water.
f. **Disinfection:** Sulfur dioxide (SO_2) or sodium thiosulfate ($Na_2S_2O_3$) can be used as disinfectants in water treatment. These compounds are effective against bacteria, viruses, and other pathogens. Sulfur dioxide can be used to disinfect water through gas-phase or liquid-phase applications, while sodium thiosulfate is often used as a neutralizing agent for chlorine residuals after disinfection.
g. **Sulfate Removal:** Sulfur compounds, such as calcium oxide (CaO) or calcium carbonate ($CaCO_3$), can be used to remove sulfate ions from water through precipitation reactions. These compounds react with sulfate ions to form insoluble calcium sulfate (gypsum) precipitates, which can be removed by filtration or sedimentation.

These applications demonstrate the versatility of sulfur and sulfur compounds in water treatment processes, where they are used to remove contaminants, adjust pH, control odors, and disinfect water, ultimately ensuring the provision of safe and clean drinking water for various purposes.

REFERENCES

1. Metcalf & Eddy Wastewater Engineering: Treatment and Reuse (4th ed.). New York: McGraw-Hill. 2003.
2. "wastewater treatment Process, History, Importance, Systems, & Technologies". Encyclopedia Britannica.
3. Kurniawan, Tonni Agustiono; Chan, Gilbert Y. S.; Lo, Wai-Hung; Babel, Sandhya (2006-05-01). "Physico–chemical treatment techniques for wastewater laden with heavy metals". Chemical Engineering Journal. 118 (1): 83–98.
4. Vigneswaran, Saravanamuthu; Ngo, Huu Hao; Chaudhary, Durgananda Singh; Hung, Yung-Tse (2005), "Physicochemical Treatment Processes for Water Reuse", Physicochemical Treatment Processes, Totowa, NJ: Humana Press, pp. 635–676
5. Singh, N. B.; Nagpal, Garima; Agrawal, Sonal; Rachna (2018-08-01). "Water purification by using Adsorbents: A Review". Environmental Technology & Innovation. 11: 187–240.
6. Saravanan, A.; Senthil Kumar, P.; Jeevanantham, S.; Karishma, S.; Tajsabreen, B.; Yaashikaa, P. R.; Reshma, B. (2021-10-01). "Effective water/wastewater treatment methodologies for toxic pollutants removal: Processes and applications towards sustainable development". Chemosphere. 280: 130595.
7. Bhalotra, Sonia R.; Diaz-Cayeros, Alberto; Miller, Grant; Miranda, Alfonso; Venkataramani, Atheendar S. (2021). "Urban Water Disinfection and Mortality Decline in Lower-Income Countries". American Economic Journal: Economic Policy. 13 (4): 490–520.

Chapter 13: From Papyrus to Print: A Journey Through the Paper Industry

HISTORY: EVOLUTION OF THE PAPER INDUSTRY

The paper industry traces its origins back to ancient China, around 100 BC during the Han Dynasty. Cai Lun, a government official, is credited with inventing paper by blending mulberry bark, hemp, and rags with water, forming thin sheets dried in the sun.

Papermaking techniques spread to Korea and Japan in the 6th and 7th centuries, then reached the Arab world in the 8th century through Chinese prisoners of war. By the 12th century, papermaking had arrived in Europe via the Arab world.

Italy became the hub of paper production in the 13th century, followed by England in the late 15th century. The industry burgeoned across Europe in subsequent centuries, aided by advancements in printing technology, notably Johannes Gutenberg's printing press.

The 19th century witnessed significant advancements, including the Fourdrinier machine for continuous paper production and the widespread use of wood pulp, supplanting rags as the primary raw material.

a. **Paper Making Process**
   - Raw Materials and Stages: Paper production entails several stages, beginning with pulping to separate fibers and eliminate impurities. Bleaching follows to enhance whiteness, and refining improves fiber quality. Papermaking involves spreading pulp on a mesh screen, pressing, and drying to form sheets.

b. **Current Landscape and Challenges**
   - Market and Consumer Base: Today, the paper industry encompasses a vast array of products, from printing and packaging to specialty

papers, serving diverse sectors like publishing, printing, and consumer needs.

- Environmental Concerns and Sustainability: Environmental challenges loom large, with concerns over raw material usage and energy consumption. Companies are increasingly adopting sustainable practices, utilizing recycled materials and renewable energy sources.
- Technological Advancements and Global Presence: Technological innovations continue to drive the industry forward, with the adoption of digital technology and ongoing research into novel paper products. The paper industry remains a global force, navigating economic and political factors to meet market demands.

The paper industry has evolved from ancient origins to a global enterprise, shaped by innovation and adaptation. With sustainability at the forefront, it strives to meet market demands while addressing environmental concerns, ensuring a future that balances tradition with progress.

APPLICATION OF SULPHUR COMPOUND IN PAPER INDUSTRY

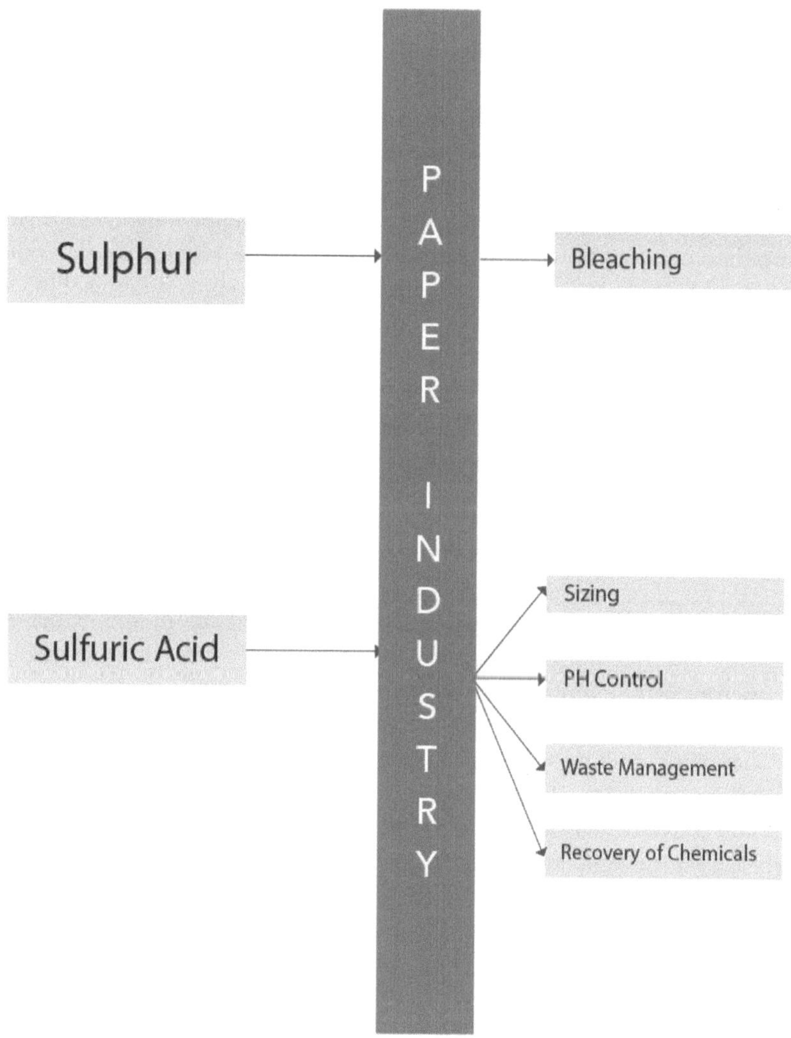

1. Utilizing Sulphur Compounds for Bleaching Purposes

While sulphur is not commonly employed in bleaching products, certain sulphur-containing compounds possess bleaching properties, notably sulphur dioxide and sodium sulfite.

Sulphur dioxide (SO_2), a colorless gas with a pungent odor, finds utility in various industrial processes, including as a bleaching agent in the pulp and paper industry. Through its reaction with lignin, the brown-coloring component in wood, SO_2 effectively breaks it down, resulting in bleached paper.

Another compound, sodium sulfite (Na_2SO_3), manifests as a white crystalline powder and serves multiple purposes in the food industry, functioning as both a preservative and antioxidant. Additionally, it acts as a bleaching agent in the manufacturing of wood pulp, textiles, and paper. Sodium sulfite targets oxygen, the culprit behind material discoloration, and eradicates it, resulting in a bleached appearance.

It is imperative to acknowledge the potential adverse environmental repercussions associated with the use of sulphur-containing compounds in bleaching processes, particularly in the pulp and paper industry. The employment of chlorine-based bleaching agents, for instance, may lead to the emission of toxic dioxins and other pollutants into the environment. Consequently, many industries have shifted their focus towards adopting more environmentally-friendly alternatives, such as hydrogen peroxide and ozone.

2. Sizing with Sulphuric Acid

While sulphuric acid is not a conventional sizing agent in the paper and textile industries, it does find application in certain niche processes within these sectors.

In the paper industry, sulphuric acid plays a role in the sulfite pulping process, where it aids in breaking down wood fibers and removing lignin. Although not directly involved in sizing, this step is crucial in preparing the pulp for subsequent treatments, including the application of sizing agents. These agents are essential for enhancing the strength and water-resistance of the paper.

Similarly, sulphuric acid contributes to the production of viscose rayon, a regenerated cellulose fiber utilized in textiles. In this process, cellulose undergoes dissolution in a solution of sodium hydroxide and carbon disulphide (CS_2). Subsequently, sulphuric acid is employed to transform the resultant solution into a viscous gel-like substance, facilitating extrusion into fibers.

It's paramount to recognize that sulphuric acid poses significant hazards due to its highly corrosive nature. Consequently, stringent regulations and safety protocols must govern its handling and usage to mitigate risks to both personnel and the environment. Adequate protective equipment and adherence to safety guidelines are imperative whenever sulphuric acid is involved in industrial processes.

3. Sulphuric Acid's Role in Paper Industry pH Control

In the realm of paper production, maintaining precise pH levels is paramount for optimal manufacturing conditions. Sulphuric acid emerges as a vital tool in this endeavor, wielding its influence across various stages of the papermaking process.

Here's how sulphuric acid contributes to pH control within the paper industry:

a. **pH Fine-Tuning:** Sulphuric acid serves as a go-to solution for adjusting the pH of water utilized in papermaking. Through careful dosing, it allows operators to finely calibrate the water's acidity to meet specific production requirements. This meticulous adjustment is instrumental in optimizing the performance of other chemicals and additives crucial for paper production.

   a. **Alkalinity Neutralization:** When faced with overly alkaline water, sulphuric acid steps in to restore balance. By administering sulphuric acid, alkalinity levels are brought down, nudging the water towards a neutral pH range. This neutralization process fosters an environment conducive to efficient fiber processing and chemical interactions.
   b. **Catalyst for Chemical Reactions:** Many chemical processes integral to paper manufacturing hinge on precise pH conditions. Sulphuric acid aids in establishing the ideal pH milieu for these reactions to unfold effectively. For instance, in the bleaching phase, maintaining pH stability is pivotal for the successful eradication of lignin and other contaminants from the pulp.
   c. **Equipment Safeguard:** By orchestrating pH control, sulphuric acid acts as a shield against equipment corrosion. Keeping pH levels within optimal parameters mitigates the risk of damage to pipelines, tanks, and other infrastructure components. This safeguarding measure extends the longevity of machinery, curtails maintenance expenses, and upholds operational efficiency.
   d. **Consistency Breeds Excellence:** The meticulous pH regulation facilitated by sulphuric acid fosters uniformity in paper attributes like strength, color, and texture. This consistency underpins the creation of premium-grade paper products that align with industry benchmarks and customer preferences.

In essence, sulphuric acid emerges as a linchpin in the pH control framework of the paper industry, fostering streamlined manufacturing processes and ensuring the delivery of superior-quality paper goods. However, prudent handling and adherence to safety guidelines are imperative to mitigate potential risks associated with its utilization.

4. Managing Waste Containing Sulphuric Acid in the Paper Industry

In the paper industry, waste containing sulphuric acid emerges as a byproduct of various manufacturing processes. To mitigate environmental impact and ensure regulatory compliance, efficient waste management strategies are imperative.

Here's a refined approach to managing waste containing sulphuric acid in the paper industry:

a. **Waste Reduction:** The foremost objective is to minimize the generation of sulphuric acid-containing waste through process optimization, enhanced efficiency, and the adoption of cleaner production techniques. By addressing root causes, such as inefficiencies and excess resource consumption, paper mills can curtail waste generation at its source.

b. **Segregation:** Rigorous segregation protocols are implemented to isolate waste containing sulphuric acid from other waste streams. This practice prevents cross-contamination and streamlines handling procedures, ensuring that sulphuric acid-containing waste receives appropriate treatment tailored to its composition.

c. **Secure Storage:** Designated storage areas equipped with robust containment measures are employed to store waste containing sulphuric acid safely. These facilities are engineered to avert spills and leaks, safeguarding environmental integrity and preserving worker safety. Clear labeling and signage facilitate swift identification and adherence to handling protocols.

d. **Tailored Treatment:** Depending on its characteristics, waste containing sulphuric acid undergoes tailored treatment processes to mitigate its environmental impact. Neutralization, chemical precipitation, and biological treatments are employed to reduce acidity and remove contaminants, rendering the waste less hazardous and suitable for disposal or potential reuse.

e. **Disposal Compliance:** Adhering to stringent regulatory guidelines, waste containing sulphuric acid is disposed of responsibly. Disposal methods, such as landfilling, incineration, or recycling, are chosen based on waste composition and local regulations. Collaborating closely with waste management authorities ensures compliance and fosters responsible disposal practices.

f. **Regulatory Adherence:** Stringent compliance with environmental regulations and permit requirements is upheld throughout the waste management process. Paper mills diligently adhere to stipulated guidelines governing storage, treatment, transportation, and disposal, with regular monitoring and reporting to demonstrate compliance and drive continual improvement.

g. **Sustainable Evolution:** Embracing a culture of innovation and sustainability, paper mills pursue ongoing improvement in waste management practices. This entails embracing emerging technologies, bolstering waste reduction initiatives, and investing in sustainable treatment and disposal methods. By fostering a culture of continuous improvement, paper mills bolster environmental stewardship and operational sustainability.

In summary, effective management of waste containing sulphuric acid in the paper industry demands a holistic approach encompassing waste reduction, meticulous handling, tailored treatment, and compliant disposal practices. By prioritizing environmental stewardship and sustainability, paper mills can mitigate risks, uphold regulatory compliance, and forge a path toward greener operations.

5. Recovering Chemicals from Sulphuric Acid in the Paper Industry

Sulphuric acid is a staple in various paper manufacturing processes like pulping, bleaching, and pH adjustment. While integral to these operations, the recovery and reuse of sulphuric acid are paramount to curbing waste and conserving resources.

Here's how chemicals can be recovered from sulphuric acid in the paper industry:

a. **Acid Regeneration:** Used sulphuric acid, particularly from processes like bleaching, can lose efficacy or become contaminated over time. Rather than disposing of it, regeneration processes can be employed to reclaim usable sulphuric acid. Treatment methods involving appropriate chemicals or processes remove impurities, restoring the acid's purity for reuse in similar processes. Acid regeneration minimizes waste and reduces the need for fresh acid.

b. **Neutralization and Precipitation:** Waste streams containing sulphuric acid can undergo neutralization and precipitation to recover valuable chemicals. Neutralizing sulphuric acid with bases like lime forms salts, which are then separated from the solution through filtration or sedimentation. Recovered salts may contain valuable chemicals or minerals suitable for reuse or sale in other applications.

c. **Recovery of Metal Salts:** Sulphuric acid is often used in processes dissolving metal salts, like aluminium sulphate for paper sizing. After the reaction, the solution may contain dissolved metal salts alongside sulphuric acid.

Techniques such as precipitation or crystallization selectively recover these salts, which are then purified for reuse in various industrial processes, reducing the need for fresh raw materials.

d. **Electrochemical Processes:** Advanced electrochemical techniques like electrodialysis and electrolysis can selectively separate and recover specific chemicals or ions from sulphuric acid-containing waste streams. These methods offer high efficiency and selectivity, enabling the recovery of valuable chemicals while minimizing waste generation.
e. **Distillation and Evaporation:** Sulphuric acid-containing waste streams can undergo distillation or evaporation to recover sulphuric acid and other volatile compounds. Through heating or water removal, volatile components are separated based on their boiling points or concentration. Both techniques recover sulphuric acid and other compounds for reuse in industrial processes.
f. **Resource Optimization:** Paper mills can optimize resource utilization by recycling process water, implementing closed-loop systems, and optimizing manufacturing processes. These measures reduce waste generation and lower consumption of raw materials, energy, and water, leading to enhanced environmental sustainability.

In conclusion, recovering chemicals from sulphuric acid in the paper industry is integral to sustainable operations and resource conservation. Through effective recovery processes and resource optimization measures, paper mills can minimize waste generation, reduce costs, and bolster environmental sustainability.

Summary

Sulfur and sulfur compounds find several applications in the paper industry, mainly in the pulp and papermaking processes.

Here are some common applications:

a. **Bleaching Agents:** Sulfur compounds, such as sodium sulfite (Na_2SO_3) and sodium bisulfite ($NaHSO_3$), are used as bleaching agents in the paper industry. These compounds are effective at removing lignin, which is the natural adhesive that binds fibers together in wood pulp. Bleaching improves the brightness and whiteness of paper products.
b. **Pulp Delignification:** Elemental sulfur (S) or sulfur dioxide (SO_2) is used in the kraft pulping process for the delignification of wood pulp. In the kraft process, wood chips are treated with a mixture of sodium hydroxide (NaOH) and sodium sulfide (Na_2S) under high temperature and pressure conditions. Sulfur compounds, such as hydrogen sulfide (H_2S) and sulfur dioxide, are

released during this process, which helps dissolve lignin and remove it from the pulp.

c. **Sulfuric Acid in Digestion:** Sulfuric acid (H_2SO_4) is used in the sulfite pulping process, where it reacts with wood chips to dissolve lignin and separate it from the cellulose fibers. In this process, wood chips are treated with a mixture of sulfur dioxide (SO_2) and sulfuric acid at elevated temperatures. The sulfuric acid helps to accelerate the delignification process, resulting in the production of high-quality pulp.
d. **Control of pH and Alkalinity:** Sulfuric acid is often used in papermaking processes to adjust pH levels and control alkalinity. Proper pH control is critical for various stages of papermaking, including pulp preparation, sizing, and coating. Sulfuric acid is used to neutralize alkaline substances and maintain the desired pH levels for optimal paper quality.
e. **Sizing Agents:** Some sulfur-containing compounds, such as alkyl ketene dimer (AKD) and alkyl succinic anhydride (ASA), are used as sizing agents in papermaking. These compounds are applied to paper surfaces to improve water resistance and ink holdout. Sizing agents containing sulfur can enhance the performance of paper products in printing and writing applications.
f. **Effluent Treatment:** Sulfur compounds are also used in the treatment of wastewater generated during the papermaking process. Sulfur-containing chemicals, such as ferrous sulfate ($FeSO_4$) or sodium hydrosulfide (NaHS), can be added to wastewater to remove contaminants, such as suspended solids and colorants, through precipitation or chemical oxidation.

These applications highlight the importance of sulfur and sulfur compounds in various stages of the papermaking process, from pulp delignification and bleaching to pH control and effluent treatment, ultimately contributing to the production of high-quality paper products.

REFERENCES:

1. Tsien, Tsuen-Hsuin (1985), Paper and Printing, Science and Civilisation in China: Chemistry and Chemical Technology, vol. 5 Part 1, Cambridge University Press, p. 58
2. Hills, Richard, "Papermaking in Britain 1488–1988", Athlone Press, 1988.
3. "Global Forest, Paper & Packaging Industry Survey: 2009 Edition - Survey of 2008 Results" (PDF). PricewaterhouseCoopers. p. 12.
4. Merck Index, 11th Edition, 9625
5. Austin, George T. (1984). Shreve's Chemical process industries (5th ed.). New York, NY: McGraw-Hill. p. 357

Chapter 14: Unveiling the Leather Industry: From Ancient Craft to Modern Innovation

HISTORY OF THE LEATHER INDUSTRY

The origins of the leather industry trace back to ancient civilizations, with evidence of leather making dating back over 5000 years BC. Early societies like the Egyptians, Greeks, and Romans mastered the art of tanning animal hides, utilizing leather for diverse purposes such as clothing, footwear, armor, and bookbinding.

During the Middle Ages, leather production flourished in Europe, marked by the rise of leather guilds and markets across many cities. The industry experienced further growth in the 16th and 17th centuries, fueled by innovations like chrome tanning, which revolutionized leather production, making it faster and more efficient.

The 19th and early 20th centuries witnessed significant transformations in the leather industry, driven by technological advancements such as the steam engine, enabling mass production of leather goods. Geographically, the industry expanded with the establishment of new tanneries and workshops in regions like North America and Australia.

Throughout the 20th century, the leather industry continued to innovate and diversify, introducing materials like synthetic leather and adopting new processing techniques. However, challenges emerged, including environmental concerns regarding tannery waste and ethical debates surrounding animal rights.

Today, the leather industry remains a vital sector, producing a wide array of products including shoes, handbags, clothing, furniture, and automotive upholstery. Despite sustainability and ethical challenges, the industry adapts to meet evolving consumer demands and technological advancements.

LEATHER GOODS & MANUFACTURING STAGES

The leather industry encompasses the production of various leather products, ranging from shoes and handbags to jackets, belts, furniture, and automotive upholstery. Key stages in leather manufacturing include leather production, tanning, finishing, and product manufacturing.

The process begins with sourcing animal hides and skins from animals such as cows, sheep, goats, and pigs. These hides undergo cleaning, dehairing, and defleshing to prepare them for tanning. Various tanning methods, including vegetable, chrome, and synthetic tanning, are employed based on the intended use and desired quality of the leather.

Following tanning, the leather undergoes finishing processes such as dyeing, embossing, and polishing to achieve the desired texture, color, and appearance. The finished leather is then ready for use in manufacturing various products.

Leather manufacturing involves cutting, sewing, and assembling leather pieces, either manually or using automated equipment. Despite facing challenges related to sustainability, environmental impact, and animal welfare, the industry adopts practices such as sustainable sourcing, recycling, and waste reduction, while exploring eco-friendly tanning methods.

Overall, the leather industry remains essential, providing high-quality and durable products while evolving to address contemporary challenges and consumer preferences.

Key Aspects of the Leather Industry

a. **Raw Materials:** Animal hides and skins, sourced from cows, sheep, goats, and pigs, serve as the primary raw materials for the leather industry.
b. **Tanning Process:** The tanning process is pivotal in leather production, involving the treatment of animal hides with chemicals to stabilize them and prevent decay. Various methods such as vegetable tanning, chrome tanning, and synthetic tanning are employed.
c. **Finishing Processes:** Following tanning, leather undergoes diverse finishing treatments like dyeing, embossing, and polishing to achieve the desired texture, color, and appearance.
d. **Manufacturing:** Leather products, including shoes, handbags, and jackets, are crafted through cutting, sewing, and assembling leather pieces. This can be accomplished manually or using automated equipment.
e. **Sustainability:** The leather industry grapples with sustainability challenges encompassing environmental impact, waste management, and animal welfare. To tackle these issues, the industry embraces sustainable practices like utilizing eco-friendly raw materials, recycling, waste reduction, and the development of environmentally friendly tanning techniques.

In conclusion, the leather industry stands as a significant sector, offering an array of high-quality and durable products. Despite confronting various challenges, including environmental concerns and evolving consumer demands, the industry persists in fostering innovation and adaptation to ensure its continued relevance and sustainability.

APPLICATION OF SULPHUR & SULPHUR COMPOUNDS IN LEATHER INDUSTRY

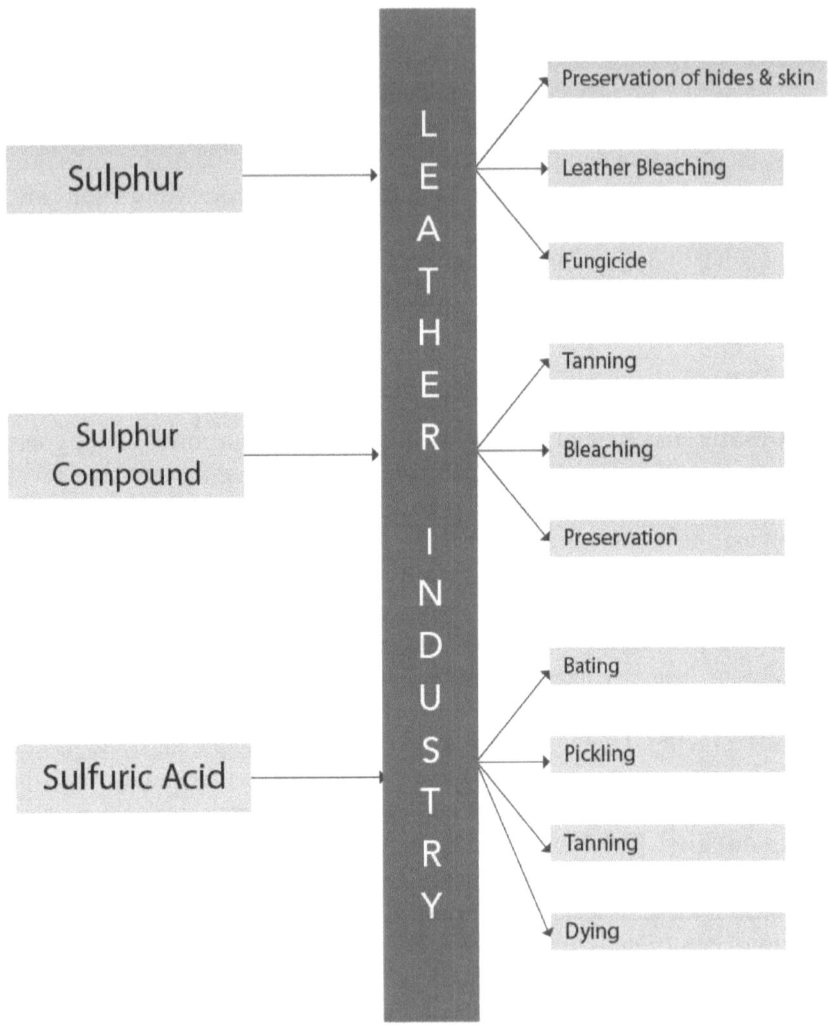

Sulfur, its compounds, and sulfuric acid are integral components of the leather industry, serving diverse functions in tanning, dyeing, and finishing processes.

Below are key applications of these chemicals in leather manufacturing

a. **Tanning:** Sulfur acts as a crucial reducing agent in the tanning process. By interacting with chromium salts, it facilitates the transformation into a more stable form, enhancing leather durability and water resistance.
b. **Dyeing:** Sulfur functions as a primary dyeing agent, particularly for achieving black and dark tones in leather production. Its effectiveness extends to suede and nubuck leather, contributing to their distinct coloration.
c. **Finishing:** Sulfuric acid finds widespread use in leather finishing stages. Through pH adjustment, it enhances leather properties such as color vibrancy, texture refinement, and overall durability.
d. The utilization of sulfur and its derivatives is pivotal across leather manufacturing stages. However, due diligence in handling is paramount, given their potential hazards to both human health and the environment if mishandled.

1. Sulphur Applications

a. **Preservation of Raw Hides and Skins:** Sulphur serves as a preservative for raw hides and skins, safeguarding them against spoilage during transportation, especially over long distances.
b. **Leather Bleaching:** Utilized in the bleaching process, sulphur contributes to achieving lighter shades in leather. Through its reaction with hydrogen peroxide in a water solution, sulphur forms sulphuric acid, which effectively bleaches the leather.
c. **Fungicide:** Sulphur acts as a potent fungicide, shielding leather goods from mold and mildew, particularly beneficial for products stored in humid environments.
d. While not traditionally employed in the tanning process, sulphur finds diverse applications within the leather industry, including preservation, bleaching, fungicidal protection, and sulphuric acid production.

2. Sulphur Compound Applications

a. **Tanning:** Sulphur compounds like sodium sulfide or sodium hydrosulfide are pivotal in the tanning process, facilitating the conversion of collagen in hides into a stable, insoluble material.

b. **Bleaching:** Sulphur dioxide finds application in the leather industry for bleaching hides to achieve lighter color variations, enhancing their aesthetic appeal.
c. **Preservation:** Utilized as preservatives, sulphur dioxide and sodium metabisulfite effectively inhibit bacterial and fungal growth on hides during storage and transportation, ensuring their integrity.

Sulphur compounds serve indispensable roles in various facets of leather production, encompassing deliming, pickling, tanning, bleaching, and preservation processes, thus contributing to the creation of diverse leather products.

3. Sulphuric Acid Applications

a. **Bating:** Sulphuric acid serves a crucial role in bating, a process utilizing enzymes to soften leather. It adjusts the pH of the solution to an optimal range, enhancing enzyme effectiveness.
b. **Pickling:** During the initial stages of leather processing, hides or skins undergo treatment with sulphuric acid to lower pH levels and eliminate unwanted substances like hair and flesh.
c. **Tanning:** In the tanning process, sulphuric acid prepares hides or skins for tanning agents by increasing fiber porosity and facilitating agent penetration, thereby aiding in the leather's preparation.
d. **Dyeing:** Sulphuric acid is integral in the dyeing process, regulating the pH of the dye bath to influence the final color of the leather product.

Summary:

Sulfur and sulfur compounds are used in various processes within the leather industry, primarily in leather tanning and finishing.

Here are some common applications:

a. **Sulfide Deliming and Bating:** Sulfur compounds, such as sodium sulfide (Na_2S) or sodium hydrosulfide (NaHS), are used in the deliming and bating stages of leather processing. Deliming involves the removal of lime and alkalinity from the hides or skins after the initial unhairing process. Sulfide-based chemicals are used to break down proteins and remove unwanted hair and tissue residues. Bating is a process that softens and opens up the fibers of the leather, improving its softness and flexibility. Sulfur compounds help in this process by promoting enzymatic degradation of the collagen fibers.

b. **Sulfide Scouring:** Sulfur compounds are used in the scouring process to remove natural fats, greases, and impurities from the hides or skins. Sodium sulfide or sodium hydrosulfide is typically added to the scouring bath to facilitate the removal of non-collagenous proteins, lipids, and other contaminants from the surface of the hides.

c. **Hair Removal:** Sulfide-based chemicals are used in the hair removal process during leather processing. Sodium sulfide or calcium hydrosulfide is added to the hair removal bath to dissolve and loosen hair roots, making it easier to mechanically remove hair from the hides or skins.

d. **Liming:** Sulfur compounds are sometimes used in the liming process, which involves the treatment of hides or skins with lime (calcium hydroxide) to swell and open up the fibers. Sodium sulfide or sodium hydrosulfide may be added to the liming bath to accelerate the loosening of hair and tissue residues.

e. **Bleaching:** Sulfur dioxide (SO_2) or sodium sulfite (Na_2SO_3) can be used as bleaching agents in the leather industry to lighten the color of hides or skins. These compounds react with natural pigments and tannins present in the leather, helping to achieve the desired color or shade.

f. **Sulfuric Acid in Pickling:** Sulfuric acid (H_2SO_4) is used in the pickling process to adjust the pH of the hides or skins and prepare them for tanning. Sulfuric acid lowers the pH of the hides, facilitating the penetration of tanning agents and ensuring uniform tanning throughout the leather.

Overall, sulfur and sulfur compounds play important roles in various stages of leather processing, from hair removal and deliming to bleaching and pickling. These compounds help achieve desired properties such as softness, flexibility, and color in the finished leather products.

REFERENCES:

1. Gillan, Kevin; Murray, Jason (May 2019). Comprehensive Guide to Leather Repair and Restoration: Leather Repair Training Manual. Advanced Leather Solutions, Inc. p. 16.
2. NIIR Board of Consultants (2011). Leather Processing & Tanning Technology Handbook. NIIR Project Consultancy Services. p. 323.
3. Heidemann, E. (1993). Fundamentals of Leather Manufacture. Eduard Roether KG. p. 211.
4. Bienkiewicz, K. (1983). Physical Chemistry of Leather Making. Robert E. Krieger. p. 226.
5. Merck Index, 11th Edition, 9625
6. Austin, George T. (1984). Shreve's Chemical process industries (5th ed.). New York, NY: McGraw-Hill. p. 357

Chapter 15: From Farm to Fork: A Journey Through the Food Industry

A HISTORICAL PERSPECTIVE

a. **Ancient Origins:** The food industry traces its roots to ancient civilizations, where agriculture and animal husbandry marked the beginnings of food production. Initially, food was cultivated and processed locally to sustain communities.
b. **Expansion and Commercialization:** As societies advanced and trade networks flourished, food production transitioned into a commercial endeavor. Ancient Rome exemplified this shift with its large-scale farming and specialized food processing facilities.
c. **Industrial Revolution:** The 18th and 19th centuries ushered in transformative changes with the advent of technologies like canning and refrigeration. These innovations facilitated food preservation and distribution on a global scale, fueling the growth of large-scale processing industries.
d. **20th Century Innovations:** The 20th century witnessed further evolution, marked by breakthroughs in food science, genetics, and biotechnology. This era saw the emergence of genetically modified crops and plant-based meat alternatives, revolutionizing food production and consumption.
e. **Modern Landscape:** Today, the food industry stands as a formidable global force, offering a vast array of products ranging from fresh produce to packaged and processed foods. Yet, it grapples with challenges such as food safety, environmental sustainability, and ethical concerns regarding animal welfare and labor practices.

Efforts to address these challenges are underway, emphasizing sustainable production methods and enhanced transparency throughout the food supply chain. The journey of the food industry from ancient roots to modern complexities reflects its enduring significance and ongoing quest for improvement.

THE FOOD INDUSTRY: AN EXPANSIVE OVERVIEW

The food industry encompasses a vast array of activities spanning the entire lifecycle of food production, processing, distribution, and retail. From the cultivation of crops to the shelves of supermarkets, it involves diverse players ranging from multinational corporations to local farmers' markets and cooperatives.

Agriculture lies at the heart of the food industry, encompassing the cultivation of crops and the rearing of livestock. This includes a spectrum of operations, from traditional family farms to technologically advanced commercial enterprises striving for optimal efficiency and yield.

Food processing transforms raw ingredients into a multitude of packaged and prepared food products through various methods such as freezing, canning, baking, and preserving, among others.

Distribution networks are integral to the food industry, ensuring the seamless movement of products from producers to retailers and eventually to consumers. This entails intricate logistics and transportation systems to guarantee timely delivery to diverse destinations.

Food retailing encompasses the final stage of the journey, involving the sale of food products through supermarkets, restaurants, and other outlets. It entails marketing strategies, inventory management, and supply chain optimization to meet consumer demand effectively.

Overall, the food industry serves as a vital cornerstone in feeding the global population. Yet, it grapples with multifaceted challenges, including food safety, environmental sustainability, and ethical considerations regarding labor practices and animal welfare. To tackle these issues, initiatives promoting transparency, sustainability, and responsible practices are underway.

THE DYNAMIC LANDSCAPE OF THE FOOD INDUSTRY

The food industry stands as a multifaceted realm encompassing diverse activities ranging from production to distribution and sale of food products. This expansive sector hosts a spectrum of players, from local family farms to expansive multinational corporations.

Key facets of the food industry include:

a. **Agriculture:** Covering the cultivation of crops and rearing of livestock, spanning from traditional methods to advanced industrial-scale operations.
b. **Food processing:** Involves the conversion of raw ingredients into packaged and prepared food products, employing techniques like canning, baking, and preservation.
c. **Food distribution:** Ensures seamless movement of products from producers to retailers and consumers, facilitated by intricate logistics and transportation systems.
d. **Food retailing:** Encompasses the sale of food products through various outlets, requiring effective marketing, inventory management, and supply chain operations.
e. **Food safety:** Emphasizes the assurance of food product safety for human consumption through rigorous testing, quality control, and regulatory compliance measures.
f. **Sustainability:** Advocates for reducing the environmental footprint of the industry, encompassing initiatives like emission reduction, waste minimization, and sustainable farming practices.
g. **Ethical concerns:** Addresses issues concerning labor practices, animal welfare, and social responsibility, increasingly pertinent to consumers and stakeholders.

Undoubtedly, the food industry holds pivotal significance in the global economy, serving as a cornerstone for providing essential goods and services worldwide. However, it grapples with challenges such as environmental sustainability and ethical considerations, driving initiatives toward enhanced transparency, sustainability, and responsible practices.

APPLICATION OF SULPHUR, SULPHUR COMPOUNDS AND SULPHURIC ACID IN FOOD INDUSTRY

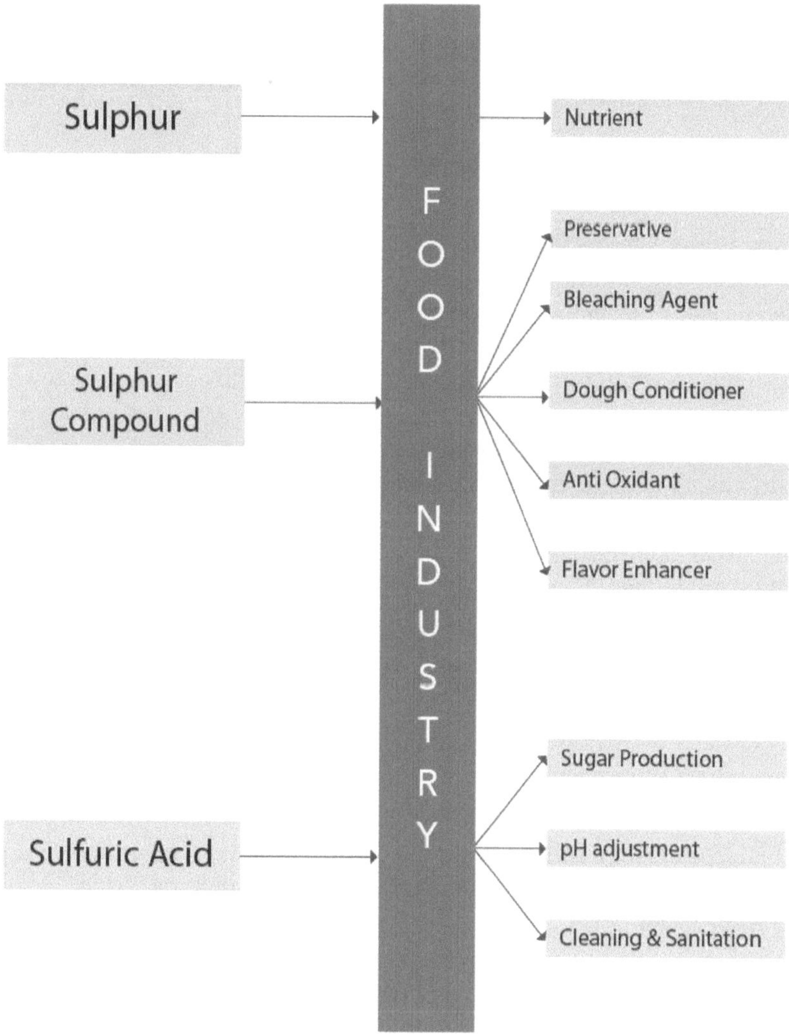

Sulphur serves as a vital nutrient for plants, commonly applied as a fertilizer to enhance plant growth and augment crop yields. Moreover, sulphur dioxide (SO_2) finds application as a preservative in dried fruits, fruit juices, and wine, effectively thwarting spoilage and microbial proliferation.

In agriculture, sulphuric acid assumes significance as a soil pH adjuster and a constituent in certain fertilizers. Although not directly incorporated into food production processes, it may indirectly participate in the cultivation of specific food items.

It's crucial to emphasize that regulatory bodies governing food safety meticulously oversee the use of sulphur-containing compounds in food production to ensure consumer safety. Strict enforcement of maximum allowable limits for sulphur-containing compounds in foodstuffs is implemented to safeguard human health.

1. Sulphur as a Nutrient

Sulphur and its derivatives serve diverse functions within the food industry, including its role as a vital nutrient for human health.

Sulphur stands as an essential nutrient necessary for human well-being. It is naturally present in various food sources like eggs, meat, and specific vegetables. These foods provide sulphur essential for the synthesis of crucial amino acids and proteins vital for bodily functions.

While sulphur is generally safe for consumption in small amounts, excessive intake can lead to adverse health effects, including headaches, nausea, and digestive issues like diarrhea. Therefore, it is imperative to exercise caution in the use of sulphur-containing food additives, ensuring moderation and adherence to recommended consumption guidelines.

2. Utilization of Sulphur Compounds

Sulphur compounds play a significant role in various aspects of the food industry, serving as preservatives, bleaching agents, dough conditioners, antioxidants, and flavor enhancers.

Sodium bisulphite, a key sulphur compound utilized in the food industry, is prepared by reacting a suitable base like sodium hydroxide with sulphur dioxide:

a. **Preservative**

   Formation of Sodium Bisulphite

- $NaOH + H_2SO_4 \rightarrow NaHSO_4 + H_2O$
- $NaHSO_4\ (aq) + NaOH\ (aq) \rightarrow Na_2SO_4\ (aq) + H_2O$
- Sodium bisulphite serves as a common preservative, inhibiting bacterial and fungal growth in various food products like fruits, vegetables, and meats.

b. **Bleaching Agent:** In certain food items such as dried fruits and vegetables, sodium bisulphite acts as a bleaching agent, enhancing their appearance and color.
c. **Dough Conditioner:** Sodium bisulphite functions as a dough conditioner in specific bread types, facilitating dough rising and improving texture.
d. **Antioxidant:** Utilized as an antioxidant in products like wine, sodium bisulphite prevents oxidation, preserving the color and flavor integrity.
e. **Flavor Enhancer:** Sodium bisulphite enhances the flavor profile of foods like jams and jellies, enriching their taste.

While sodium bisulphite enhances food quality and shelf life, individuals with sensitivities may experience allergic reactions. Hence, prudent usage and adherence to safety guidelines are imperative.

3. Utilization of Sulphuric Acid

a. While not directly incorporated into food products, sulphuric acid serves several pivotal roles as a processing aid in food manufacturing processes. Its applications span across various stages, from sugar production to cleaning and sanitation within food processing facilities.
b. **Sugar Production:** Sulphuric acid plays a vital role in sugar production, particularly from sugarcane or sugar beets. Through hydrolysis, it facilitates the breakdown of sucrose into glucose and fructose, essential components of sugar.
c. **pH Adjustment:** In certain food products, such as fruit juices or canned fruits and vegetables, sulphuric acid serves as a pH adjuster. Its controlled addition helps achieve the desired acidity level, enhancing both flavor and preservation.
d. **Cleaning and Sanitation:** Within food processing facilities, sulphuric acid functions as a potent cleaning and sanitation agent. Its efficacy in dissolving mineral deposits, rust, and other residues ensures the maintenance of hygienic standards across equipment and surfaces. While sulphuric acid offers indispensable utility in food processing, stringent safety measures must be observed:
    - Handling Precautions: Proper handling protocols, including the use of protective gear, are essential to prevent skin and eye contact.
    - Dilution: Dilution of sulphuric acid solutions is crucial to mitigate its corrosive effects and ensure safe application.

Sulphuric acid, though not a direct component of food products, plays a pivotal role as a processing aid in the food industry. From facilitating sugar production to aiding in pH adjustment and sanitation, its multifaceted applications underscore its significance in ensuring the efficiency and safety of food manufacturing processes.

Summary: Application of Sulphur and Sulphur Compounds

Sulfur and sulfur compounds have several applications in the food industry, primarily as food additives, preservatives, and processing aids.

Here are some common applications:

a. **Preservatives:** Sulfur dioxide (SO_2) and sulfating agents, such as sodium sulfite (Na_2SO_3) and potassium metabisulfite ($K_2S_2O_5$), are commonly used as preservatives in the food industry. They inhibit the growth of bacteria, yeasts, and molds, thereby extending the shelf life of various food products, including dried fruits, wines, and fruit juices.

b. **Antioxidants:** Some sulfur-containing compounds, such as sulfur dioxide and its derivatives, act as antioxidants in food products. They prevent the oxidation of fats and oils, thereby inhibiting rancidity and extending the shelf life of processed foods containing fats and oils, such as fried foods, snacks, and baked goods.

c. **Dough Conditioners:** Sulfur-containing compounds, such as cysteine and its derivatives (e.g., L-cysteine hydrochloride), are used as dough conditioners in bakery products. They help improve dough elasticity, texture, and volume by breaking down disulfide bonds in gluten proteins, thereby enhancing dough handling properties and producing softer, more uniform bread products.

d. **Acidulants:** Sulfuric acid (H_2SO_4) and sulfurous acid (H_2SO_3) are used as acidulants and pH regulators in food processing. They are added to acidic foods and beverages, such as soft drinks, fruit juices, and pickles, to adjust acidity levels, enhance flavor, and improve microbial stability.

e. **Processing Aids:** Sulfur-containing compounds, such as sulfuric acid and sulfur dioxide, are used as processing aids in various food manufacturing processes. For example, sulfuric acid is used in the production of high-fructose corn syrup (HFCS), a sweetener used in many processed foods and beverages. Sulfur dioxide is used in the production of sugar from sugarcane or sugar beets to inhibit microbial growth and clarify sugar solutions.

f. **Color Retention:** Sulfur-containing compounds, such as sulfites and sulfur dioxide, are used to prevent browning and discoloration in fruits and vegetables. They are commonly used in the processing and preservation of dried fruits, canned fruits, and fruit juices to maintain their natural color and appearance.

g. **Flavor Enhancers:** Certain sulfur-containing compounds, such as sulfur-containing amino acids (e.g., cysteine and methionine), contribute to the flavor profile of foods. They are naturally present in many food sources, including meats, dairy products, and vegetables, and can enhance the savory or umami taste of foods.

While sulfur and sulfur compounds have various beneficial applications in the food industry, it's important to use them in compliance with regulatory guidelines and to consider potential sensitivities or allergies in consumers. Additionally, labeling requirements ensure transparency and informed consumer choices regarding the presence of sulfur compounds in food products.

REFERENCES:

1. Sadler, Christina R.; Grassby, Terri; Hart, Kathryn; Raats, Monique; Soković, Milka; Timotijevic, Lada (1 June 2021). "Processed food classification: Conceptualisation and challenges". Trends in Food Science & Technology. 112: 149–162.
2. Nestle, Marion (2013) [2002]. Food Politics: How the Food Industry Influences Nutrition and Health. University of California Press. pp. 36–37.
3. Marcus, Jacqueline B. (2013). "Protein Basics: Animal and Vegetable Proteins in Food and Health". Culinary Nutrition. Elsevier. pp. 189–230.
4. Merck Index, 11th Edition, 9625
5. Austin, George T. (1984). Shreve's Chemical process industries (5th ed.). New York, NY: McGraw-Hill. p. 357

Chapter 16: Exploring the Depths: An Overview of the Mining Industry

HISTORICAL EVOLUTION

The mining industry boasts a rich and ancient history, stretching back millennia to the earliest human civilizations. Initially centered around the extraction of precious metals like gold and silver for decorative and monetary purposes, mining later expanded to encompass base metals such as copper, iron, and lead to meet growing demands. From rudimentary tools and manual labor in ancient times, the industry advanced with the advent of steam power in the 18th and 19th centuries, paving the way for mechanized extraction methods.

Technological Advancements

The early 20th century witnessed significant breakthroughs in mineral exploration and extraction techniques, leading to the discovery of vast mineral reserves worldwide and the rise of large-scale mining operations. Technological innovations have remained pivotal, facilitating efficient resource extraction while reducing environmental impact and enhancing safety standards.

Environmental and Social Impacts

Despite its historical significance and economic contributions, the mining industry has grappled with environmental and social challenges. Issues such as deforestation, pollution, and community displacement have underscored the need for sustainable practices and responsible engagement with local stakeholders.

Industry Description

Today, the mining industry stands as a global juggernaut, encompassing diverse activities ranging from exploration to closure. It spans multiple sub-sectors, including exploration, development, production, and site reclamation. From artisanal endeavors in developing nations to large-scale industrial operations, mining plays a pivotal role in supplying raw materials essential for myriad products, from electronics to infrastructure.

Adapting to Challenges

Acknowledging its environmental and social footprint, the mining industry has embraced initiatives to enhance sustainability and community involvement. Implementing stringent regulations and standards, coupled with technological innovations, the industry strives to balance economic imperatives with environmental and social responsibilities.

In summary, the mining industry stands as a testament to human ingenuity and resourcefulness, driving economic growth while navigating complex environmental and social landscapes. Embracing sustainability and innovation, the industry charts a course toward a more equitable and environmentally conscious future.

OVERVIEW OF THE MINING INDUSTRY

The mining sector encompasses the extraction, processing, and distribution of vital natural resources like metals, minerals, and fossil fuels, serving as a linchpin in the global economy's machinery.

a. **Diverse Sub-Sectors and Operations**: Within the mining industry, distinct segments such as exploration, development, production, and closure cater to a broad spectrum of operations, ranging from small-scale artisanal endeavors to expansive industrial undertakings.
b. **Environmental and Social Challenges**: Despite its economic significance, the mining industry faces pressing environmental and social issues, including pollution, habitat degradation, and community displacement, necessitating concerted efforts for mitigation.
c. **Sustainable Solutions and Community Engagement**: To tackle these challenges head-on, the mining sector has adopted sustainable practices and fostered closer ties with local communities, aiming to strike a balance between economic development and environmental stewardship.
d. **Regulatory Framework and Technological Advancements:** Stringent regulations and standards govern mining operations, ensuring safety and responsible practices. Moreover, ongoing technological innovations and a heightened emphasis on sustainability are shaping the industry's trajectory towards a more responsible future.

APPLICATION OF SULPHUR & SULPHUR COMPOUNDS IN MINING INDUSTRY

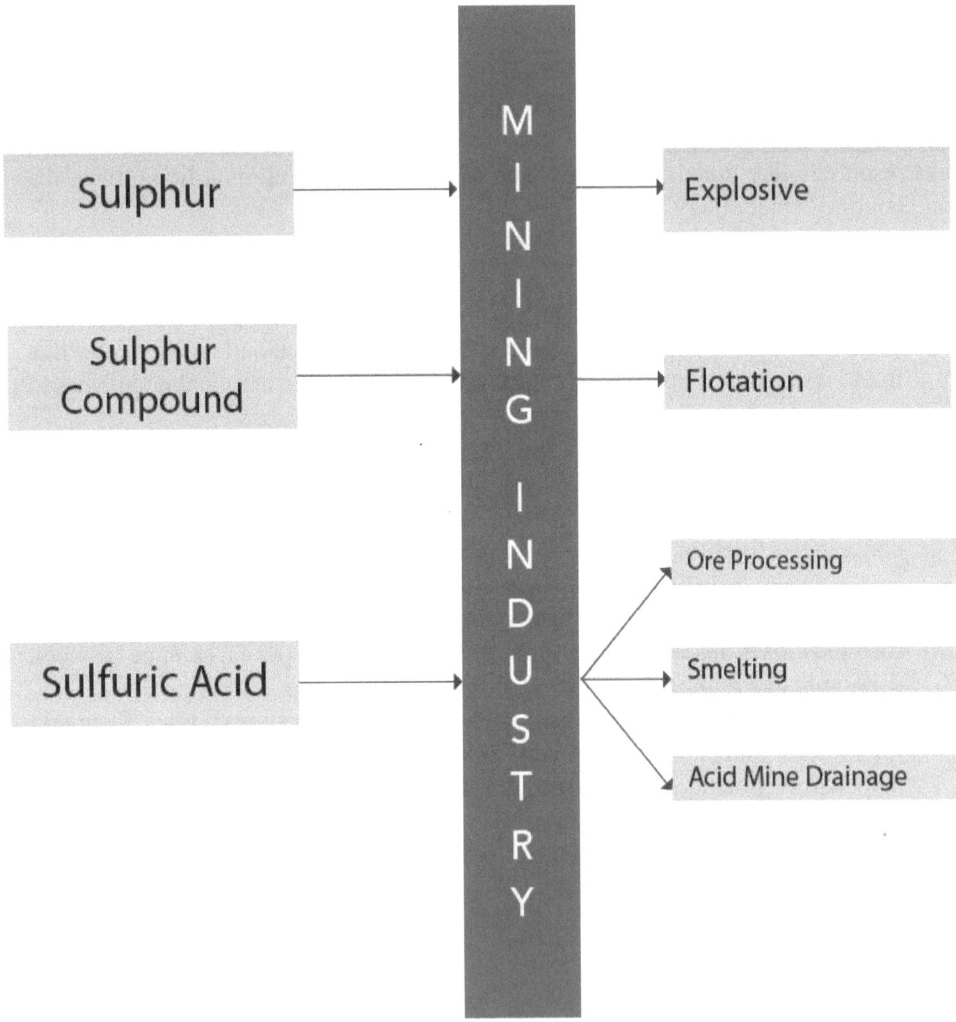

Sulphuric acid finds extensive use in mineral processing operations to extract metals like copper, nickel, zinc, and gold from their ores. It facilitates the dissolution of metal ores, aiding in the separation of valuable minerals from waste materials.

- **Sulphur in Explosives:** Sulphur serves as a crucial component in explosive compounds such as gunpowder and dynamite, vital for fragmenting rock and ore in mining operations.
- **Heap Leaching with Sulphuric Acid:** Heap leaching, employed for extracting metals like copper from low-grade ores, utilizes dilute sulphuric acid sprayed onto crushed ore heaps. This process aids in leaching metals from the ore, facilitating their extraction.
- **Flotation with Sulphuric Acid:** Sulphuric acid acts as a pH regulator in the flotation process, enhancing the separation of minerals from their ores by controlling the acidity of the pulp.
- **Acid Mine Drainage Management:** In addressing acid mine drainage concerns, sulphuric acid plays a dual role. While naturally produced acid can be detrimental to the environment, sulphuric acid can be employed to neutralize acidity and mitigate environmental impact.

a. Use of Sulphur

- **Sulphur in Explosives Production:** Explosives crucial for mining activities often incorporate sulphur, amalgamated with other compounds like nitrates, to effectively fragment rock and extract minerals.
- **Sulphur's Role in Ore Processing:** Sulphur finds application in ore processing, particularly in leaching copper from ores, leading to the formation of copper sulphate solutions that aid metal extraction.
- **Sulphur for Water Treatment:** In water treatment during mining operations, sulphur is utilized to eliminate impurities and neutralize acidity, safeguarding against environmental degradation.

b. Use of Sulphur Compounds

- **Flotation Agents:** Sulphur-based compounds like sodium hydrosulphide serve as flotation agents, facilitating the separation of minerals from ores by rendering them hydrophobic.
- **Acid Mine Drainage Treatment:** Compounds like calcium carbonate and calcium oxide, derived from sulphur-based sources, aid in neutralizing acid mine drainage, preventing harmful metal release into the environment.
- **Cyanide Detoxification:** Sulphur-based compounds like sodium metabisulphite are instrumental in detoxifying cyanide in mining operations, forming harmless compounds and averting environmental contamination.

- **Smelting By-Products:** Sulphur dioxide, a by-product of smelting, is captured and converted into sulphuric acid, serving various industrial processes.
- **Sulphur in Explosives Production:** Sulphur's inclusion in explosive formulations, alongside compounds like potassium nitrate and charcoal, contributes to their efficacy as blasting agents.

Sulphur, sulphur compounds, and sulphuric acid constitute indispensable elements in the mining industry, playing multifaceted roles from ore processing and water treatment to explosives production and environmental management. Their judicious use, accompanied by stringent safety measures, is imperative to mitigate environmental and safety risks associated with mining operations.

1. Sulphuric Acid in Explosive Production

Sulphur-based compounds play a crucial role in the manufacturing of explosive materials utilized in the mining industry.

Here are several notable examples:

a. **Black Powder:** Black powder, a historical explosive and propellant, is crafted by blending sulphur with charcoal and potassium nitrate, offering a combustible composition vital in firearms and mining.

b. **TNT (Trinitrotoluene):** TNT, recognized for its potency, is a staple explosive in mining and construction sectors. Its production involves nitrating toluene with a blend of nitric and sulphuric acid, resulting in a highly effective explosive compound.

c. **Sulphuric Acid Fuel:** Sulphuric acid serves as a potent oxidizer in rocket propulsion systems, reacting with various fuels like hydrazine to induce controlled explosions, crucial for high-performance applications in the mining domain.

d. **Gunpowder:** Historically significant, gunpowder combines saltpeter (potassium nitrate), charcoal, and sulphur, forming an explosive mixture utilized in military and industrial settings.

e. **Nitroglycerin:** Nitroglycerin, a liquid explosive, finds extensive use in dynamite manufacturing and high-powered explosive formulations. Its production involves nitrating glycerin with a blend of nitric and sulphuric acid.

These examples underscore the integral role of sulphur-based compounds in explosive production for mining operations. Given the inherent risks associated with explosives, stringent regulations govern their handling and storage, emphasizing safety as paramount.

2. Sulphuric Acid in Flotation Products

Sulphur-based compounds play a pivotal role in the froth flotation process, facilitating the separation of minerals from ores based on their hydrophobic properties.

Here are key examples of flotation products leveraging sulphur:

a. **Xanthates:** Xanthates serve as collector chemicals in flotation, selectively binding to mineral particles and rendering them hydrophobic for separation. They are synthesized by reacting an alcohol with carbon disulphide in the presence of sodium hydroxide.
b. **Dithiophosphates:** Dithiophosphates, another class of collector chemicals, are integral to the flotation process. Synthesized by reacting phosphorus oxychloride with an alcohol, followed by reaction with a sulphide, they aid in mineral separation.
c. **Thionocarbamates:** Thionocarbamates function as depressant chemicals, impeding the attachment of certain minerals to air bubbles and their ascent to the surface during flotation. Synthesized by reacting a primary amine with carbon disulphide in the presence of an alkali, they optimize separation efficiency.
d. **Sulphuric Acid:** Sulphuric acid finds utility in adjusting the pH of the flotation slurry, optimizing conditions for mineral separation. Additionally, it can activate specific minerals, enhancing their amenability to separation processes.
e. These examples underscore the significance of sulphur-based compounds in flotation products essential for efficient mineral extraction. Given their crucial role, meticulous handling and storage are imperative to ensure both safety and efficacy in the flotation process.

3. Application of Sulphuric Acid in Ore Processing

Sulphuric acid (H_2SO_4) stands as a cornerstone within the ore processing operations of the mining industry. Its multifaceted properties render it invaluable across numerous stages of mineral extraction and refinement.

a. **Acid Leaching**

- Metal Dissolution: Employed in leaching processes, sulphuric acid dissolves metals like copper, nickel, zinc, and uranium from their respective ores. This reaction yields soluble metal sulphate complexes.
  $M_{(s)} + H_2SO_{4(aq)} \rightarrow MSO_{4(aq)} + H_{2(g)}$

- pH Optimization: By regulating pH levels, sulphuric acid ensures optimal conditions for leaching, enhancing metal extraction efficiency while averting the precipitation of insoluble metal compounds.

b. pH Adjustment

- Controlled Acidity: Sulphuric acid is instrumental in modulating pH levels within ore processing solutions, fostering an acidic milieu conducive to targeted mineral reactions and averting undesired precipitation or alkalinity.

  $H_2SO_{4(aq)} + H2O_{(l)} \leftrightharpoons H_3O + HSO^{+}_{(aq)} + HSO^{-}_{4(aq)}$
- Efficiency Enhancement: Through pH regulation, sulphuric acid facilitates superior separation of valuable minerals from gangue materials, leading to heightened metal recovery rates.

c. Solvent Extraction

- Metal Retrieval: Within solvent extraction processes, sulphuric acid aids in selectively extracting metals like copper and nickel from pregnant leach solutions, effectively segregating metal ions from the aqueous phase.

  $MSO_{4(aq)} + 2RH_{(org)} \leftrightharpoons M_{(org)} + (HSO_4)^{-}_{(aq)} + R_2SO_4$
- Purity Refinement: Sulphuric acid contributes to purifying extracted metals, ensuring the production of high-quality final products for subsequent processing.

d. Tailings Management:

- pH Control: Applied in tailings treatment, sulphuric acid neutralizes alkaline residues and adjusts pH levels, thereby stabilizing tailings and mitigating the risk of environmental contamination.

  $H_2SO_{4(aq)} + Ca(OH)_{2(aq)} \rightarrow CaSO_{4(s)} + 2H_2O_{(l)}$
- Environmental Responsibility: Through alkaline component neutralization, sulphuric acid aids in minimizing the environmental impact of tailings, ensuring adherence to regulatory standards.

Sulphuric acid emerges as a versatile agent across various ore processing stages within the mining sector, facilitating efficient metal extraction, pH regulation, solvent extraction, and tailings management. Its widespread application underscores its pivotal role in augmenting productivity and sustainability within mining operations.

4. Application of Sulphuric Acid in Smelting Processes

Sulphuric acid (H_2SO_4) stands as a crucial element in smelting operations within the mining industry. Its adaptability and effectiveness make it an indispensable reagent across various stages of metal extraction and refinement.

a. **Ore Concentration**

- Enhanced Leaching: Sulphuric acid boosts leaching efficiency in ore concentration processes, facilitating the dissolution of metal-bearing minerals and aiding in the extraction of valuable metals from ores.
  $M_{(s)} + H_2SO_{4(aq)} \rightarrow MSO_{4(aq)} + H_{2(g)}$
- pH Optimization: By adjusting pH levels, sulphuric acid ensures favorable conditions for leaching reactions, maximizing metal recovery while preventing the formation of insoluble metal compounds.

b. **Smelting Operations**

- Catalytic Role: Acting as a catalyst, sulphuric acid promotes oxidation reactions in smelting processes. It facilitates the conversion of metal sulfide ores into metal oxides, which are subsequently reduced to yield pure metals.
  $MS_{(s)} + 2H_2SO_{4(aq)} + O_{2(g)} \rightarrow MO_{(s)} + 2HSO^{-}_{4(aq)}$
- Impurity Removal: Through its oxidizing properties, sulphuric acid aids in eliminating sulfide impurities from metal ores, ensuring the production of high-purity metal concentrates.

c. **Environmental Considerations**

- Emissions Control: Sulphuric acid can be employed to curb emissions of sulfur dioxide (SO_2) during smelting, converting it into less harmful sulfuric acid mist, thus minimizing environmental impact.
  $2SO_{2(g)} + O_{2(g)} + 2H_2O_{(l)} \rightarrow 2H_2SO_{4(aq)}$

d. **Waste Management:** Utilizing sulphuric acid in effluent treatment processes can neutralize acidic waste streams generated during smelting, ensuring compliance with environmental regulations and mitigating ecological damage.

Sulphuric acid stands as a linchpin in smelting processes within the mining sector, facilitating ore concentration, oxidation reactions, and environmental stewardship. Its pivotal role in augmenting metal extraction efficiency and minimizing environmental impact underscores its significance in modern mining operations.

5. Application of Sulphuric Acid in Mining

Sulphuric acid serves as a versatile chemical with diverse applications across the mining and mineral processing industry. Below are key examples of ore processing products employing sulphuric acid:

a. **Leaching of Metal Ores:** Sulphuric acid, renowned for its potency as a mineral acid, plays a pivotal role in leaching processes to extract metals like copper, zinc, and uranium from ores. By introducing sulphuric acid to a slurry of crushed ore, metals dissolve, enabling their separation and extraction.

b. **pH Adjustment:** Utilizing sulphuric acid for pH adjustment in mineral processing slurries enhances ore processing efficiency and metal recovery rates. By carefully controlling pH levels, optimal conditions for ore processing can be maintained, ensuring efficient extraction processes.
c. **Mineral Beneficiation:** Sulphuric acid facilitates mineral beneficiation by eliminating undesired impurities from minerals. Through sulphuric acid treatment, impurities such as silica are removed, yielding a more concentrated and purified mineral product.
d. **Electro-winning:** In the electro-winning process, sulphuric acid plays a crucial role in producing metals like copper and nickel. Through the application of electrical current, metal ions in solution are reduced, leading to the deposition of pure metals.
e. **Refining:** Sulphuric acid is instrumental in refining processes for metals such as zinc, copper, and nickel. By dissolving and eliminating impurities from crude metals, sulphuric acid contributes to refining operations, ensuring the production of high-quality metal products.

These applications underscore the indispensable role of sulphuric acid in ore processing within the mining industry. Given its significance, meticulous handling and storage practices are essential to uphold safety standards and optimize process efficiency.

6. Application of Sulphuric Acid in Acid Mine Drainage (AMD) Treatment

Acid Mine Drainage (AMD) poses a significant environmental challenge in mining activities. The reaction of sulphide minerals in rocks and soils with air and water during mining generates acidic solutions containing heavy metals and contaminants. Sulphuric acid-derived chemicals play a crucial role in treating AMD, neutralizing it, and mitigating its environmental impact.

Here are key examples of sulphuric acid applications in AMD treatment:

a. **Lime Production**: Sulphuric acid is utilized in lime production, an essential step in AMD treatment. Lime serves to neutralize the acidic water by reacting with it to form calcium sulphate, which is less harmful to the environment. Incorporating lime into water treatment systems raises pH levels, effectively neutralizing acidity.
b. **Gypsum Generation:** Treating AMD with sulphuric acid often yields gypsum as a by-product. Gypsum, a calcium sulphate compound, finds diverse

applications, including cement, plaster, and drywall production. Moreover, gypsum acts as a soil conditioner, enhancing soil structure and mitigating erosion.

c. **Metal Recovery:** Sulphuric acid aids in recovering valuable metals from AMD-contaminated water. Dissolving metal ions in water, sulphuric acid enables their recovery via precipitation, ion exchange, or solvent extraction methods. Metal recovery not only reduces AMD's environmental impact but also offers a valuable resource.
d. **Sulphate Reduction:** Utilizing sulphuric acid facilitates sulphate ion concentration reduction in water, a crucial step in AMD treatment. High sulphate levels can compromise water quality, making reduction essential. Methods like biological and chemical treatments effectively achieve sulphate reduction.

These applications underscore the critical role of sulphuric acid in AMD treatment. However, meticulous handling and disposal procedures are imperative to ensure safety and efficacy in AMD treatment processes.

Summary

Sulfur and sulfur compounds find several applications in the mining industry, particularly in mineral processing and ore beneficiation processes.

Here are some common applications:

a. **Flotation:** Sulfur compounds, such as sodium sulfide (Na_2S) or potassium ethyl xanthate (KEX), are widely used as flotation reagents in the froth flotation process. In froth flotation, mineral ores are treated with flotation reagents to selectively separate valuable minerals from gangue minerals based on their hydrophobicity. Sulfur compounds act as collectors or frothers, promoting the attachment of collector-coated bubbles to the surface of valuable mineral particles, thereby facilitating their separation from gangue minerals.
b. **Sulfide Mineral Processing**: Sulfur compounds play a crucial role in the processing of sulfide minerals, such as copper, lead, zinc, and nickel ores. Sulfide minerals are often concentrated through froth flotation, where sulfur-containing collectors selectively bind to the surface of sulfide minerals, enabling their recovery from ore slurry.
c. **Smelting and Refining:** Sulfur compounds are used in smelting and refining processes to extract metals from sulfide ores and to remove impurities. During smelting, sulfur dioxide (SO_2) is generated as a byproduct and can be captured and converted into sulfuric acid for further industrial applications. In refining processes, sulfur-containing fluxes, such as calcium sulfate (gypsum) or

sodium sulfate (salt cake), are used to remove impurities, such as iron oxide, silica, and alumina, from metal concentrates.

d. **Environmental Control:** Sulfur compounds are also used in environmental control measures to mitigate the environmental impact of mining operations. For example, lime or limestone (calcium carbonate) is often added to tailings ponds or wastewater treatment facilities to neutralize acidic mine drainage, which may contain sulfuric acid and heavy metals leached from sulfide minerals. Additionally, biological processes, such as bioremediation, may be employed to oxidize or reduce sulfur compounds in contaminated soils or water bodies.

e. **Cyanide Detoxification:** Sulfur compounds, such as sodium metabisulfite ($Na_2S_2O_5$) or sodium sulfite (Na_2SO_3), are used in cyanide detoxification processes to neutralize residual cyanide in mine tailings or process effluents. These compounds react with cyanide to form less toxic cyanate or thiocyanate species, reducing the environmental impact of cyanide-containing wastewater discharged from mining operations.

Overall, sulfur and sulfur compounds play critical roles in various stages of mineral processing, from flotation and concentration to smelting, refining, and environmental management in the mining industry. These applications contribute to the efficient extraction of valuable metals from ores while minimizing environmental impacts and ensuring regulatory compliance.

REFERENCES

1. Hartman, Howard L. SME Mining Engineering Handbook, Society for Mining, Metallurgy, and Exploration Inc, 1992, p. 3.
2. Leotaud, V. R. "Scientists develop technique to reduce cost, environmental impact of mining precious metal", by Valentina Ruiz
3. "Enhanced Landfill Mining: Material recovery, energy utilization and economics in the EU (Directive) perspective" (PDF). Enhanced Landfill Mining.
4. Zvereva, V. P.; Frolov, K. R.; Lysenko, A. I. (2021-10-13). "Chemical reactions and conditions of mineral formation at tailings storage facilities of the Russian Far East". Gornye Nauki I Tekhnologii = Mining Science and Technology (Russia). 6 (3): 181–191.
5. Merck Index, 11th Edition, 9625
6. Austin, George T. (1984). Shreve's Chemical process industries (5th ed.). New York, NY: McGraw-Hill. p. 357

Index

A

Accelerators, 285
Acetaldehyde, products and uses, 146–147
Acetic acid, synthesis of, 137–138
Acid-catalyzed hydration of ethylene, 214, 215
Acid-catalyzed vulcanization, 280–281
Acid dyes, 263
Acidizing
  products and applications, 273
  with sulphuric acid, 272–273
Acid mine drainage (AMD), 348–349
Acid sulphite oxidation of sludge, 55
Acidulants, 338
Acrylic fibers, 212
Acrylonitrile (C_3H_3N)
  applications, 212–213
  production, 212
Activators, 285
Agriculture industry, sulphur and sulphur compounds in, 290–299
Algaecide, 143
Alkyl aryl sulfonates (AAS), 229–230
  applications, 230
  sulfonation, 229, 230
Alkylated aromatics, 272
Alkylation
  products and applications, 272
  from sulphuric acid, 272, 273
Alkylation gasoline
  reaction, 191
  uses of, 191–192
Alkyl ketene dimer (AKD), 325
Alkyl succinic anhydride (ASA), 325
Alkyl sulphites formation, 79
Alum, 125–126, 193
  industrial applications, 126
  production, 126
Aluminum sulphate, 273
  applications of, 193
Amidosulfuric acid, 107
  production, 108
  uses of, 108
Amidosulphuric acid (ASA)
  production process, 238
  utilization of, 238
Amino acids, sulphur, 294
Ammonium sulphate (AMS), 271, 292
  applications, 130
  chemical synthesis, 129
  as fertilizer, 129–130
  production, 297
Ammonium sulphate fertilizers, 261
Amorphous sulphur, 13–14
Andrussow process, 211
Aniline
  applications, 171–172
  production, 171
Anthraquinone, 158–160
  applications of, 158–159
  based dyes, 159–160
  production, 158
Anthraquinone-based dyes
  applications of, 159–160
  production of, 159
  types of, 159
Antibiotics, 260, 304, 305
Anticoagulants, 260–261
Antifungals, 260, 305
Anti-inflammatory agents, 305
Antioxidants, 34, 285, 338
Antiozonants, 285
Antivirals, 305
Arc furnace, 44

Arc ignition, 44
Aspirin, 304
Automotive components, 144
Azo dyes, 156, 157–158

B

Bating, 330
Battery electrolyte, 77
Battery-grade sulphuric acid, 11
Beet sugar juice, 71–72
  decolorization process, 71
  filtration and separation, 72
  pH adjustment, 72
  reaction with color compounds, 72
  sulphur dioxide addition, 72
Benzene, 147–148
  products and applications, 147–148
  from sulphuric acid, 147
Benzenesulfonic acid, 148
Beverage alcohol, 145
Beverage preservation, 85
Biological production, 246
Biological symbiosis, 298–299
Biological treatment, 22, 312, 315
Biopesticides, 294–295
Birkeland-Eyde process, 44
Black powder, 344
Bleaching, sulphite pulping, 58–59
Bleaching agent
  sodium sulfite, 331
  sulphur dioxide, 59, 331
Blue vitriol, 143
Boric acid, 160–163
  applications of, 160–161
  detergents with, 162–163
  in glassmaking, 161–162
  in herbicides, 162
  production of, 160
Borosilicate glass
  production of, 161
  uses of, 161–162
Brimstone see Sulphur
Bromine (Br_2), 132–133
  applications and uses, 132
  production, 132
Butyl acetate, 136–137

C

Calcium oxide (CaO), 22–23
Calcium sulphate, 273
  production, 142
  products and uses, 142
Calcium sulphide (CaS), 22
Carbolic acid see Phenol
Carbon disulphide (CS_2), 43–47, 270
  in carbon tetrachloride, 47
  in cellophane, 46–47
  in soil and grain fumigants, 45
  in viscose rayon, 46
Carbon tetrachloride, 47
Carbonylation of methanol, 137, 138
Carbonyl sulphide (COS), 23, 33, 250–251
Carboxylic anhydride production, 77
Casein production
  in food and industry, 66
  sulphur dioxide in, 66–67
Catalytic cracking, 273
Catalytic oxidation, 21–22
  of sulphuric acid, 28
Cellophane, 46–47
Cellulose acetate
  applications of, 169–170
  fibre and plastics from, 170
  production process, 169
Cellulose nitrate see Nitrocellulose
Chamber process, 227
Chelating agents, 305
Chemical feedstock, 275
Chemical processing, 24
Chemical production, 246

Chemical treatment, 23
Chloral, 163–164
  applications of, 163–164
  production process of, 163
Chloral hydrate, 163–164
Chlor-alkali process, 199
Chlorine, 199–201, 315
  applications of, 200
  dechlorination, 315
  production process, 199
  in pulp and paper manufacturing,200–201
Chloroform, 133–135, 201
  adverse effects, 135
  applications, 134–135
  production, 133–134
Chloroformic acid, 134
Chlorophenol production, from sulphuryl chloride, 61–62
Chlorosilanes
  production, 113
  uses, 113
Chlorosulfonic acid ($ClSO_3H$), 11, 229
  industrial applications, 229
Chlorosulfonic acid esters see Haloalkyl chlorosulphates
Chlorosulphuric acid (HSO_3Cl), 80
Chlorothymol production, from sulphuryl chloride, 60–61
Chrome liquor
  incorporation of sulphur dioxide, 70
  synthesis process, 70
Chromic oxide, 42–43
  applications of, 139
  in metal treatment processes, 139–140
  production, 139
  from sulphuric acid, 139–140
Citric acid
  applications and uses, 131–132
  production, 131
Claus process, 23, 24, 25, 26, 54, 248, 271, 274
Coal, sulphur from, 23–24
Coal preparation, 23
Coal tar, 201
Coal washing, 23
Collection, 17, 18, 37, 45
Compounding, 38
Concentrated superphosphate see Triple superphosphate (TSP)
Contact process, 33, 55, 86, 93, 123, 227, 263
Copper sulphate, 142–143
  production process, 143
  products and uses, 143
Corn oil production, sulphur dioxide in
  extraction, 63
  health benefits of, 63
  high, 63
  steeping, 62–63
Corrosion control, 274
Corrosion inhibitors, 275
Crafts process, 44
Cresols, 201–203
  applications, 202
  phenolic resins from, 202–203
  production process, 201–202
Cross-linking with sulphur, 39
Crown-Eckroth process, 24
Crushing and grinding, 18, 20
Crystalline sulphur, 13
Cumene process, 147, 272
Curing process, 39
Cyclic sulphates, 199

D

DDT (dichlorodiphenyltrichloroethane), 164
Dechlorination, 77, 315
Decorative coatings, 143
Deliming, 330
Desalting, 25
Desulphurization, 21, 23, 53, 268
  of sludge/flue gases, 55

from sulphuric acid, 271, 273–274
Detergents
organic sulphonates, 76
from sodium dodecylbenzenesulfonate,99
Dialkyl phthalates (DEP), 151–153
applications, 151–152
plasticizers, 152–153
production via sulphuric acid, 151
Diammonium phosphate (DAP)
applications, 127
from phosphoric acid, 127–129
Dibasic calcium phosphate see Dicalcium phosphate
Dibutyl ether (DBE), 176
Dicalcium phosphate, 154–155
in animal feed, 155
industrial applications, 154
production, 154
Dichlorodiphenyl sulfone (DCDPS)
production, 112
uses, 112–113
Dichlorodiphenyltrichloroethane (DDT), 232–233
applications, 233
Diethyl ether, 175
Diethyl sulphate (DES), 234–236
applications, 235
production process, 234
resins production and uses from, 235–236
Diisopropyl ether (DIPE), 176
Dimercaprol (BAL), 305
Dimercaptosuccinic acid (DMSA), 305
Dimethyl disulfide (DMDS), 269
Dimethyl sulfochloride (DMSC), 250
Dimethyl sulfoxide (DMSO), 75, 249–250
as refrigerant, 75
solvent production from, 249–250
synthesis, 249
Dimethyl sulphate (DMS)
in dye manufacturing, 116
in pharmaceuticals, 116–117
production, 115
uses, 115
see also Methylene sulphate
Dimethyl sulphide (DMS), 34
Dimethyl sulphoxide (DMSO), 34
Direct dyes, 42, 159
Direct sulfation, 101
Dishwashing detergents, 262
Disinfectants from chloro thymol and chlorophenol, 63
Disinfection, 315
Distillation, 25
Disulphur trioxide, 32–33
Dithiocarbamates, 260, 280
Dithionous acid
applications of, 195–196
synthesis, 195
Dithiophosphates, 345
Dome deposits, 16–17
Dough conditioners, 337, 338
Drilling, 17
Drying, 43
Dry scrubbing, 21
Dyeing, 329
Dyes, 233–234
industrial applications, 234
from oleum, 233–234
Dynamite
applications, 207
from nitroglycerin, 206–207

E

Electrolysis, 246
Electronic components, 144
Electroplating, 143–144
Elemental sulphur, 24, 48, 257, 270, 271, 292
in fertilizer production, 270
Enzyme inhibition, 34

Epoxy resins, 259
Epoxy sulphates see Cyclic sulphates
Epsom salt, 184
Ester synthesis, 77
Ethanol (C_2H_5OH), 214–215
Ethers, 175–176
    diethyl ether, 175
    ethylene glycol, 175
    ethyl ether, 175
    solvents derived from, 175–176
Ethionic acid
    applications, 102
    production process, 102
    sodium ethylene sulfonate from, 102–103
Ethyl acetate
    lacquers, 173–174
    synthesis of, 173
    uses of, 173
    vinyl acetate, 174
Ethyl alcohol/ethanol, 145–147
    products and uses, 145–146
Ethylbenzene, 29
Ethyl chloride, 201
Ethylene dibromide (EDB)
    applications and uses, 133
    from bromine, 133
    production, 133
Ethylene glycol, 175
    applications of, 190–191
    production process, 190
Ethylene sulphate, 100–101
    production process, 100
    uses, 100–101
Ethyl ether, 175
Evaporative basin deposits, 17–19
Exothermic reaction, 33, 53, 186, 234
Exploration, 17, 19–20
Explosives, 172
Extraction process, 23

F

Fatty acids
    applications of, 192
    esterification reaction, 192
Fatty alcohols, 213–214
    applications, 213–214
    Ziegler process, 213, 214
Ferrous sulphate
    industrial applications, 125
    production, 125
Fertilization, sulphur, 293
Fertilizer production, 261–262, 263, 270, 291
    sulphur for, 292–293
    sulphuric acid in, 296–297
Filtration, 43
Finishing, 329
Flavour enhancement, 34
Flotation, 18, 20–21
Flue gas, 21–22
Flue-gas desulphurization (FGD), 21, 54
Food industry
    facets of, 334
    overview, 333
    sulphur applications, 336
    sulphur compounds applications, 336–337
    sulphuric acid applications, 337
Food preservation, 85
Fool's gold see Iron pyrite
Formic acid
    in leather tanning, 177
    production process, 176
    products and uses, 176–177
    in textile dyeing, 178
Frasch process, 1, 16, 16–17, 25–26
Free machining steel, 35–36
Free vulcanization, 38
Fuel additives, 275
Fumigants
    applications of, 72–73
    environmental and health considerations, 73

sulphur dioxide in, 72
Fuming sulphuric acid, 109, 224, 227, 239
see also Oleum
Fungicides, 28, 143, 291, 298

G

Gasoline, 272
Gelatine
preservative function, 81
production process, 81
regulatory considerations, 81
sulphur dioxide in, 81
Germination control, 34
Glass production, sulphur dioxide in, 70–71
batch adjustment, 71
decolorization, 71
oxidation state control, 71
as refining agent, 71
Glauber's salt see Sodium sulphate
Glue, 82–83
color and consistency maintenance, 82
preservative function, 82
production process, 82–83
Glutamic acid, 135–136
applications, 135
production, 135
Glutathione, 303
Grain fumigant, 45
Grain preservation, 34
Guanidines, 280
Gunpowder, 344
Gypsum, 22–23, 142, 154, 271, 293

H

Halides, 31–32
Haloalkyl chlorosulphates
production, 111
uses, 111–112
Halogenation, 31–32
Heating process, 35
Heavy metals, reduction of, 315
Herbicide, 143
High-purity sulphur, 14
Hydrochloric acid (HCl), 204–205
applications, 204–205
in metal industry, 205
production, 204
Hydrodesulphurization (HDS), 268
Hydrofluoric acid (HF), 210–211
in aluminum industry, 211
industrial and commercial applications, 210
production, 210
Hydrogen cyanide (HCN), 211–213
applications, 212
Hydrogen peroxide
applications, 208
production, 208
Hydrogen sulphide (H_2S), 54, 270, 273
chemical reactions, 242
generation in Athabasca, Canada, 245
industrial applications, 242
occurrence, 241
from organic sulphur coal, 244–245
from petroleum refinery streams, 245
physical properties, 242
production methods, 246
properties, 241
removal from sour natural gas, 244
safety precautions, 242
to sulphur conversion, 24, 26
Hydrogen sulphide (H_2S), uses of
analytical chemistry, 252
chemical manufacturing, 251
dimethyl sulfoxide, 249–250
healthcare and biomedical research, 252
mining and metallurgy, 251–252
oil and gas industry, 251
sulphur, 248
thioaldehyde, 249

thioketones, 250
thiol, 248
thiophene, 251
thiophosgene, 250–251
wastewater treatment, 252
water treatment, 252
Hydrotreating, 25
2-hydroxyethanesulfonic acid see Ethionic acid
Hydroxytoluene see Cresols

I

Indirect sulfation, 101
Industrial alcohol/ethanol, 214–215
Industrial detergents, 262
Industrial preservatives, 85
Inorganic sulphur compounds, 293
Insecticide, 28, 291
Insoluble sulphur, 11
Iron and steel pickling, 198
Iron(II) sulphate see Ferrous sulphate
Iron pyrite, 54–55
Iron pyrite, burning of, 55
Iron sulphate, 273
Isopropanol see Isopropyl alcohol
Isopropyl alcohol
applications of, 189–190
production process, 189

K

Kilning, 34
Kraft process, 31, 324–325

L

Lacquers
synthesis and composition, 173
uses of, 173–174
Lactic acid
applications of, 138–139
sulphuric acid as catalyst, 138–139
Laundry detergents, 262
Lead-acid batteries, 215–216
Lead chamber process, 55
Leather industry
key aspects of, 327
manufacturing stages, 326–327
origins, 326
production of leather goods, 326
sulphur applications, 329
sulphur compound applications, 329–330
sulphuric acid applications, 330
Lime-sulphur, 292
Liming, 331
Linear alkylbenzene (LAB), 272
Liquid sulphur dioxide (SO_2), 11
Lithium carbonate
applications of, 188–189
production process, 188
Livestock feed, sulphur supplements, 291, 293–294
Ludox process, 18

M

Magnesium metallurgy
control of reaction temperature, 84
efficiency enhancement, 84
emission control and environmental considerations, 84
sulphur dioxide as reducing agent, 84–85
Magnesium sulphate (Epsom salt), 271
applications, 184
Malt, 34–35
Malting process, 34
Mannheim process, 204
Melting, 37
Mercaptobenzothiazole (MBT), 280
Methylcyclopentadienyl manganese tricarbonyl (MMT), 269
Methylene chloride, 201

Methylene sulphate, 101
Methyl hydrogen sulphate (CH_3HSO_4)
  production, 114
  uses, 114–115
Mining, 17, 18, 19, 20, 37
Mining industry
  adapting to challenges, 341
  description, 340
  environmental and social impacts, 340
  and ore processing, 264
  overview, 341
  sulphur application, 343
  sulphur compounds uses, 343–344
  sulphuric acid application, 344–349
  technological advancements, 340
Monoammonium phosphate (MAP)
  applications, 129
  from phosphoric acid, 128–129
  preparation of, 129
Monoclinic sulphur, 13
Monosodium glutamate (MSG), 135, 136
  applications, 136
  from glutamic acid, 136
  production, 136
MSG see Monosodium glutamate (MSG)

N

Naphthalene, 153–154
  applications, 153–154
  production method, 153
B-Naphthol, 155–158
  azo dyes, 156, 157–158
  dyes synthesis, 157–158
  production process, 155–156
  rubber additives, 156–157
Natural dyes, 263
Natural gas sweetening, 244, 246
Natural resin, 259
Natural seepage, 19
N-chloramines production, 79
Neutralisation, 43
Nitration, 230, 231
Nitrobenzene, 148, 170–172
  in aniline production, 171–172
  applications, 171
  production process, 171
Nitrocellulose, 172, 230–231
  nitration process, 230
  uses of, 230–231
Nitrogen-based fertilizers, 261
Nitroglycerin, 172, 206–207, 344
  applications, 206
  dynamite from, 206–207
  production, 206
Nitrotoluene, 149
N-Methylmorpholine N-oxide (NMMO), 46
Nordhausen acid see Oleum
Nutrient synergism, 298

O

Odor control, 315
Oilfield chemicals
  organic sulphonates, 76
Oil refining, 274
Oleic acid
  production of, 168
  uses of, 168–169
Oleum, 109
  physio-chemical properties, 224–225
  production pathway of, 227
  production process, 105
  uses of, 105
Oleum, uses of
  acid catalyst, 239
  alkyl aryl sulfonates, 229–230
  amidosulphuric acid, 238
  battery manufacturing, 239
  chemical synthesis, 239
  chlorosulfonic acid, 229
  DDT, 232–233

diethyl sulphate, 234–236
dyes, 233–234
laboratory reagent, 239
nitrocellulose, 230–231
petroleum refining, 239
purification, 239
sodium dodecyl benzene sulfonate, 236–237
sulfonating agent, 239
sulfonating agents, 237
treflan, 232
trinitrotoluene, 231
Open-pit mining, 18, 20
Organic chlorides, applications and uses, 201
Organic sulphonates, 75–76
applications, 76
Organic sulphur compounds, 293
Orthorhombic sulphur, 13
Oxalic acid
applications, 137
production, 137
Oxidation, 23, 43
Oxyhalides, 31–32

P

Paper industry
current landscape and challenges, 317–318
making process, 317
origins, 317
sulphur compounds in, 320
sulphuric acid in, 320–324
Personal care products
organic sulphonates, 76
preservatives, 85
Pesticides, 292, 298
Petroleum refining, 25, 264, 270
PH adjustment, 315
Pharmaceutical industry
business stages, 301
key features of, 300–301
sulphur and sulphur compounds in, 303–305
Pharmaceutical preservation, 85
Phenol, 203–204
applications, 203–204
production process, 203
Phenolic resins, 259
applications, 202–203
from cresols, 202–203
Phosgene method, 250
Phosphoric acid, 126–127, 154
applications, 126–127
monoammonium phosphate from, 128–129
production of, 296–297
triple superphosphate fertilizer from, 128
Physical separation, 23
Pickle liquor, 198
Pickling, sulphuric acid in, 330, 331
Pigment dyes, 263
Pigments, 42–43, 159
Plaster of Paris, 142
Plasticizers, 152–153
Polymeric sulphur nitride, 14
Polymerization, 39, 264
Polysulfide resins, 259
Polysulfone (PSU), 260
Poly-sulphide elastomer, 39–40
Poly sulphuryl chloride (SxOyClz)
industrial applications, 96
production process, 96
properties, 96
from sulphur trioxide, 96
vs. thionyl chloride, 96
Polyvinyl chloride (PVC), 260
Potassium aluminum sulphate see Alum
Potassium sulphate fertilizers, 261
Preservatives, 34
beverage, 85
cosmetic and personal care products, 85

food, 85
industrial, 85
pharmaceutical, 85
sulphur dioxide as, 85–86, 338
Pressure vulcanization, 38
Processing and refining, 18, 20
Protective coatings, 144
P-toluenesulfonic acid, 149
Pulp delignification, 324–325
Pumping, 17
Purification, 25, 37
Pyridine, 193–194
applications of, 194
production, 194
Pyritic sulphur in coal, 23–24
Pyrosulfuric acid see Oleum
Pyrosulphuric acid (fuming sulphuric acid/ oleum)
production, 109
uses of, 109

R

Radiopharmaceuticals, 261, 305
Reaction chamber, 44
Recovery of chemicals from sulphuric acid, 323–324, 325
Refining, 17
Refrigerants
dimethyl sulfoxide, 75
sulfolane, 75
Resins production and uses from diethyl sulphate, 235–236
Ring-opened sulphates see Cyclic sulphates
Roasting, 21, 23
Rubber
additives, 156
applications of, 157
vulcanization of, 38–39, 279
Rubber industry, 276–277

S

Salicylic acid
applications, 182
production process, 182
Scrubbing, 248
Seed treatment, 299
Single superphosphate (SSP), 130–131
applications, 131
see also Single superphosphate
Sizing agents, 320–321, 325
Sludge acid, 53
Smelting of ores, 54
S-methylmethionine (SMM), 34
Soda ash, 186
Sodium bisulphate, 144–145
industrial applications, 144–145
production, 144
Sodium bisulphite, 57, 315
as dough conditioner, 337
Sodium dichromate, 178–180
applications, 178–179
as pigment, 179–180
production from sulphuric acid, 178
Sodium dodecyl benzene sulfonate (SDBS), 236–237
applications, 236
detergents from, 99
production, 98, 236
uses of, 99
Sodium ethylene sulfonate (SES)
from ethionic acid, 102–103
Sodium hydrogen sulphate, 186
Sodium hydrosulfide (NaHS)
in deliming and bating, 330
effluent treatment, 325
leather tanning, 329
liming, 331
Sodium polysulphide, 41–42
Sodium sulphate, 186–188
applications, 186

filler in detergent formulations, 186–187
manufacturing sodium hydrogen sulphate, 186
production and variants, 186
sodium sulphide from, 187–188
Sodium sulphide (Na_2S), 30–31
applications of, 187–188
in deliming and bating, 330
leather tanning, 329
odor control, 315
reduction of heavy metals, 315
from sodium sulfate, 187–188
Sodium sulphite (Na_2SO_3), 303, 315
bleaching agent, 320, 331
synthesis, 65
thiosulphate from, 65–66
uses of, 65
Sohio process, 212, 213
Soil amendment, by sulphur, 291, 298
Soil conditioner, 28
Soil fumigants, 45, 294
Soil sterilization
integrated approach to, 295
for sulphur compounds, 294–295
Solid sulphur, 44
Soybean protein preservation, sulphur dioxide in, 67
Starch production, sulphur dioxide in, 63–64
Stearic acid, 281, 285
production of, 166
uses of, 166–167
Steroids, 304
Storage batteries, 215–216
Styrene-butadiene rubber (SBR), 260
Styrene monomer, 29
Sugar refining, sulphur dioxide in
antimicrobial protection and preservation, 59–60
bleaching enhancement, 59
Sulfamic acid (H_3NSO_3), 238
production process, 106, 107
uses of, 106, 107
Sulfate removal, 315
Sulfide ores
roasting, 93
smelting of, 55
Sulfide scouring, 331
Sulfite pulping process, 320, 325
Sulfolane, 73–74
alternative production method, 73
applications of, 73–74
production process, 73
as refrigerant, 75
Sulfonamide antibiotics, 303, 305
Sulfonating agents, 99–100
from oleum, 237
production and application of, 237
Sulfonation, 43, 201–202, 229, 230, 233, 236
Sulfones
applications of, 77
mechanism, 76
from sulphur dioxide, 76–77
variations and optimization, 77
Sulfonic acids
applications, 183
derivatives, 79
production process, 182–183
Sulfonylurea drugs, 303
Sulphate of soda see Sodium sulphate
Sulphate process, 31
Sulphates, 23, 47
Sulphide ore, 20–21
Sulphide process, 30
Sulphides, 22, 47–48
Sulphite pulping
bleaching, 58–59
cooking, 58
Sulphonyl chloride (SO_2Cl_2)
applications of, 79
chlorinating sulphonating agent from, 80
chloro sulphuric acid from, 80
from sulphur dioxide, 79–80

synthesis, 77
Sulphur, 1, 53
amino acids, 294
based fertilizers, 297
carbon, reaction with, 45
extraction process, 23
from flue gas, 21–22
from gypsum, 22–23
heat effect on, 14
from hydrogen sulphide, 24, 248
occurrence, 13
from petroleum refinery, 25
production methods, 25–26
properties, 13–14
from pyritic sulphur in coal, 23–24
in rubber vulcanization, 279
from sulphide ore, 20–21
Sulphur applications
antibiotics, 305
carbon disulphide, 43–47
carbonyl sulphide, 33
catalysts in refining, 269–270, 274
chromic oxide, 42–43
disulphur trioxide, 32–33
drugs from, 303
elemental sulphur, 257
in explosives, 343
fertilizers, 257, 292–293
free machining steel, 35–36
in fungicide, 28, 329
halides, 31–32
infused pesticides, 257
in insecticide, 28
leather bleaching, 329
livestock feed, 293–294
malt, 34–35
as nutrient, 336
oxyhalides, 31–32
pesticides, 292
pigments, 42–43
for plant growth, 257
poly-sulphide elastomer, 39–40
preservation of hides and skins, 329
sodium polysulphide, 41–42
in sodium sulphide, 30–31
soil amendment by, 291
in soil conditioner, 28
styrene monomer, 29
sulfonamides, 303, 305
sulphur cement, 36–37
sulphur dioxide, 47, 257
sulphur dyes, 41–42
sulphur monoxide, 37–38
synthetic rubber vulcanization, 257–258
tetrasulphur tetranitride, 40
treflan, 40
ultramarine blue pigment, 35
vulcanization of rubber, 38–39
wood pulp, 31
Sulphur black, 43
Sulphur browns, 43
Sulphur burning, 26, 37
Sulphur cement (SC), 36–37
Sulphur-coated urea, 293
Sulphur combustion, 93
Sulphur compounds applications
accelerators, 285
activators, 285
as additives, 269, 270
antifungals, 305
antioxidant/reducing agent from, 303–304, 337
antioxidants, 285
antiozonants, 285
biological treatment, 315
bleaching, 330
in bleaching, 320, 324, 337
catalysts in refining, 274
chemical feedstock, 275
chemical industry, 270
chemical synthesis, 264
color retention, 338

corrosion inhibition, 275
as crude oil desulphurization, 268
in crude oil refining, 269–270
cyanide detoxification, 350
dechlorination, 315
disinfection, 311–312, 315
dough conditioner, 337
eliminating odors in water, 312–313
environmental control, 350
in fertilizers production, 270
flavor enhancers, 337, 338–339
flotation, 343, 349
fuel additives, 275
hydrodesulphurization, 268, 274
iron and manganese, removal of, 310–311
liming, 331
in metal production, 271
mining and ore processing, 264
odor control, 315
in petroleum industry, 268–275
petroleum refining, 264, 270
in pharmaceuticals, 260–261
in plastic production, 259–260
polymerization, 264
preservation, 330
preservatives, 336
processing aids, 286, 338
recovery, 270–271, 274
reduction of heavy metals, 315
in resin production, 259
in rubber industry, 279–286
smelting and refining, 349
for soil sterilization, 294–295
sulfate removal, 315
sulfide deliming and bating, 330
sulfide mineral processing, 349
sulfide scouring, 331
tanning, 329
in textile dyeing, 258, 264
vitamins and coenzymes, 305
vulcanization, 285
as vulcanization accelerators, 279–280
water treatment, 264
Sulphur-containing additives, 260, 269
Sulphur-containing supplements, 303
Sulphur deposits
from dome deposits, 16–17
from evaporative basin deposits, 17–19
from volcanic deposits, 19, 19–20
Sulphur dichloride (SCl_2), 32
Sulphur dioxide (SO_2), 22–23, 24, 34, 47, 50–51, 257, 270, 292, 294, 304
addition, 72
antioxidants, 338
bleaching agent, 320
chemical characteristics, 50
from hydrogen sulphide, 54
from iron pyrite, 54–55
oxidation, 65, 78
production methods, 55
from sludge acid, 53
from smelting of ores, 54
from sulphur, 53
Sulphur dioxide (SO_2), uses
beet sugar juice, decolorization of, 71–72
casein, 66–67
chrome liquor, 70
corn oil, 62–63
corrosion control in water treatment, 314
environmental applications, 87
fumigants, 72–73
gelatine, 81
glass, 70–71
glue, 82–83
industrial applications, 87
laboratory applications, 87–88
magnesium metallurgy, 84–85
organic sulphonates, 75–76
pH adjustment of water, 313–314
preservative, 85–86
processing aids, 338
refrigerants, 75

sodium bisulphite, 57
sodium sulphite, 64–66
soybean protein preservation, 67
starch, 63–64
sugar refining, 59–60
sulfolane, 73–74
sulfones, 76–77
sulphite pulp, 58–59
sulphonyl chloride, 79–80
sulphur trioxide and oleum production, 86–87
sulphuryl chloride and derivatives, 60–62
thionyl chloride, 77–79
vanillin, 74–75
wine, 83
zinc hydrosulphite, 67–68
Sulphur dissolution, 43
Sulphur dyes, 41–42
Sulphur green, 43
Sulphuric acid ($H2SO_4$), 21, 47
acidizing with, 272–273
alkylation from, 272
chemical properties, 121
contact process, 123
desulphurization of, 271, 273–274
esters, 284
fuel, 344
history, 120
neutralization, 78
physical properties, 120–121
production, 263
production methods, 123
in refining process, 273–274
wet sulphuric acid process, 123
Sulphuric acid anhydride see Oleum
Sulphuric acid applications
acetic acid, 137–138
acidulants, 338
alkylation gasoline, 191–192
alum, 125–126
aluminum sulphate, 193
in AMD treatment, 348–349
ammonium sulphate, 129–130
anthraquinone, 158–160
bating, 330
battery manufacturing, 218
benzene, 147–148
boric acid, 160–163
bromine, 132–133
butyl acetate, 136–137
calcium sulphate, 142
cellulose acetate, 169–170
chemical manufacturing, 217
chloral, 163–164
chlorine, 199–201
chloroform, 133–135
chromic acid, 139–140
citric acid, 131–132
for cleaning, 282–283
copper sulphate, 142–143
cresols, 201–203
cyclic sulphates, 199
in dehydration processes, 281–282
in detergent production, 262
dialkyl phthalates, 151–153
diammonium phosphate, 127–129
dicalcium phosphate, 154–155
dithionous acid, 195–196
dyeing, 330
in dye manufacturing, 262–263
electroplating, 143–144
ethers, 175–176
ethyl acetate, 173–174
ethyl alcohol, 145–147
ethylene glycol, 190–191
explosive, 172
in explosives, 344
fatty acid, 192
fatty alcohols, 213–214
ferrous sulphate, 125
fertilizer production, 217, 261–262, 263, 296–297

flotation products, 345
food processing, 218
formic acid, 176–178
glutamic acid, 135–136
hydrochloric acid, 204–205
hydrofluoric acid, 210–211
hydrogen cyanide, 211–213
hydrogen peroxide, 208
industrial alcohol, 214–215
iron and steel pickling, 198
isopropyl alcohol, 189–190
lactic acid, 138–139
lithium carbonate, 188–189
magnesium sulphate, 184
mineral processing, 218
in mining, 347–348
naphthalene, 153–154
for neutralization, 283, 284
nitrobenzene, 170–172
nitroglycerin, 206–207
oleic acid, 168–169
ore processing, 345–346
oxalic acid, 137
petroleum refining, 218
pH adjustment, 337
pharmaceutical products from, 304
pH control and water treatment, 218, 315, 321–322, 325
phenol, 203–204
phosphoric acid, 126–127
pickling, 330, 331
processing aids, 338
in production of rubber chemicals, 284–285
pyridine, 193–194
recovery of chemicals from, 323–324, 325
salicylic acid, 182
single superphosphate, 130–131
sizing with, 320–321, 325
smelting processes, 346–347
sodium bisulphate, 144–145
sodium dichromate, 178–180
sodium sulphate, 186–188
stearic acid, 166–167
storage batteries, 215–216
sugar production, 337
sulfite pulping process, 325
sulfonic acid, 182–183
tall acids, 216–217
tanning, 330
tartaric acid, 167
tetryl, 207
textile industry, 218
titanium dioxide, 183–184
toluene, 148–150
vanillin, 194–195
waste management, 322–323, 325
water quality, 295–296
xylene, 150–151
zinc sulphate, 185
β-Naphthol, 155–158
Sulphurized oils, 259
Sulphur-modified resins, 259
Sulphur monochloride (S_2Cl_2), 32
Sulphur monoxide (SO)
applications of, 38
environmental considerations, 38
from sulphur, 37–38
Sulphur preparation, 44
Sulphur recovery, 21–23
applications, 270–271
during petroleum refining, 270–271, 274
Sulphur Recovery Unit (SRU), 18, 26
Sulphur trioxide (SO_3), 11, 37
chemical properties, 118
contact process, 93
history, 90
industrial uses, 91
physical properties, 90–91
primary application, 90
production methods, 93, 118

reactions with metal oxides, 93
roasting sulfide ores, 93
sulphur combustion, 93
Sulphur trioxide (SO_3), uses of, 118
amido sulphuric acid, 108
catalysts and chemical synthesis, 118
chlorosilanes, 113
desiccant and dehydrating agent, 118
dichlorodiphenyl sulfone, 112–113
dimethyl sulphate, 115–117
ethionic acid, 102–103
ethylene sulphate, 100–101
haloalkyl chlorosulphates, 111–112
laboratory reagent, 118
methylene sulphate, 101
methyl hydrogen sulphate, 114–115
oleum, 105
poly sulphuryl chloride, 96
pyrosulphuric acid, 109
sodium dodecylbenzenesulfonate, 98–99
sulfamic acid, 106–107
sulfonating agents, 99–100
sulfonation reactions, 118
sulphuric acid production, 118
sulphur trioxide-dioxane, 97–98
sulphur trioxide pyridine complex, 110
thionyl chloride, 95
trimethylamine, 97
Sulphur trioxide (SO_3) and oleum, 86–87
Sulphur trioxide-dioxane, 97–98
applications, 98
production process, 98
Sulphur trioxide pyridine (pyridine complex of sulphur trioxide/pyridine-sulphur trioxide complex)
production, 110
uses, 110
Sulphur trioxide trimethylamine
industrial applications, 97
production process, 97
Sulphur vaporization, 44, 60
Sulphur yellows, 42
Sulphuryl chloride (SO_2Cl_2)
chlorophenol from, 61–62
chloro thymol from, 60–61
disinfectants from chloro thymol and chlorophenol, 62
Sulphuryl dioxane see Sulphur trioxide-dioxane
Superheating, 17
Superphosphate fertilizers, 261
Surfactants, 262
Synthetic dyes, 263
Synthetic resin, 259
Synthetic rubber vulcanization with sulphur, 257–258

T

Tall acids, 216–217
Tall oil fatty acids (TOFA), 216–217
Tanning, 329
Tartaric acid
applications of, 167
production, 167
Tert-butyl methyl ether (TBME), 176
Tetrasulphur tetranitride, 40
Tetryl, 207
Textile industry, 264
Textiles, organic sulphonates, 76
Thermal decomposition, 246
Thermoplastic elastomers (TPEs), 260
Thioaldehydes
from hydrogen sulphide, 249
synthesis, 249
Thioketone synthesis, 250
Thiokol/thiothene elastomers, 39
Thiolation, 248
Thiols, 303–304
from hydrogen sulphide, 248
see also Hydrogen sulphide
Thionocarbamates, 345

Thionyl bromide ($SOBr_2$)
  bromination reaction, 79
  preparation of thionyl chloride, 78–79
  quality control, 79
  separation and purification, 79
  from thionyl chloride, 78–79
Thionyl chloride ($SOCl_2$)
  chlorinating agent from, 78
  hydrolysis, 78
  industrial applications, 95
  preparation of, 95
  production process, 95
  separation and purification, 95
  storage and distribution, 95
  sulfur dichloride, preparation of, 95
  from sulphur dioxide, 77–79
  thionyl bromide from, 78–79
  uses of, 77–78
Thiophene synthesis, 251
Thiophosgene, 250–251
Thiosulphate production, from sodium sulphite, 65–66
Thioureas, 280
Thiurams, 260, 280
Titanium dioxide (TiO_2)
  applications, 183–184
  production process, 183
TMS see Sulphur trioxide trimethylamine
TNT see Trinitrotoluene (TNT)
TNT (Trinitrotoluene), 344
Toluene, 148–150
  applications, 149
  chemical formula and physical characteristics, 148–149
  derivatives, 149
  from sulfuric acid, 148
  versatility and applications, 149–150
Topical sulphur preparations, 303
Treflan, 40, 232
  in weed management, 232
Trichloroethylene, 201
Trichloromethane ($CHCl_3$) see Chloroform
Trifluralin, 40, 232
Trimethylamine sulfonate see Sulphur trioxide trimethylamine
Trinitrotoluene (TNT)
  nitration process, 231
  uses of, 231
Triple superphosphate (TSP) fertilizer
  applications and uses, 128
  from phosphoric acid, 128
  production, 128

U

Ultra-low sulphur diesel (ULSD), 269
Ultramarine blue pigment, 35
Underground mining, 18, 20
Utilisation, 20

V

Vaccines, 304
Vanillin, 194–195
  applications and significance, 74–75, 195
  commercial production methods, 74
  synthesis from sulphur dioxide, 74, 194
VAT dyes, 42, 159
Vinyl acetate ($C_4H_6O_2$)
  applications of, 174
  synthesis and composition, 174
Vinyl chloride, 201
Viscose rayon, 46
Vitamins and coenzymes, 305
Volcanic deposits, 19, 19–20
Vulcanization, 279, 285
  accelerators, 279–280
  of rubber, 38–39

W

Waste management, 322–323, 325
Water quality, sulphuric acid in, 295–296

Water treatment, 264
  components in, 307–308
  origins, 307
  process, 307
  sulphur compounds in, 310–313
  sulphuric acid in, 313–314
Wellman-Lord process, 53
Wet scrubbing, 21
Wet sulphuric acid process, 123
Wettable sulphur, 292
Wine, 83
Wood preservation, 143
Wood pulp, 31

X

Xanthates, 345
Xylene
  applications, 150–151
  industrial scale and significance, 150
  production via sulphuric acid, 150

Z

Ziegler process, 213, 214
Zinc hydrosulphite, sulphur dioxide in, 67–68
  production process, 68
  in sodium hydrosulphite production, 68
Zinc oxide, 279, 281, 285
Zinc sulphate
  applications, 185
  forms, 185
  production, 185

About the Authors:

Mahendra and Aditya Sheth form a dynamic father-son team driven by a shared passion for disseminating value adding knowledge. Their collaborative efforts and shared vision to give back, have culminated in the creation of "Versatilities of Sulphur," a comprehensive exploration of the multifaceted sources and applications of sulphur and its derivatives across various industries.

Mahendra Sheth, boasts over 40 years of experience as a Chemical Engineer. Educated in the United States, he holds dual Bachelor's Degrees in Chemical and Industrial Engineering, as well as a Master's Degree in Chemical Engineering. As the Chairman of Galiakotwala Engineering Company, a leading manufacturer of sulfuric acid equipment, Mahendra's profound expertise in sulphur chemistry has been pivotal in shaping the content of this book.

Aditya Sheth, following in his father's esteemed footsteps, earned his Chemical Engineering degree from the prestigious Columbia University (New York, USA) and Economics Degree from the prestigious Denison University (Ohio, USA). Despite his youth, Aditya has demonstrated a remarkable aptitude for engineering. His contributions to the book have brought forth fresh perspectives and innovative ideas.

Together, Mahendra Sheth and Aditya Sheth have synergized their knowledge, creativity, and passion to craft a book that not only delves into the intriguing realm of sulphur products but also aims to ignite curiosity and foster appreciation for the marvels of chemistry across readers of all ages.